Building Scalable Deep Learning Pipelines on AWS

Develop, Train, and Deploy Deep Learning Models

Abdelaziz Testas

Apress®

Building Scalable Deep Learning Pipelines on AWS: Develop, Train, and Deploy Deep Learning Models

Abdelaziz Testas
Fremont, CA, USA

ISBN-13 (pbk): 979-8-8688-1016-9 ISBN-13 (electronic): 979-8-8688-1017-6
https://doi.org/10.1007/979-8-8688-1017-6

Copyright © 2024 by Abdelaziz Testas

This work is subject to copyright. All rights are reserved by the Publisher, whether the whole or part of the material is concerned, specifically the rights of translation, reprinting, reuse of illustrations, recitation, broadcasting, reproduction on microfilms or in any other physical way, and transmission or information storage and retrieval, electronic adaptation, computer software, or by similar or dissimilar methodology now known or hereafter developed.

Trademarked names, logos, and images may appear in this book. Rather than use a trademark symbol with every occurrence of a trademarked name, logo, or image we use the names, logos, and images only in an editorial fashion and to the benefit of the trademark owner, with no intention of infringement of the trademark.

The use in this publication of trade names, trademarks, service marks, and similar terms, even if they are not identified as such, is not to be taken as an expression of opinion as to whether or not they are subject to proprietary rights.

While the advice and information in this book are believed to be true and accurate at the date of publication, neither the authors nor the editors nor the publisher can accept any legal responsibility for any errors or omissions that may be made. The publisher makes no warranty, express or implied, with respect to the material contained herein.

> Managing Director, Apress Media LLC: Welmoed Spahr
> Acquisitions Editor: Celestin Suresh John
> Development Editor: Laura Berendson
> Coordinating Editor: Kripa Joseph

Cover designed by eStudioCalamar

Cover image designed by Pixabay

Distributed to the book trade worldwide by Springer Science+Business Media New York, 233 Spring Street, 6th Floor, New York, NY 10013. Phone 1-800-SPRINGER, fax (201) 348-4505, e-mail orders-ny@springer-sbm.com, or visit www.springeronline.com. Apress Media, LLC is a California LLC and the sole member (owner) is Springer Science + Business Media Finance Inc (SSBM Finance Inc). SSBM Finance Inc is a **Delaware** corporation.

For information on translations, please e-mail booktranslations@springernature.com; for reprint, paperback, or audio rights, please e-mail bookpermissions@springernature.com.

Apress titles may be purchased in bulk for academic, corporate, or promotional use. eBook versions and licenses are also available for most titles. For more information, reference our Print and eBook Bulk Sales web page at http://www.apress.com/bulk-sales.

Any source code or other supplementary material referenced by the author in this book can be found here: https://www.apress.com/gp/services/source-code.

If disposing of this product, please recycle the paper

To my managers, who entrusted me with meaningful projects that sharpened my data science skills.

And to my fellow data scientists, for embracing my first book, "Distributed Machine Learning with PySpark: Migrating Effortlessly from Pandas and Scikit-Learn."

Table of Contents

About the Author xi

About the Technical Reviewer xiii

Acknowledgments xv

Introduction xvii

Chapter 1: Overview of Scalable Deep Learning Pipelines on AWS 1

Components of a Deep Learning Workflow on AWS 2

    Data Source (S3) 3

    Data Preprocessing (PySpark) 7

    Model Building (PyTorch and TensorFlow) 12

    Model Training (EC2) 19

    Model Evaluation (EC2) 28

    Model Deployment (Airflow) 30

    Project Directory Structure 49

    Virtual Environment 54

    Development Environment 55

Summary 56

Chapter 2: Setting Up a Deep Learning Environment on AWS 57

Creating an AWS Account 59

Provisioning Amazon EC2 Instances 60

Setting Up Amazon S3 70

Creating a Project Directory 75

TABLE OF CONTENTS

Creating a Virtual Environment 80

Installing and Configuring Dependencies 83

    Installing PySpark..83

    Installing PyTorch 85

    Installing TensorFlow 85

Installing Boto3 86

Installing Airflow 87

    Standalone Installation..87

    Docker-Based Setup 95

Installing JupyterLab...100

Setting Up a Databricks Account and Workspace 102

Summary...113

Chapter 3: Data Preparation with PySpark for Deep Learning 115

The Dataset...116

PySpark's Parallel Processing 137

    Data Preparation with PySpark for PyTorch..149

    Data Preparation with PySpark for TensorFlow 174

Bringing It All Together..191

    Data Exploration with PySpark 191

    Parallel Processing in PySpark..195

    Data Preparation with PySpark for PyTorch..199

    Data Preparation with PySpark for Tensorflow 205

Summary...210

Chapter 4: Deep Learning with PyTorch for Regression 213

The Dataset...214

Predicting Tesla Stock Price with PyTorch 223

    Bringing It All Together 257

TABLE OF CONTENTS

Exploring the Tesla Stock Dataset 258
Predicting Tesla Stock Price with PySpark and PyTorch............................... 260
Comparing and Plotting Actual and Predicted Values................................... 269

Summary... 272

Chapter 5: Deep Learning with TensorFlow for Regression 275

The Dataset 276
Predicting Tesla Stock Price with TensorFlow 277
TensorFlow vs. PyTorch.. 305
Bringing It All Together 309
    Layers and Activation Functions 309
    environment.yml... 310
    TensorFlow Regression Code 310
Summary... 319

Chapter 6: Deep Learning with PyTorch for Classification 321

The Dataset 322
Predicting Diabetes with PyTorch 359
    Enhancing Model Evaluation with Cross-Validation 391
Bringing It All Together 395
    Exploring the Pima Diabetes Dataset 396
    Model Building, Training, and Evaluation with PyTorch.......................... 403
    Diabetes Classification Without K-Fold Cross-Validation 403
    Diabetes Classification With K-Fold Cross-Validation............................ 415
Summary... 428

Chapter 7: Deep Learning with TensorFlow for Classification 431

The Dataset 432
Predicting Diabetes with TensorFlow 438

vii

TABLE OF CONTENTS

TensorFlow vs. PyTorch 463
    Optimizing Model Performance with Hyperparameter Tuning 466

Bringing It All Together 472
    Building and Training a TensorFlow Model with Fixed Hyperparameters 472
    Optimizing the TensorFlow Model with Hyperparameter Tuning 479

Summary 488

Chapter 8: Scalable Deep Learning Pipelines with Apache Airflow 489

An Airflow Pipeline for Tesla Stock Price Prediction 490

Tesla Stock Price Prediction Without Airflow DAG 492

Tesla Stock Price Prediction with Airflow DAG 511

An Airflow Pipeline for Diabetes Prediction 528
    Diabetes Prediction Without Airflow DAG 529
    Diabetes Prediction with Airflow DAG 546

Bringing It All Together 558
    Tesla Stock Price Prediction Without Airflow DAG 558
    Tesla Stock Price Prediction with Airflow DAG 567
    Diabetes Prediction Without Airflow DAG 573
    Diabetes Prediction with Airflow DAG 582

Summary 584

Chapter 9: Techniques for Improving Model Performance 585

The Baseline Model 586

Early Stopping 603

Dropout 607

L1 and L2 Regularization 610

Learning Rate 616

Model Capacity 619

viii

TABLE OF CONTENTS

Automating Hyperparameter Optimization with Keras Tuner 629
Bringing It All Together 634
    Baseline Model 635
    Early Stopping 643
    Dropout 651
    L1 Regularization 658
    L2 Regularization 666
    Learning Rate 674
    Model Capacity 682
    Automating Hyperparameter Tuning Using Keras Tuner 690
Summary 701

Chapter 10: Deploying and Monitoring Deep Learning Models 703

Steps in Deploying and Monitoring Deep Learning Models 704
    Step 1: Setting Up the Environment 706
    Step 2: Developing the DAG 716
    Step 3: Uploading to S3 719
    Step 4: Configuring the MWAA Environment 732
    Step 5: Triggering DAG Execution 733
    Step 6: Monitoring Execution 734
Summary 738

Index 739

ix

About the Author

Abdelaziz Testas, Ph.D., is a seasoned data scientist with over a decade of experience in data analysis and machine learning. He earned his Ph.D. in Economics from the University of Leeds in England and holds a master's degree in the same field from the University of Glasgow in Scotland. Additionally, he has earned several certifications in computer science and data science in the United States.

For over ten years, Abdelaziz served as a lead data scientist at Nielsen, where he played a pivotal role in enhancing the company's audience measurement capabilities. He was instrumental in planning, initiating, and executing end-to-end data science projects and developing methodologies that advanced Nielsen's digital ad and content rating products. His expertise in media measurement and data science drove the creation of innovative solutions.

Recently, Abdelaziz transitioned to the public sector, joining the State of California's Department of Health Care Access and Information (HCAI). In his new role, he leverages his coding and data science leadership skills to make a meaningful impact, supporting HCAI's mission to ensure quality, equitable, and affordable health care for all Californians.

Abdelaziz is also the author of *Distributed Machine Learning with PySpark: Migrating Effortlessly from Pandas and Scikit-Learn*, published by Apress in November 2023.

About the Technical Reviewer

Abhik Choudhury is a senior analytics managing consultant and data scientist with 12 years of experience in designing and implementing scalable data solutions for organizations across various industries. Throughout his career, he has developed a strong understanding of AI/ML, cloud computing, database management systems, data modeling, ETL processes, and big data technologies. His expertise lies in leading cross-functional teams and collaborating with stakeholders at all levels to drive data-driven decision making in longitudinal pharmacy and medical claims and wholesale drug distribution areas. Abhik's technical skills encompass a wide range of areas, including architecting cloud solutions, business intelligence, data visualization, machine learning algorithms, probability and statistics, data analysis, data warehousing, data quality, linear algebra, cloud computing, big data, data governance, and relational databases. Abhik has hands-on experience with languages such as Python, R, SQL, UNIX, and DAX, as well as packages like Scikit-Learn, Pandas, SciPy, NumPy, Matplotlib, XGBoost, NLTK, ggplot2, dplyr, and tidyverse. In addition, he has worked with tools and technologies such as Google Cloud Platform, IBM Watson, Microsoft PowerBI, Jupyter Notebook, GIT, Tensor, PyTorch, MySQL, SAP BusinessObjects, IBM DB2, SAP HANA, Hadoop, and Microsoft Excel. Abhik has completed his M.S. in Analytics from the Georgia Institute of Technology, GA, USA, in December 2022. Prior to that, he obtained a

ABOUT THE TECHNICAL REVIEWER

B.Tech. in Electronics and Communications Engineering from the West Bengal University of Technology, WB, India, in May 2011. He has been certified as IBM DB2 database developer, SAP BusinessObjects BI tools developer, and Google Cloud professional data engineer. In his current role as a senior analytics managing consultant at IBM Corporation, USA, since August 2022, Abhik has been responsible for designing and implementing scalable data architecture solutions for healthcare and pharma clients. He has also played a key role in architecting, building, and managing legal and compliance data platforms using the Google Cloud Platform. Additionally, he has led the investigation and adaptation of beneficial data architecture technologies and tools, as well as the development and implementation of an enterprise data model and strategy for operational data stores. Previously, as a senior data scientist at IBM, USA, from September 2021 to August 2022, Abhik leveraged Python to clean and analyze raw data for predicting opioid prescribers based on wholesale drug distribution patient data. He also conducted market basket analysis and association rule mining on a large customer dataset using R and Tableau. Moreover, he performed exploratory data analysis, developed machine learning models for predicting employee churn rate, conducted customer segmentation analysis, and performed time series analysis for predicting stock prices. Before that, as a lead business intelligence engineer at IBM, USA, from July 2019 to August 2021, Abhik created scalable data models, built efficient ETL pipelines, developed regression and classification models using BigQuery ML, and designed Tensor-based machine learning models in Vertex AI. Abhik also played a significant role in deploying inventory optimization applications and creating dashboards in PowerBI. During his time as a business intelligence analyst at IBM, from March 2012 to July 2018, Abhik implemented analytical reports across the supply chain spectrum using SAP BusinessObjects BI reporting tools. Abhik also worked on building reports and universes and utilized SQL, data warehousing programs, Tableau, and other visualization tools for data intelligence.

Acknowledgments

I would like to extend my sincere gratitude to everyone at Apress who contributed to the creation of this book. Specifically, I wish to thank Celestin John, the executive editor, for giving me the opportunity to write both this second book and my first, *Distributed Machine Learning with PySpark: Migrating Effortlessly from Pandas and Scikit-Learn*. Celestin's invaluable guidance and support have been instrumental throughout both projects.

Special thanks to Kripa Joseph, the coordinating editor, whose diligent oversight ensured that every detail was carefully addressed. I also appreciate Laura Berendson, the development editor, for her expertise in shaping the content into its final form.

I would like to acknowledge Abhik Choudhury, the technical reviewer, whose thorough review and expert insights greatly enhanced the technical accuracy and depth of the material presented.

Many thanks to the entire team at Apress for their commitment to excellence and collaborative spirit, which has made the publishing process a rewarding experience. Your combined efforts have been instrumental in bringing this book to life, and I am immensely grateful for your contributions.

Introduction

With the exponential growth of data and the increasing demand for sophisticated AI applications, traditional on-premises infrastructure often falls short in providing the computational power and scalability required for modern deep learning workflows. Consequently, cloud-based deep learning is becoming increasingly important as organizations seek scalable and efficient solutions for their machine learning tasks. By leveraging the flexibility, scalability, and cost-effectiveness of cloud services, organizations can overcome hardware limitations and accelerate the development and deployment of advanced machine learning algorithms.

Integrating PySpark, PyTorch, TensorFlow, and Apache Airflow with Amazon Web Services (AWS), including Amazon Simple Storage Service (S3) and Amazon Elastic Compute Cloud (EC2), offers powerful tools for building and deploying deep learning models at scale. AWS, as the most popular cloud service provider with the largest market share, provides a robust infrastructure for executing these tasks efficiently.

This book is designed for data scientists, machine learning engineers, and AI practitioners interested in building scalable deep learning pipelines on AWS or seeking to enhance their skills in integrating cloud services with deep learning frameworks. It provides a comprehensive guide to constructing a deep learning pipeline on AWS using a combination of PySpark, PyTorch, TensorFlow, and Airflow. We illustrate the entire workflow, starting with data ingestion and preprocessing using S3 and PySpark and concluding with scheduling the model to run on Apache Airflow. Each stage of the pipeline is dissected, offering clear explanations, hands-on examples, and insights drawn from real-world applications.

INTRODUCTION

This pipeline offers several advantages. Firstly, it capitalizes on the scalability of PySpark for large-scale data processing and the compute capacity of EC2 instances for model training, ensuring efficient utilization of resources. Secondly, a cloud-based approach provides flexibility, allowing resource scaling based on workload demands and easy provisioning of resources. Thirdly, PySpark and Apache Airflow simplify data processing tasks with their distributed computing capabilities, enabling efficient preprocessing, transformation, and workflow orchestration. Fourthly, the large storage capacity of Amazon S3 facilitates uninterrupted data storage and retrieval, supporting the processing of massive datasets. Lastly, the separation of concerns among preprocessing (PySpark), model development, training, and evaluation (PyTorch, TensorFlow), model orchestration (Apache Airflow), storage (S3), and compute (EC2) simplifies the architecture and enhances modularity, facilitating streamlined workflow management.

The book also examines JupyterLab and Databricks to further enhance the development and deployment process. JupyterLab provides an interactive computing environment ideal for data exploration and prototyping, while Databricks offers a collaborative platform for big data processing and advanced analytics.

This integration of multiple cloud services and data science tools fosters an end-to-end workflow, promoting collaboration and efficiency across the deep learning pipeline.

The book is divided into ten chapters. In Chapter 1, readers are introduced to the fundamentals of deep learning pipelines, emphasizing the importance of scalability in such frameworks. Additionally, an overview of AWS as a platform for scalability is provided, laying the foundation for subsequent chapters that delve deeper into setting up, configuring, and optimizing deep learning environments on the AWS cloud.

INTRODUCTION

Chapter 2 guides readers through configuring AWS services tailored for deep learning tasks, establishing development environments, and securing the AWS environment for deep learning pipelines. This chapter serves as a crucial starting point for readers to effectively leverage AWS resources for scalable deep learning solutions.

Chapter 3 focuses on data preprocessing and feature engineering with PySpark, while Chapters 4 and 5 explore deep learning with PyTorch and TensorFlow for regression tasks, covering the frameworks, model building, and training techniques. Chapters 6 and 7 shift focus to deep learning for classification tasks using PyTorch and TensorFlow.

In Chapter 8, we demonstrate how to build scalable data processing pipelines with Apache Airflow. We create entire AWS workflows that include preprocessing data using PySpark, developing deep learning models using PyTorch and TensorFlow, and running the code using Airflow.

Chapter 9 demonstrates advanced techniques for enhancing the performance of deep learning models, including handling overfitting and underfitting, implementing regularization techniques, early stopping strategies, and fine-tuning models for improved accuracy and efficiency.

Finally, in Chapter 10, we delve into the deployment process by showcasing how to deploy deep learning models using Amazon Managed Workflows for Apache Airflow (MWAA), providing insights into this managed option for workflow orchestration.

Together, these chapters provide a comprehensive guide for building end-to-end scalable deep learning pipelines on AWS, from data preprocessing and model training to deployment and monitoring in production environments. This book complements our first book, *Distributed Machine Learning with PySpark: Migrating Effortlessly from Pandas and Scikit-Learn* (Apress, November 2023), which focuses on transitioning from small-scale tools like Pandas and Scikit-Learn to PySpark. While the first book emphasizes scalable machine learning

INTRODUCTION

workflows, this book expands on that foundation by focusing on building scalable deep learning workflows in the cloud. There is a small overlap between the two books, specifically in the use of the Pima dataset for diabetes classification, which we revisit here through the lens of PyTorch and TensorFlow. This allows us to showcase the enhanced flexibility and scalability these platforms offer for deep learning workflows.

CHAPTER 1

Overview of Scalable Deep Learning Pipelines on AWS

As data volumes surge and the demand for sophisticated AI applications increases, traditional on-premises infrastructure often falls short in providing the computational power and scalability required for modern deep learning workflows. Consequently, cloud-based deep learning is becoming increasingly vital.

In this chapter, we introduce a deep learning pipeline that leverages these cloud-based solutions effectively by integrating key technologies and services. This pipeline, which will be detailed throughout the book, incorporates popular big data frameworks—PySpark, PyTorch, TensorFlow, and Apache Airflow (as a standalone deployment, Docker containerized setup, or managed service through Amazon Managed Workflows for Apache Airflow (MWAA))—with Amazon Web Services (AWS) cloud solutions, including Amazon Simple Storage Service (S3) and Amazon Elastic Compute Cloud (EC2). AWS, the leading cloud service provider with the largest market share, provides a robust infrastructure for efficiently executing these tasks.

The first section of this chapter provides a detailed examination of the main components of this deep learning workflow on AWS. It covers the

CHAPTER 1   OVERVIEW OF SCALABLE DEEP LEARNING PIPELINES ON AWS

entire pipeline, from data ingestion using S3 to scheduling model runs with Apache Airflow. Subsequent chapters will delve deeper into each stage of this pipeline, offering clear explanations, hands-on examples, and insights from real-world applications.

Components of a Deep Learning Workflow on AWS

Figure 1-1 illustrates a comprehensive deep learning pipeline leveraging the power of Amazon Web Services (AWS) to handle large-scale data processing, model training, evaluation, and deployment tasks. At the heart of this pipeline lies Amazon S3, serving as the centralized data source for storing raw datasets, preprocessed data, and model artifacts. Data preprocessing is facilitated by PySpark, enabling distributed data processing on large datasets stored in S3. The pipeline utilizes both PyTorch and TensorFlow for building and training deep learning models, harnessing their flexibility and scalability. Amazon EC2 instances provide the necessary computational resources for model training, ensuring efficient utilization of cloud resources. Apache Airflow, as a standalone deployment, Docker containerized setup, or managed service through Amazon MWAA, orchestrates the deployment of trained models onto production environments, streamlining the deployment process and ensuring smooth execution of tasks.

Let's break down the figure in detail, starting with the data source.

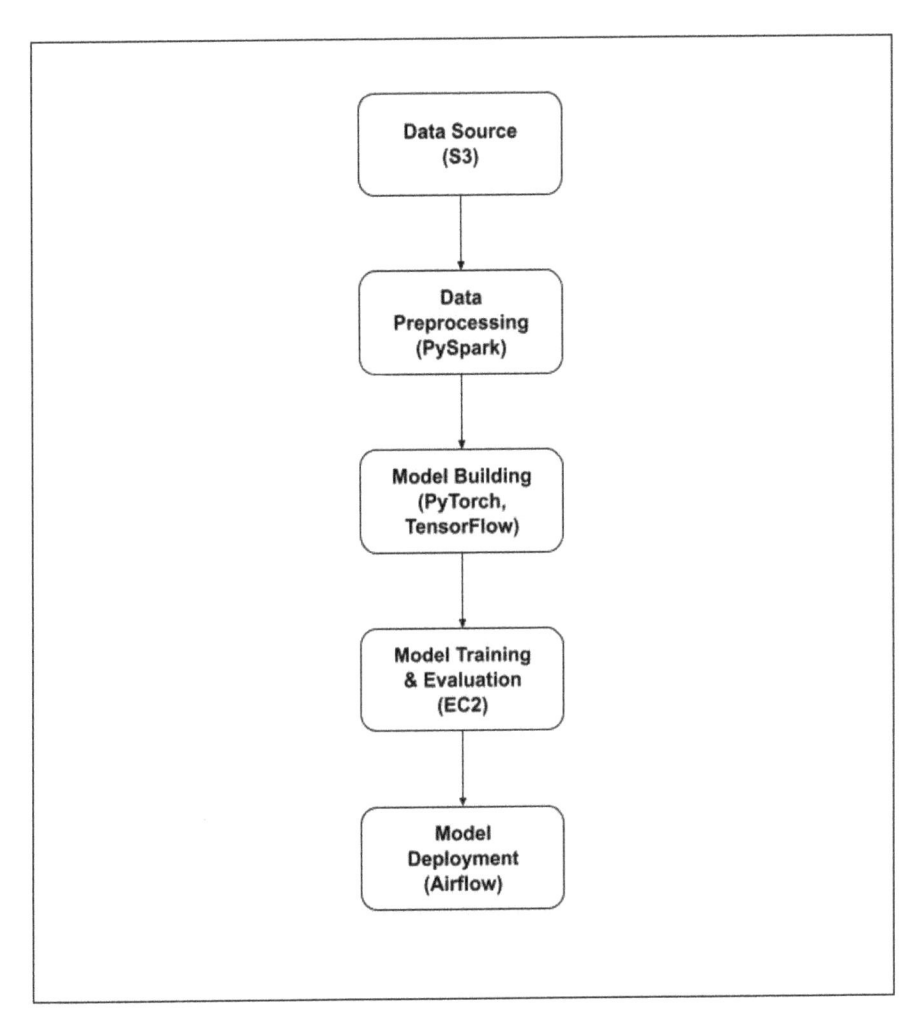

CHAPTER 1　OVERVIEW OF SCALABLE DEEP LEARNING PIPELINES ON AWS

Figure 1-1. Components of a Deep Learning Workflow on AWS

Data Source (S3)

This represents the initial point where data is sourced from Amazon S3, which is a cloud-based storage service provided by AWS. S3 is commonly used for storing various types of data, including raw datasets, preprocessed data, and model checkpoints (such as weights). Data stored in S3 can be

3

CHAPTER 1 OVERVIEW OF SCALABLE DEEP LEARNING PIPELINES ON AWS

accessed and utilized by other components in the pipeline including data preprocessing, model training, model evaluation, model deployment, and monitoring and maintenance.

Source data in S3 could be copied to a local directory on EC2 and then fed into the deep learning model, or it can be copied as part of the pipeline, meaning that the retrieval and integration of data from S3 into various stages of the machine learning (ML) pipeline are automated for efficiency and scalability.

Below are two examples (using AWS Command Line Interface (CLI) with subprocess and Boto3) of how a CSV file named "TSLA_stock.csv" (which will be used in Chapters 4 and 5 to predict Tesla stock prices with PyTorch and TensorFlow, respectively) is copied from an S3 bucket named "instance1bucket" to a local directory on EC2 named "/home/ubuntu/airflow/dags".

Using AWS CLI with subprocess:

```
[In]: import subprocess

[In]: try:
         subprocess.run([
         "aws", "s3", "cp",
         "s3://instance1bucket/TSLA_stock.csv",
         "/home/ubuntu/airflow/dags/"
         ], check=True)
         print("Download successful.")
      except subprocess.CalledProcessError as e:
         print(f"Error occurred: {e}")

[Out]: Download successful.
```

This code utilizes the AWS Command Line Interface (CLI) through the subprocess module in Python to copy the file "TSLA_stock.csv" from the S3 bucket "instance1bucket" to the local directory

CHAPTER 1   OVERVIEW OF SCALABLE DEEP LEARNING PIPELINES ON AWS

"/home/ubuntu/airflow/dags/" on EC2 instance. The subprocess.run() function executes the AWS CLI command "aws s3 cp" with the specified source and destination paths.

The try/except block is used to handle potential errors that might occur during the execution of the command. If the command is successful, the try block prints "Download successful." If the command fails and raises a subprocess.CalledProcessError, the except block catches the error and prints a message indicating that an error occurred, along with the error details.

The check=True parameter in subprocess.run() ensures that a CalledProcessError is raised if the command returns a non-zero exit status, triggering the except block. In this example, the output indicates that the file has been successfully downloaded from the S3 bucket to the local directory.

After running this code, we can use standard file system commands (e.g., "ls" in Linux/MacOS) to check whether the file exists in the directory:

```
[In]: ls /home/ubuntu/airflow/dags/
```

Using Boto3:

```
[In]: import boto3

[In]: try:
         s3_client = boto3.client('s3')
         bucket_name = 'instance1bucket'
         key = 'TSLA_stock.csv'
         local_path = '/home/ubuntu/airflow/dags/TSLA_stock.csv'
         s3_client.download_file(bucket_name, key, local_path)
         print("Download successful.")
     except Exception as e:
         print(f"An error occurred: {e}")

[Out]: Download successful.
```

CHAPTER 1 OVERVIEW OF SCALABLE DEEP LEARNING PIPELINES ON AWS

In this code, Boto3, the official AWS SDK (Software Development Kit) for Python, is used to interact with the Amazon S3 service. The s3_client.download_file() method is used to download the file "TSLA_stock.csv" from the S3 bucket "instance1bucket" to the local directory "/home/ubuntu/airflow/dags/" on EC2 instance. After running this code, the file "TSLA_stock.csv" should be present in the specified local directory.

The process of copying files from the directory to S3 bucket is essentially the reverse of the process of copying files from S3 to the directory.

Using AWS CLI with subprocess:

```
[In]: import subprocess

[In]: try:
          subprocess.run([
          "aws", "s3", "cp",
          "/home/ubuntu/airflow/dags/TSLA_stock.csv",
          "s3://instance1bucket/"
          ], check=True)
          print("File uploaded successfully.")
      except Exception as e:
          print(f"An error occurred: {e}")

[Out]: File uploaded successfully.
```

In this code, we reversed the order of the source and destination paths compared with the code for downloading from S3.

Using Boto3:

```
[In]: import boto3

[In]: try:
          s3_client = boto3.client('s3')
          bucket_name = 'instance1bucket'
          key = 'TSLA_stock.csv'
```

```
        local_path = '/home/ubuntu/airflow/dags/TSLA_
        stock.csv'
        s3_client.upload_file(local_path, bucket_name, key)
        print("File uploaded successfully.")
    except Exception as e:
        print(f"An error occurred: {e}")
```

[Out]: File uploaded successfully.

In this code, we reversed the order of the parameters compared with the code for downloading from S3. In other words, for downloading, the parameters were

```
s3_client.download_file(bucket_name, key, local_path)
```

For uploading, the parameters are

```
s3_client.upload_file(local_path, bucket_name, key)
```

It is worth noting that, if the file already exists in the destination location (S3 bucket in this case) and we attempt to upload a file with the same key (filename), it will typically overwrite the existing file with the new one.

Data Preprocessing (PySpark)

Once data is retrieved from S3, as explained in the preceding step, it undergoes preprocessing using PySpark, a distributed data processing framework that enables efficient data cleaning, transformation, and feature engineering on large datasets.

Leveraging PySpark for preprocessing tasks offers distinct advantages over traditional libraries like Pandas and Scikit-Learn. Firstly, its distributed nature allows PySpark to handle big data processing tasks with superior scalability and performance compared with single-node

CHAPTER 1   OVERVIEW OF SCALABLE DEEP LEARNING PIPELINES ON AWS

processing tools. Secondly, by utilizing PySpark's distributed processing capabilities, preprocessing tasks can be parallelized across multiple nodes, reducing processing times and enabling efficient handling of large-scale datasets. Additionally, PySpark seamlessly integrates with other components of the machine learning pipeline, facilitating a cohesive and streamlined workflow from data ingestion to model deployment.

In *Distributed Machine Learning with PySpark: Migrating Effortlessly from Pandas and Scikit-Learn* (Apress, November 2023), we demonstrated how transitioning from small data libraries like Pandas and Scikit-Learn to PySpark can be relatively straightforward. Although Pandas and Scikit-Learn are primarily designed for small data processing and analysis, transitioning is made easier by the similarity in syntax in many instances. PySpark offers functionality similar to Pandas and Scikit-Learn, including data preprocessing and feature engineering. The presence of these familiar functionalities in PySpark facilitates a smoother transition for data scientists accustomed to working with Pandas and Scikit-Learn.

In this section, we provide a simple example of creating a DataFrame using PySpark, followed by printing it on an EC2 instance, serving as an initial introduction to its capabilities. In Chapter 3, we'll delve deeply into exploring data preparation techniques using PySpark. This includes more advanced topics such as data exploration, preprocessing, feature engineering, and formatting data for PyTorch and TensorFlow:

```
[In]: from pyspark.sql import SparkSession

[In]: try:
          spark = (SparkSession.builder
                  .appName("TacosDataFrame")
                  .getOrCreate())
          data = [("Taco 1", "Beef", 2.50),
                  ("Taco 2", "Chicken", 2.00),
                  ("Taco 3", "Vegetarian", 2.00),
                  ("Taco 4", "Fish", 3.00)]
```

CHAPTER 1   OVERVIEW OF SCALABLE DEEP LEARNING PIPELINES ON AWS

```
        schema = ["Taco Name", "Type", "Price"]
        tacos_df = spark.createDataFrame(data, schema)

        # Cache the DataFrame to optimize performance
        tacos_df.cache()

        tacos_df.show()
    except Exception as e:
        print(f"An error occurred: {e}")
    finally:
        spark.stop()
```

[Out]:

Taco Name	Type	Price
Taco 1	Beef	2.5
Taco 2	Chicken	2.0
Taco 3	Vegetarian	2.0
Taco 4	Fish	3.0

The code above creates a PySpark DataFrame with taco-related data (taco name, type, and price) and then prints the DataFrame using the show() method.

Here's an explanation of each line of code:

- from pyspark.sql import SparkSession: This statement imports the SparkSession class from the pyspark.sql module in the PySpark library. The SparkSession serves as the primary interface for programming Spark using the DataFrame application programming interface (API). It facilitates interaction with Spark and enables the creation of DataFrames, which are essential for distributed data processing tasks.

9

- spark = SparkSession.builder: This line creates a new SparkSession using the SparkSession.builder method.
- .appName("TacosDataFrame"): This line sets the application name for the Spark job to "TacosDataFrame".
- .getOrCreate(): This line retrieves an existing SparkSession if one exists or creates a new one if none is found.
- data = [("Taco 1", "Beef", 2.50), ...]: This line defines a sample dataset of taco-related data. Each tuple in the list represents a row of data, with three values: taco name, type, and price.
- schema = ["Taco Name", "Type", "Price"]: This line defines the schema for the DataFrame. It specifies the column names as "Taco Name", "Type", and "Price".
- tacos_df = spark.createDataFrame(data, schema): This line creates a DataFrame called tacos_df using the createDataFrame() method of the SparkSession. It takes two parameters: the data (defined in data) and the schema (defined in schema).
- tacos_df.cache(): Caches the DataFrame in memory to enhance performance for subsequent operations that access the DataFrame multiple times. Caching is particularly advantageous when handling large datasets or during computationally intensive operations to prevent unnecessary recomputation. This method stores the intermediate DataFrame in memory following its initial computation, triggered by the first action such as a transformation or action on

the DataFrame. This approach is especially beneficial in scenarios involving iterative algorithms or multiple actions on the same DataFrame, as it eliminates the need to recompute the DataFrame from scratch for each operation. On the other hand, tacos_df.persist() offers more detailed control over the storage level, allowing choices between memory only, disk only, or a combination of both, depending on the dataset's size and available system resources.

- tacos_df.show(): This line prints the contents of the DataFrame tacos_df using the show() method (displays the first 20 rows of the DataFrame by default).

- try/except Block:

  - Inside the try Block: Performs operations like displaying the DataFrame with tacos_df.show().

  - except Exception as e: Catches any exceptions that might occur during the operations, handling runtime errors such as data processing issues or Spark connection problems.

- finally Block:

  - spark.stop(): Stops the SparkSession, releasing the resources associated with it. Using finally ensures this cleanup happens whether the try block completes successfully or an exception occurs.

CHAPTER 1   OVERVIEW OF SCALABLE DEEP LEARNING PIPELINES ON AWS

To run this code, PySpark needs to be installed and configured correctly for Python 3 (we will cover how to do this in the next chapter). Save the PySpark code in a file, e.g., "tacos_dataframe.py", on the EC2 instance, and then execute the script using one of the two options below:

1. Running as a Python Script:

    You can SSH into your EC2 instance and execute the script using the Python interpreter as follows:

    [In]: python3 tacos_dataframe.py

    This approach utilizes the local resources of your EC2 instance and the Spark installation to create and process the DataFrame.

2. Running with spark-submit:

    Alternatively, you can leverage Spark's distributed computing capabilities by submitting the script to the local Spark instance using "spark-submit" as follows:

    [In]: spark-submit tacos_dataframe.py

    This method allows your script to be executed in a distributed manner, utilizing the computing resources provided by Spark.

Model Building (PyTorch and TensorFlow)

In this step of the pipeline, the preprocessed data from PySpark is used for building deep learning models using both PyTorch and TensorFlow frameworks. These are popular deep learning frameworks known for their flexibility and scalability in constructing and training neural network models.

PyTorch was developed by Facebook's AI Research lab (FAIR) as an open source machine learning framework. It was released in October 2016, building upon the Torch library and providing a more Pythonic interface. The framework was created with a focus on flexibility, ease of use, and dynamic computation graphs. One of its key features is dynamic computation, which allows for intuitive model construction and debugging. PyTorch gained popularity among researchers and developers due to its user-friendly API, extensive documentation, and support for advanced features like automatic differentiation. Over time, PyTorch has become one of the most widely used frameworks in the deep learning community, powering research projects, production systems, and educational initiatives.

TensorFlow, developed by Google Brain, was first released as an open source machine learning library in November 2015. This framework was designed to be scalable, flexible, and efficient, capable of handling both research and production workloads. It introduced the concept of computational graphs, allowing users to define complex neural network architectures and execute them efficiently on CPUs, GPUs, or TPUs. TensorFlow's initial release provided a low-level API for building and training deep learning models. However, to make deep learning more accessible, Google introduced Keras as a high-level API for TensorFlow in March 2017. Keras simplified the process of building and training neural networks, offering a more user-friendly interface while retaining TensorFlow's performance and scalability. The integration of Keras into TensorFlow contributed to its widespread adoption across various industries and academic institutions, making it one of the most popular deep learning frameworks in the world.

In this section, we provide an overview of the common steps between the two frameworks, while in Chapters 4–7, we delve deeper into the specific implementation details and demonstrate how to execute each step using both frameworks:

CHAPTER 1   OVERVIEW OF SCALABLE DEEP LEARNING PIPELINES ON AWS

- Data Loading and Preprocessing: Both PyTorch and TensorFlow require data loading and preprocessing steps to prepare the data for training. This may involve tasks such as reading data from files, normalization, and splitting the data into training and validation sets.

- Model Building: In both frameworks, we define the architecture of our neural network model. This includes specifying the number of layers, the type of layers (e.g., dense, convolutional, recurrent), activation functions, and any other relevant parameters.

- Defining Loss Function and Optimizer: Both PyTorch and TensorFlow allow us to choose a loss function based on the task at hand (e.g., Mean Squared Error (MSE) for regression, categorical cross-entropy for classification) and an optimizer for optimizing the model's parameters (e.g., stochastic gradient descent (SGD), Adam).

- Model Training: The training loop in both frameworks involves iterating over batches of data, passing them through the model to obtain predictions, calculating the loss, backpropagating the gradients, and updating the model parameters using the optimizer.

- Model Evaluation: After training, we evaluate the performance of the model on a separate validation or test dataset using appropriate metrics such as accuracy, precision, recall, or Mean Squared Error, depending on the task.

- Inference: Once the model is trained and evaluated, we can use it for making predictions on new, unseen data. This involves passing new data through the trained model and obtaining predictions.

CHAPTER 1   OVERVIEW OF SCALABLE DEEP LEARNING PIPELINES ON AWS

While the general workflow is similar between PyTorch and TensorFlow, there are differences in their APIs, syntax, and specific implementation details. For example:

- In PyTorch, we define models using the nn.Module class and build the forward pass in the forward method, whereas in TensorFlow, we use the tf.keras.Model class and define the forward pass in the call method.

- PyTorch emphasizes dynamic computation graphs, allowing for more flexibility and easier debugging, while TensorFlow originally used static computation graphs but has adopted a more dynamic approach with TensorFlow 2.0 and later versions.

- TensorFlow has a high-level API, Keras, which provides a simpler and more user-friendly interface for building and training neural network models, whereas PyTorch's API is more low-level and provides greater flexibility.

Below are examples of defining a neural network model in both PyTorch and TensorFlow/Keras to illustrate the differences in their APIs and syntax. We begin by defining a neural network model using the nn.Module in PyTorch, which specifies a feedforward network with one hidden layer (we examine how to use more layers in other chapters of the book):

```
[In]: import torch
[In]: import torch.nn as nn

[In]: class PyTorchModel(nn.Module):
        def __init__(self, input_size, hidden_size,
        output_size):
            super(PyTorchModel, self).__init__()
```

CHAPTER 1   OVERVIEW OF SCALABLE DEEP LEARNING PIPELINES ON AWS

```
# Input layer: Fully connected layer
self.fc1 = nn.Linear(input_size, hidden_size)

# Activation layer: ReLU function
self.relu = nn.ReLU()

# Output layer: Fully connected layer
self.fc2 = nn.Linear(hidden_size, output_size)

def forward(self, x):
# Applying the input layer
x = self.fc1(x)

# Applying the activation layer
x = self.relu(x)

# Applying the output layer
x = self.fc2(x)
return x
```

The equivalent code using tf.keras.Model in TensorFlow is as follows:

```
[In]: import tensorflow as tf
[In]: from tensorflow.keras import layers

[In]: class TensorFlowModel(tf.keras.Model):
        def __init__(self, hidden_size, output_size):
        super(TensorFlowModel, self).__init__()

        # Input layer: Fully connected (dense) layer
        self.fc1 = layers.Dense(hidden_size,
        activation='relu')

        # Output layer: Fully connected (dense) layer
        self.fc2 = layers.Dense(output_size)

        def call(self, inputs):
```

16

CHAPTER 1 OVERVIEW OF SCALABLE DEEP LEARNING PIPELINES ON AWS

```
# Applying the input layer
x = self.fc1(inputs)

# Applying the activation layer (ReLU)
x = tf.nn.relu(x)

# Applying the output layer
x = self.fc2(x)
return x
```

In the two examples, we can see the differences in how the neural network model is defined and used in PyTorch vs. TensorFlow/Keras:

- In PyTorch, we subclass nn.Module to define the model and override the forward method to specify the forward pass computation, while in TensorFlow/Keras, we subclass tf.keras.Model to define the model and implement the forward pass computation within the call method.

- PyTorch uses modules like nn.Linear and nn.ReLU to define layers, while TensorFlow/Keras uses tf.keras.layers.Dense and activation functions like tf.nn.relu.

We can see that the overall workflow and usage of the models are similar, but there are syntactic differences in how the models are defined and called.

This type of neural network architecture can be used for both regression and classification tasks (which are the focus of this book), but there will be some modifications needed based on the specific task requirements:

- Output Layer Activation Function: For regression tasks, the output layer should typically not have an activation function, or it can use a linear activation function to allow the network to output continuous values. For classification tasks, the output layer may use different activation functions depending on the number of classes. For binary classification, we might use a sigmoid activation function. For multi-class classification, we might use a softmax activation function.

- Loss Function: For regression tasks, we typically use loss functions like Mean Squared Error (MSE) or Mean Absolute Error (MAE). For classification tasks, we typically use cross-entropy loss (or Log-Loss) for binary classification or categorical cross-entropy loss for multi-class classification.

- Evaluation Metrics: The evaluation metrics used would depend on the task. For regression, we might use metrics like Mean Absolute Error (MAE) or Mean Squared Error (MSE). For classification, we might use metrics like accuracy, precision, recall, or F1 score.

Throughout this book, we will demonstrate how to use deep learning to predict continuous variables such as stock prices and binary outcomes such as diabetes diagnosis. These types of models are commonly used in finance and healthcare for their ability to analyze complex data and make accurate predictions. Moreover, it's worth noting that the same models can easily be modified to handle cases beyond finance and healthcare, demonstrating their versatility and applicability across various domains.

Understanding how to use deep learning for such prediction tasks provides practitioners (data scientists, machine learning engineers, researchers, and analysts) with foundational knowledge of neural networks. This includes concepts such as feedforward networks (the basic architecture of neural networks), backpropagation (the algorithm used to train neural networks), and gradient descent (the optimization algorithm used to minimize the loss function). It also serves as a solid basis for understanding specialized models designed for specific data types. For example, Convolutional Neural Networks (CNNs) are commonly used for image data in computer vision tasks, while Recurrent Neural Networks (RNNs) are used for sequential data such as time series or text. Despite their differences, these models share many underlying concepts and principles, including activation functions, weight initialization, and optimization techniques.

Model Training (EC2)

Once we have built the deep learning models using PyTorch or TensorFlow, the next step is to train them on computational resources provided by Amazon EC2 instances. EC2 (Elastic Compute Cloud) is a web service that provides resizable compute capacity in the cloud, allowing for efficient training of deep learning models.

EC2 instances offer a range of instance types optimized for different workloads, including those tailored for deep learning tasks. The instance types that EC2 provides can be grouped into a number of categories: general purpose, memory optimized, storage optimized, compute optimized, and accelerated computing. For all machine learning and deep learning applications, accelerated computing types of instances are the preferred choice.

Let's discuss each type in some detail:

General-Purpose Type:

General-purpose instances offer a blend of computing power, memory capacity, and network capabilities, catering to a wide array of workloads. They are particularly well-suited for tasks that demand a balanced utilization of these resources, such as hosting web servers and managing code repositories.

CHAPTER 1   OVERVIEW OF SCALABLE DEEP LEARNING PIPELINES ON AWS

The most prominent of these instances include

- M4: Fourth generation of general-purpose instances offered by Amazon EC2.

- M5, T3, T2: These instances belong to the "M5" and "T3" generations, which are the fifth and sixth generations, respectively. The "T2" instances are also part of the sixth generation, introduced alongside the "T3" instances.

- M6g, T4g: These instances belong to the "M6g" and "T4g" generations, respectively. They are the latest generation of general-purpose instances and are powered by AWS Graviton2 processors, offering improved performance and efficiency compared with previous generations.

Memory-Optimized Type:

Memory-optimized instances are designed to deliver fast performance for workloads that process large datasets in memory. They can be divided into two categories based on their memory capacity and performance characteristics:

- Instances with moderate to large memory sizes and balanced performance characteristics suitable for general-purpose workloads—examples include

  - R4
  - R5
  - R6g

CHAPTER 1 OVERVIEW OF SCALABLE DEEP LEARNING PIPELINES ON AWS

- Instances with large memory sizes and high-performance capabilities, ideal for memory-intensive applications and demanding workloads—examples include

  - R7g
  - R8g
  - X1e
  - X2gd

Storage-Optimized Type:

Storage-optimized instances are crafted for tasks demanding extensive, sequential access to huge datasets stored locally. Engineered to furnish tens of thousands of low-latency, random I/O operations per second (IOPS), these instances cater to applications with stringent performance requirements. They can be divided into the following categories:

- General Storage–Optimized Instances (designed for tasks requiring high-performance access to large datasets stored locally):

  - I4g
  - I4i
  - I3
  - D2

CHAPTER 1   OVERVIEW OF SCALABLE DEEP LEARNING PIPELINES ON AWS

- Enhanced Storage–Optimized Instances (offering enhanced performance and capabilities for demanding storage-intensive workloads):
  - Im4gn
  - Is4gen
  - I3en
  - D3
  - D3n
  - H1

Compute-Optimized Type:

Compute-optimized instances excel in powering compute-intensive applications, leveraging high-performance processors to deliver exceptional speed and efficiency. These instances are tailor-made for tasks that heavily rely on computational prowess, including batch processing, media transcoding, scientific modeling, HPC (High Performance Computing), dedicated gaming servers, ad server engines, machine learning inference, and other demanding compute-intensive workloads.

These instances can be grouped into the following categories:

- Advanced Compute-Optimized Instances (with cutting-edge technologies and high-performance processors optimized for demanding compute-intensive workloads):
  - C7g
  - C7gn
  - C7i
  - C7a

CHAPTER 1   OVERVIEW OF SCALABLE DEEP LEARNING PIPELINES ON AWS

- Standard Compute–Optimized Instances (offering a balance of performance and cost-effectiveness for compute-intensive applications):
  - C6g
  - C6gn
  - C6i
  - C6in
  - C6a
- Legacy Compute–Optimized Instances (from previous generations, still offering reliable performance for certain compute-intensive tasks):
  - C5
  - C5n
  - C5a
  - C4

Accelerated Computing Type:

Accelerated computing instances harness the power of hardware accelerators or co-processors to execute tasks such as floating-point number calculations, graphics processing, or data pattern matching with heightened efficiency compared with software running on conventional CPUs. For all machine learning and deep learning applications, these types of instances are the preferred choice. Renowned for their exceptional speed and precision, they are well-suited to handling complex computational tasks. These instances can be categorized into the following types:

CHAPTER 1   OVERVIEW OF SCALABLE DEEP LEARNING PIPELINES ON AWS

- Graphics Processing Units (GPU)-Based Instances (equipped with GPU accelerators, optimized for parallel processing tasks such as machine learning, deep learning, and graphics processing):
  - P2
  - P3
  - P4
  - P5
  - G5
  - G5g
  - G4dn
  - G4ad
  - G3
- Tensor Processing Units (TPU) Instances (featuring TPUs, specialized hardware accelerators developed by Google for machine learning tasks):
  - Trn1
- Inference Accelerator Instances (optimized for inference tasks in machine learning models, offering high throughput and low latency):
  - Inf2
  - Inf1

CHAPTER 1   OVERVIEW OF SCALABLE DEEP LEARNING PIPELINES ON AWS

- Deep Learning Accelerator Instances (tailored specifically for deep learning workloads, providing optimized performance for training and inference tasks):
  - DL1
  - DL2q
- FPGA-Based Instances (instances featuring FPGAs, customizable hardware accelerators suitable for various computational tasks):
  - F1
- Video Transcoding Instances (optimized for video transcoding applications, offering efficient processing of multimedia content):
  - VT1

When choosing the appropriate EC2 instance type for deep learning models built using PyTorch or TensorFlow, several factors should be considered to ensure optimal performance and efficiency. For deep learning tasks, especially those involving training and evaluation of complex models, accelerated computing instances are often preferred due to their exceptional speed and precision. These instances leverage specialized hardware accelerators or co-processors to execute tasks such as floating-point number calculations, graphics processing, and data pattern matching with heightened efficiency compared with traditional CPUs. Instances equipped with GPUs (Graphics Processing Units), such as the NVIDIA Tesla V100, are well-suited for accelerating deep learning training due to their parallel processing capabilities.

Additionally, instances with TPUs (Tensor Processing Units), such as Trn1, are available for even faster training of certain types of models. GPU-based instances, such as P2, P3, P4, P5, G5, and others, are particularly

CHAPTER 1   OVERVIEW OF SCALABLE DEEP LEARNING PIPELINES ON AWS

well-suited for parallel processing tasks inherent in deep learning, including training and inference. Therefore, for deep learning models using PyTorch or TensorFlow, opting for GPU-based or TPU instances ensures accelerated performance and efficient training and evaluation processes on Amazon EC2.

Training deep learning models on EC2 instances offers several advantages over training on traditional on-premises servers. Firstly, EC2 instances can be scaled up or down based on workload demands, allowing users to adjust compute resources as needed. This scalability ensures that models of varying complexities and sizes can be trained efficiently without being constrained by hardware limitations.

Furthermore, EC2 instances provide access to a wide range of pre-configured machine images (Amazon Machine Images (AMIs)) and deep learning frameworks, simplifying the setup and configuration process for training environments. Users can choose from popular deep learning frameworks such as TensorFlow and PyTorch and leverage pre-built AMIs optimized for these frameworks.

To set up or provision an EC2 instance for your deep learning pipeline, you need to configure the instance to meet specific requirements, such as selecting the instance type, configuring storage, setting up networking, and applying security settings.

Below are the key steps in EC2 instance provisioning:

- Choosing an AMI (Amazon Machine Image):

    - The AMI is a pre-configured template that contains the operating system, application server, and applications required to launch an instance. You can choose from a variety of public AMIs (such as Amazon Linux, Ubuntu, Windows) or create your own custom AMI.

- Selecting an Instance Type:
  - The instance type determines the hardware of the host computer used for your instance. Different instance types offer varying combinations of CPU, memory, storage, and networking capacity. For example, t2.micro is a general-purpose instance type, while p3.2xlarge is optimized for machine learning and GPU workloads.

- Configuring Storage:
  - You configure the storage for your instance by specifying the type and size of the attached Elastic Block Store (EBS) volumes. For instance, you might choose a 16 GB gp3 volume for a general-purpose workload.

- Setting Up Networking:
  - You must configure the networking for your instance, including selecting a Virtual Private Cloud (VPC), choosing a subnet, and deciding whether the instance should have a public IP address. The instance must also be placed in an appropriate security group that defines inbound and outbound traffic rules.

- Applying Security Settings:
  - Security settings include assigning a key pair for SSH access to the instance and applying security group rules to control traffic to and from the instance. The security group acts as a virtual firewall.

CHAPTER 1   OVERVIEW OF SCALABLE DEEP LEARNING PIPELINES ON AWS

- Launching the Instance:
  - Once all configurations are specified, the EC2 instance is launched. This process involves allocating the necessary resources, deploying the operating system, and setting up the networking and security configurations.
- Post-launch Configuration:
  - After the instance is running, you would need to perform additional setup tasks, such as installing software, configuring applications, and connecting to other AWS services.

In the next chapter, we explore how to provision an EC2 instance both manually and through Amazon CloudFormation, as well as how to install and configure the necessary software and applications.

Model Evaluation (EC2)

Trained models are evaluated to assess their performance and generalization ability using validation datasets. Evaluation metrics are used to measure how well the models have learned from the data and generalize to unseen examples.

For regression tasks, common measures include the Mean Squared Error (MSE) and R-squared. The MSE measures the average squared difference between the predicted values and the actual values. R-squared measures the proportion of the variance in the dependent variable that is predictable from the independent variables, providing an indication of the model's goodness of fit.

For classification, the following evaluation metrics are commonly used:

- Accuracy: This measures the proportion of correctly classified instances (both true positives and true negatives) out of the total instances.

- Precision: This measures the proportion of true positive predictions out of all positive predictions, focusing on the accuracy of positive predictions.

- Recall: This measures the proportion of true positive predictions out of all actual positive instances, focusing on the model's ability to identify all positive instances.

- F1 Score: This is the harmonic or weighted mean of precision and recall, providing a balance between precision and recall, especially when dealing with imbalanced classes.

- ROC-AUC: Receiver Operating Characteristic Area Under the Curve (ROC-AUC) measures the area under the ROC curve, which plots the true positive rate (sensitivity) against the false positive rate (1 – specificity). It provides an aggregate measure of the model's ability to discriminate between positive and negative classes across different thresholds.

- Matthews Correlation Coefficient (MCC): MCC is a correlation coefficient between the observed and predicted binary classifications, considering all four confusion matrix values (true positives, true negatives, false positives, false negatives). It ranges from –1 to 1, with 1 indicating perfect predictions, 0 indicating random predictions, and –1 indicating total disagreement between predictions and observations.

CHAPTER 1   OVERVIEW OF SCALABLE DEEP LEARNING PIPELINES ON AWS

The confusion matrix provides the necessary components for calculating metrics like MCC. It is a tabular representation of actual vs. predicted class labels, allowing for a detailed analysis of the model's performance.

We will be using some of these metrics in subsequent chapters. More specifically, in Chapters 4 and 6, we will compute MSE and R-squared to evaluate deep learning regression models, while in Chapters 5 and 7, we will compute accuracy, precision, recall, and F1 to evaluate deep learning classification models.

Model Deployment (Airflow)

Apache Airflow is utilized to orchestrate the deployment of trained models onto production environments, scheduling and managing tasks related to model deployment to ensure a smooth and reliable deployment process. These tasks may include versioning models, setting up APIs for inference, and monitoring model performance in production.

Originally developed by Airbnb in 2014 to manage complex workflows and data pipelines, Apache Airflow was later open-sourced in 2015, enabling contributions from developers worldwide. Since then, it has become a prominent project within the Apache Software Foundation. It gained popularity due to its flexibility, scalability, and robustness in managing workflows and pipelines. Its features, such as task scheduling, dependency management, and extensibility, appealed to a wide range of industries and use cases beyond just data engineering.

Additionally, the active community around Airflow contributed to its growth by continually adding new features, improving documentation, and providing support. As organizations increasingly adopted data-driven approaches and machine learning technologies, the need for reliable orchestration tools like Airflow became more pronounced, further contributing to its popularity.

CHAPTER 1   OVERVIEW OF SCALABLE DEEP LEARNING PIPELINES ON AWS

In this section, we show the reader how to orchestrate their first Directed Acyclic Graph (DAG). In Chapter 8, we will create an entire AWS pipeline that includes preprocessing data using PySpark; building, training, and evaluating models using PyTorch and TensorFlow; and deploying the models using Airflow.

Before orchestrating the DAG, let's first demonstrate how Python code is typically executed. For illustration, we'll use PySpark to create a DataFrame, perform operations on it, and calculate the average price of tacos:

```
[In]: from pyspark.sql import SparkSession
[In]: from pyspark.sql.types import (
        StructType, StructField, StringType, FloatType
      )
[In]: from pyspark.sql.functions import avg
[In]: def create_taco_dataframe(Spark):
        """
        Create a DataFrame containing taco data.

        Parameters:
            spark (SparkSession): The SparkSession object.

        Returns:
            DataFrame: DataFrame containing taco data.
        """
        schema = StructType([
            StructField("Taco", StringType(), True),
            StructField("Price", FloatType(), True)
        ])

        data = [("Chicken Taco", 2.50),
                ("Beef Taco", 3.00),
                ("Fish Taco", 3.50),
                ("Vegetarian Taco", 2.00)]
```

```
        taco_df = spark.createDataFrame(data, schema)
        return taco_df
```

[In]:
```
def calculate_average_price(taco_df):
    """
    Calculate the average price of tacos.

    Parameters:
        taco_df (DataFrame): DataFrame containing
        taco data.

    Returns:
        float: Average price of tacos.
    """
    avg_price = taco_df.select(avg("Price")).first()[0]
    return avg_price
```

[In]:
```
def main():
    """
    Main function to demonstrate PySpark DataFrame
    operations.
    """
    spark = (SparkSession.builder
            .appName("TacoPrices")
            .getOrCreate())

    taco_df = create_taco_dataframe(spark)

    taco_df.show()

    avg_price = calculate_average_price(taco_df)
    print("Average Price of Tacos:", avg_price)

    spark.stop()
```

CHAPTER 1   OVERVIEW OF SCALABLE DEEP LEARNING PIPELINES ON AWS

```
[In]: if __name__ == "__main__":
          main()
[Out]:
```

Taco	Price
Chicken Taco	2.5
Beef Taco	3.0
Fish Taco	3.5
Vegetarian Taco	2.0

[Out]: Average Price of Tacos: 2.75

The preceding code goes through a number of steps: importing necessary modules, defining functions, creating a main function, and conditional execution. Below is an explanation of what the code does at each step:

- Importing Necessary Modules: The code begins by importing required modules from PySpark, including SparkSession for creating a Spark application; StructType, StructField, StringType, and FloatType for defining the schema of the DataFrame; and avg function from pyspark.sql.functions for calculating the average price.

- Defining Functions: Two functions are defined:
  - create_taco_dataframe(spark): This function creates a DataFrame containing taco names and their prices. It defines a schema with two fields: "Taco" of type StringType and "Price" of type FloatType. Then, it creates a list of tuples with taco data and creates a DataFrame using the createDataFrame method of SparkSession.

- calculate_average_price(taco_df): This function calculates the average price of tacos by selecting the "Price" column from the DataFrame taco_df, applying the avg function, and retrieving the result.
- Main Function: The main() function is defined to execute the main logic of the program. Inside the main function
  - SparkSession is initialized with the application name "TacoPrices".
  - The create_taco_dataframe() function is called to create a DataFrame containing taco data.
  - The show() method is called on the DataFrame to display its contents.
  - The calculate_average_price() function is called to calculate the average price of tacos.
  - The average price is printed.
  - Finally, SparkSession is stopped to release the resources.
- Conditional Execution: The code block if __name__ == "__main__" ensures that the main() function is executed only if the script is run directly (not imported as a module).

Now, let's migrate this code to Apache Airflow. The code below sets up an Apache Airflow DAG to automate the process of creating a DataFrame containing taco data with PySpark, calculating the average price of tacos, and saving the result to a Parquet file. The DAG is configured with default arguments, task dependencies, and other parameters to define the workflow's behavior:

CHAPTER 1 OVERVIEW OF SCALABLE DEEP LEARNING PIPELINES ON AWS

```
[In]: from airflow import DAG
[In]: from airflow.operators.python_operator import
      PythonOperator
[In]: from datetime import datetime
[In]: from pyspark.sql import SparkSession
[In]: from pyspark.sql.types import (
          StructType, StructField, StringType, FloatType
      )
[In]: from pyspark.sql.functions import avg

[In]: def create_taco_dataframe(**context):
          """
          Create a DataFrame containing taco data and save it
              to a Parquet file.

          Parameters:
              context (dict):
                  The context dictionary containing task
                  metadata.

          Returns:
              None
          """
          schema = StructType([
              StructField("Taco", StringType(), True),
              StructField("Price", FloatType(), True)
          ])

          data = [("Chicken Taco", 2.50),
                  ("Beef Taco", 3.00),
                  ("Fish Taco", 3.50),
                  ("Vegetarian Taco", 2.00)]
```

CHAPTER 1   OVERVIEW OF SCALABLE DEEP LEARNING PIPELINES ON AWS

```
        spark = SparkSession.builder.getOrCreate()
        taco_df = spark.createDataFrame(data, schema)

        taco_df.show()

        taco_df.write.parquet(
            "/home/ubuntu/airflow/dags/taco_prices.parquet",
            mode="overwrite"
        )
[In]: def calculate_average_price(**context):
        """
        Calculate the average price of tacos from the
        DataFrame.

        Parameters:
            context (dict):
                The context dictionary containing task
                metadata.

        Returns:
            None
        """
        spark = SparkSession.builder.getOrCreate()
        taco_df = spark.read.parquet(
            "/home/ubuntu/airflow/dags/taco_prices.parquet"
        )
        avg_price = taco_df.select(avg("Price")).first()[0]
        print("Average Price of Tacos:", avg_price)
[In]: default_args = {
        'owner': 'airflow',
        'start_date': datetime(2024, 3, 20),
        'retries': 1
    }
```

```
[In]: dag = DAG('taco_prices_dag',
            default_args=default_args,
            description='DAG to calculate average taco
            prices',
            schedule_interval=None,
            catchup=False
      )

[In]: create_taco_dataframe_task = PythonOperator(
          task_id='create_taco_dataframe_task',
          python_callable=create_taco_dataframe,
          provide_context=True,
          dag=dag
      )

[In]: calculate_average_price_task = PythonOperator(
          task_id='calculate_average_price_task',
          python_callable=calculate_average_price,
          provide_context=True,
          dag=dag
      )

[In]: create_taco_dataframe_task >> calculate_average_
      price_task
```

To run the DAG, we follow the steps below:

1. Activate the Virtual Environment: Before running the DAG, activate the virtual environment where Apache Airflow and its dependencies are installed. This ensures that the correct Python environment is used for executing the DAG tasks. In this project, we have created an environment named "myenv":

    ```
    [In]: source /home/ubuntu/myenv/bin/activate
    ```

CHAPTER 1   OVERVIEW OF SCALABLE DEEP LEARNING PIPELINES ON AWS

This command activates a Python virtual environment named "myenv" located at /home/ubuntu/myenv. When we activate a virtual environment, it modifies the shell environment to use the Python interpreter and packages associated with that specific virtual environment. This ensures that any Python commands executed in that shell session will use the version of Python and the installed packages within the virtual environment, rather than the global system Python environment.

2. Navigate to the DAG Directory: After activating the virtual environment, navigate to the directory where your DAG script is located. This is typically the directory where your Apache Airflow DAG files are stored. In this project, we are using the airflow/dags directory:

```
[In]: cd /home/ubuntu/airflow/dags
```

This command changes the current directory to the directory where the DAG script (taco_prices_dag.py) is saved.

3. Run Apache Airflow Web Server and Scheduler: Once we are in the directory containing the DAG script, we start the Apache Airflow web server and scheduler to initialize the Airflow environment:

```
[In]: airflow webserver &
[In]: airflow scheduler &
```

These commands start the Airflow web server and scheduler processes in the background. The web server provides the Airflow user interface, while the scheduler orchestrates the execution of DAGs and tasks.

CHAPTER 1   OVERVIEW OF SCALABLE DEEP LEARNING PIPELINES ON AWS

4. Trigger the DAG: After the web server and scheduler are running, trigger the DAG using the Airflow command-line interface (CLI):

   [In]: airflow dags trigger taco_prices_dag

   This command triggers the execution of the taco_prices_dag DAG.

Before diving deeper into the code and explaining each component of the DAG in more detail, it's important to note that Apache Airflow offers several deployment options, each suited to different environments and requirements. The one discussed here is the standalone deployment option, where Airflow, as will be examined in more detail in the next chapter ("Setting Up a Deep Learning Environment on AWS"), is installed and run directly on a single machine. This setup is straightforward and ideal for testing. It is popular among individual developers and small teams working on less complex DAGs. Although standalone Airflow is easy to set up and requires fewer resources, it comes with limitations such as limited scalability, making it less suitable for larger production environments.

In the next chapter, we also examine the second option—a Docker containerized setup, where Airflow runs in isolated Docker containers. This method provides more flexibility and scalability than a standalone deployment. Using Docker, Airflow and its dependencies are packaged into containers, ensuring consistent environments across development, staging, and production. This setup is ideal for teams that need to manage dependencies, avoid conflicts, and easily scale their Airflow environment by adding more containers.

In Chapter 10, we explore a third option: Amazon Managed Workflows for Apache Airflow (MWAA), a fully managed service provided by AWS. MWAA takes care of the setup, scaling, and maintenance of Airflow, allowing users to focus solely on writing and running their DAGs. This option is best suited for production environments that require high availability, reliability, and seamless integration with other AWS services.

CHAPTER 1   OVERVIEW OF SCALABLE DEEP LEARNING PIPELINES ON AWS

Let's now dive deeper into the code and explain each component of the DAG in more detail:

1. Import Statements:

    ```
    [In]: from airflow import DAG
    [In]: from airflow.operators.python_operator import
              PythonOperator
    [In]: from datetime import datetime
    [In]: from pyspark.sql import SparkSession
    [In]: from pyspark.sql.types import (
              StructType, StructField, StringType,
              FloatType
    )
    [In]: from pyspark.sql.functions import avg
    ```

    These import statements bring in necessary modules and classes for creating Apache Airflow DAGs, defining PythonOperator tasks, handling datetime objects, and working with PySpark. When using the `from airflow.operators.python_operator import PythonOperator` statement, ensure that your Airflow version is compatible with the PythonOperator import. The import path for PythonOperator may differ depending on your Airflow version. For example, in Airflow 2.x, the import path has changed from `airflow.operators.python_operator` to `airflow.operators.python`. If you are using Airflow 2.x or later, update the import statement as follows:

CHAPTER 1   OVERVIEW OF SCALABLE DEEP LEARNING PIPELINES ON AWS

```
[In]: from airflow.operators.python import
PythonOperator
```

2. Task Functions:

```
[In]: def create_taco_dataframe(**context):
          """
          Create a DataFrame containing taco data and
          save it to a Parquet file.

          Parameters:
              context (dict):
                  The context dictionary containing
                  task metadata.

          Returns:
              None
          """
          schema = StructType([
              StructField("Taco", StringType(), True),
              StructField("Price", FloatType(), True)
          ])

          data = [("Chicken Taco", 2.50),
                  ("Beef Taco", 3.00),
                  ("Fish Taco", 3.50),
                  ("Vegetarian Taco", 2.00)]

          spark = SparkSession.builder.getOrCreate()
          taco_df = spark.createDataFrame(data, schema)

          taco_df.show()
```

```
         taco_df.write.parquet(
             "/home/ubuntu/airflow/dags/taco_prices.
             parquet",
              mode="overwrite"
         )
[In]: def calculate_average_price(**context):
      """
      Calculate the average price of tacos from the
      DataFrame.

      Parameters:
          context (dict):
              The context dictionary containing
              task metadata.

      Returns:
          None
      """
      spark = SparkSession.builder.getOrCreate()
      taco_df = spark.read.parquet(
          "/home/ubuntu/airflow/dags/taco_prices.
          parquet"
      )
      avg_price = taco_df.select(avg("Price")).
      first()[0]
      print("Average Price of Tacos:", avg_price)
```

CHAPTER 1   OVERVIEW OF SCALABLE DEEP LEARNING PIPELINES ON AWS

The code creates two task functions:

- create_taco_dataframe: This function creates a PySpark DataFrame containing taco data with two columns, "Taco" and "Price". It then saves this DataFrame to a Parquet file. Parquet is a columnar storage format that is highly efficient for storing and querying structured data. It offers benefits such as compression, efficient encoding, and native support for nested data structures. Storing data in Parquet format is advantageous for big data processing tasks as it improves query performance, reduces storage costs, and facilitates compatibility with various data processing frameworks like Apache Spark and Hadoop.

- calculate_average_price: This function calculates the average price of tacos from the DataFrame saved in the Parquet file.

3. Default Arguments:

```
[In]: default_args = {
          'owner': 'airflow',
          'start_date': datetime(2024, 3, 20),
          'retries': 1
      }
```

default_args defines default configuration parameters for the DAG, including the owner of the DAG, the start date for the DAG runs, and the number of retries for each task.

CHAPTER 1   OVERVIEW OF SCALABLE DEEP LEARNING PIPELINES ON AWS

4. DAG Definition:

```
[In]: dag = DAG('taco_prices_dag',
                default_args=default_args,
                description='DAG to calculate average taco prices',
                schedule_interval=None,
                catchup=False
            )
```

This section defines a new DAG named 'taco_prices_dag' with the specified default_args, description, schedule interval (None indicates the DAG is not scheduled and should be triggered manually), and catchup setting (set to False to prevent backfilling missed DAG runs).

We can schedule the DAG to run automatically, by changing schedule_interval from None to '@daily' as follows:

```
[In]: dag = DAG('taco_prices_dag',
                default_args=default_args,
                description='DAG to calculate average taco prices',
                schedule_interval='@daily',
                catchup=False
            )
```

The @daily cron expression specifies that the DAG should run once every day. So, instead of requiring manual triggering, the DAG will now run automatically according to the specified schedule. If

CHAPTER 1   OVERVIEW OF SCALABLE DEEP LEARNING PIPELINES ON AWS

we were to run it weekly, monthly, or quarterly, the cron expression would be modified to @weekly, @monthly, or @quarterly, respectively.

5. Task Operators:

   Two PythonOperator tasks are defined within the DAG:

   - create_taco_dataframe_task: This task calls the create_taco_dataframe function.

   - calculate_average_price_task: This task calls the calculate_average_price function.

6. Provide Context:

   [In]: provide_context=True

   provide_context=True is set for both tasks to pass the Airflow context dictionary containing task metadata to the task functions. This allows access to information such as the DAG run ID and task execution date.

7. Task Dependencies:

   [In]: create_taco_dataframe_task >>
           calculate_average_price_task

   The dependency between tasks is established using the >> operator, indicating that calculate_average_price_task should run after create_taco_dataframe_task.

Once the DAG taco_prices_dag runs successfully, we would typically observe the following outputs on the Airflow user interface (UI):

45

CHAPTER 1   OVERVIEW OF SCALABLE DEEP LEARNING PIPELINES ON AWS

Task Execution Logs:

- You would see logs generated by each task, indicating its execution status, any errors encountered, and any output generated during execution.

- For example, the create_taco_dataframe_task would display logs showing the creation of the DataFrame containing taco data and the process of saving it to a Parquet file.

- Similarly, the calculate_average_price_task would display logs showing the calculation of the average price of tacos from the DataFrame saved in the Parquet file.

Task Status:

- The status of each task would be updated in the Apache Airflow UI, indicating whether the task was executed successfully (success), encountered an error (failed), or is currently running (running).

- You would be able to monitor the progress of each task and identify any issues that may have occurred during execution.

Output Data:

- If the tasks were executed successfully, you would have output data generated by the DAG's tasks.

- For example, after the create_taco_dataframe_task runs, you would have a Parquet file (taco_prices.parquet) containing the DataFrame with taco data saved to the specified location (/home/ubuntu/airflow/dags/).

CHAPTER 1 OVERVIEW OF SCALABLE DEEP LEARNING PIPELINES ON AWS

- The calculate_average_price_task would output the average price of tacos, which would be printed to the task logs and visible in the Apache Airflow UI.

Execution Metadata:

- Apache Airflow maintains metadata about each DAG run, including details such as the start time, end time, duration, and any relevant metadata associated with each task execution.

- You would be able to access this metadata in the Apache Airflow UI or through Airflow's backend database.

Leveraging Apache Airflow provides a more scalable and manageable approach to workflow orchestration, made possible by replacing the main() function in the standalone script with the Airflow DAG definition.

Let's discuss how this replacement happens:

- Task Functions: Each logical step or operation performed within the main() function becomes a separate task in the Airflow DAG. For example, the create_taco_dataframe function and calculate_average_price function would become individual tasks in the DAG.

- DAG Definition: Instead of having a main() function, we define a DAG object in Airflow. This DAG object encapsulates the workflow, including task dependencies, scheduling, and other parameters. The DAG definition provides a structured way to represent and manage the workflow.

CHAPTER 1   OVERVIEW OF SCALABLE DEEP LEARNING PIPELINES ON AWS

- Task Dependencies: In the standalone script, the sequence of function calls within the main() function determines the execution order. In the Airflow DAG, task dependencies are explicitly defined using operators like >> to indicate which tasks depend on others. This ensures that tasks are executed in the correct order.

- Execution Context: In the standalone script, functions are executed sequentially within the context of the main() function. In Airflow, each task is executed independently within the context of a DAG run. Airflow manages task execution, scheduling, retries, and error handling automatically based on the DAG definition.

- Triggering Execution: While the main() function is typically triggered by directly running the script, the execution of an Airflow DAG can be triggered manually, based on a schedule, or in response to an external event. This triggering is done using Airflow's command-line interface (CLI) or web interface or programmatically through its API.

The above indicates that the replacement of the main() function with an Airflow DAG provides a more scalable and manageable approach to workflow orchestration. Airflow's DAG-based approach allows for better monitoring, scheduling, and management of complex workflows, especially in data processing pipelines and ETL (Extract, Transform, Load) tasks.

Overall, the entire pipeline (starting with the initial steps of data ingestion using S3 and concluding with scheduling the model to run on Apache Airflow) offers several advantages:

- Firstly, it capitalizes on the scalability of PySpark for large-scale data processing and the compute capacity of EC2 instances for model training, ensuring efficient resource utilization.

- Secondly, a cloud-based approach provides flexibility, allowing resource scaling based on workload demands and easy provisioning of resources.

- Thirdly, PySpark and Apache Airflow simplify data processing tasks with their distributed computing capabilities, enabling efficient preprocessing, transformation, and workflow orchestration.

- Fourthly, the large storage capacity of Amazon S3 facilitates uninterrupted data storage and retrieval, supporting the processing of massive datasets.

- Lastly, the separation of concerns among preprocessing (PySpark); model development, training, and evaluation (PyTorch, TensorFlow); model orchestration (Apache Airflow); storage (S3); and compute (EC2) simplifies the architecture and enhances modularity, facilitating streamlined workflow management.

Project Directory Structure

When starting a new project, one of the first considerations is how to organize the files and directories that will make up the project. A thoughtfully designed project directory structure offers several benefits, including organization, collaboration, reproducibility, and documentation:

CHAPTER 1   OVERVIEW OF SCALABLE DEEP LEARNING PIPELINES ON AWS

- Organization: A clear and logical structure helps keep track of all project files, making it easier to find what is needed and avoid clutter.

- Collaboration: When working with a team, a standardized directory layout ensures consistency across contributors and reduces confusion about where to find specific files.

- Reproducibility: A well-organized project directory makes it easier to reproduce experiments or analyses, as all necessary data, code, and outputs are neatly organized and documented.

- Documentation: The directory structure itself serves as a form of documentation, providing insights into the project's architecture, components, and workflow.

The ideal directory structure should possess the following attributes:

- Intuitive and easy to navigate.

- It avoids unnecessary nesting of directories or overly complex hierarchies.

- It should be divided into logical sections based on the type of content.

- It uses meaningful names for directories and files that reflect their contents and purpose.

- It includes README files or documentation within directories to provide additional context and guidance on how to use the files contained within.

Figure 1-2 is an example of how a project predicting Tesla stock price might be structured.

CHAPTER 1   OVERVIEW OF SCALABLE DEEP LEARNING PIPELINES ON AWS

```
Project Directory Structure
├── data
│   └── TSLA_stock.csv
├── logs
├── output
│   ├── TSLA_stock.parquet
│   │   ├── _SUCCESS
│   │   └── part-00000-2c717885-1e92-4091-b73e-c07fb2f42ba8-c000.snappy.parquet
│   └── TSLA_stock_train.parquet
│       ├── _SUCCESS
│       └── part-00000-3931d223-097a-4b7f-a6df-6ddc8abfcb5c-c000.snappy.parquet
├── src
│   ├── pyspark_code
│   │   └── data_preprocessing.py
│   ├── pytorch_code
│   │   ├── model.py
│   │   └── model_training.py
│   └── main.py
├── tests
│   └── test.py
├── visualizations
│   └── plot.png
├── README.md
├── config.yaml
└── requirements.txt
```

Figure 1-2. Project Directory Structure for Predicting Tesla Stock Price

To understand the organization of the Tesla stock prediction project, let's examine the directory structure in detail:

- data: This directory contains the input data for the project. In this case, it's a CSV file named TSLA_stock.csv, containing historical stock data for Tesla.

- logs: This directory might contain log files generated during the execution of the project. It's common practice to separate logs from other project files for easier management.

- output: This directory stores the output files generated by the project. It contains two subdirectories:
    - TSLA_stock.parquet: This directory holds Parquet files, representing processed data related to Tesla stock.
    - TSLA_stock_train.parquet: Similar to the previous directory, this one contains Parquet files representing training data for a machine learning model.
- src: This directory contains the source code for the project. It's further organized into subdirectories:
    - pyspark_code: Contains Python files related to PySpark data preprocessing used for data cleaning or transformation.
    - pytorch_code: Contains Python files for PyTorch model development and training.
    - main.py: This is the main entry point for the project, where different components of the project are orchestrated.
- tests: This directory holds test files for the project. In this case, there's a Python file named test.py, which contains unit tests for the project's components.
- visualizations: This directory stores visualization outputs generated by the project. It contains a PNG file named plot.png, which is a plot or chart related to the Tesla stock analysis.

CHAPTER 1   OVERVIEW OF SCALABLE DEEP LEARNING PIPELINES ON AWS

- README.md: This is a markdown file providing information about the project. It typically includes instructions for setting up the project, an overview of its purpose, and other relevant details.

- requirements.txt: This file lists the dependencies required to run the project. It typically includes Python packages and their versions needed for the project's execution.

- config.yaml: This file contains configuration settings for the project. It could include parameters for data preprocessing, model training, or other aspects of the project. Below is a basic example of a config.yaml file that could be added to the Tesla project for representation and overview of the structure:

```
[In]: project:
         name: Tesla Stock Prediction
         version: 1.0
         author: Abdelaziz Testas

[In]: data:
         input_path: data/TSLA_stock.csv
         output_path: output/TSLA_stock.parquet
         train_output_path: output/TSLA_stock_train.parquet

[In]: preprocessing:
         missing_values_strategy: fillna
         fillna_value: 0
         scaling_method: standard_scaler

[In]: training:
```

```
          model_type: pytorch
          batch_size: 32
          epochs: 50
          learning_rate: 0.001
          train_split: 0.8
          random_seed: 42
[In]: logging:
          log_level: INFO
          log_file_path: logs/project.log
[In]: visualization:
          plot_output_path: visualizations/plot.png
          plot_type: line
          x_axis: Date
          y_axis: Close
[In]: aws_s3:
          bucket_name: instance1bucket
          region: us-west-2
          input_file_key: TSLA_stock.csv
          output_file_key: TSLA_stock_output.parquet
```

Virtual Environment

In addition to creating a well-defined project structure, it's also recommended to create a virtual environment (e.g., myenv) for the project to manage dependencies and ensure reproducibility across different environments. This isolates the project's dependencies from other Python projects on the system, preventing conflicts and ensuring that the project can be run in a consistent environment regardless of the system's configuration. The next chapter explains how to create such an environment, as well as the project directory structure shown above.

CHAPTER 1   OVERVIEW OF SCALABLE DEEP LEARNING PIPELINES ON AWS

Development Environment

After examining the components of a complete deep learning workflow on AWS and discussing the importance of designing a project directory structure and setting up a virtual environment on EC2, let's delve into where we actually write the code and develop our machine learning models.

For the purpose of this book's project, we will primarily utilize JupyterLab. However, there are other popular options for deep learning development environments, including Jupyter Notebook, PyCharm, and Visual Studio Code (VS Code). Databricks is another notable option, particularly well-suited for Spark-based workflows and large-scale data processing. It provides a unified analytics platform that includes collaborative notebooks, data engineering tools, and machine learning capabilities, all within a scalable cloud environment. While Databricks excels in big data processing tasks, it may not be the first choice for deep learning–specific tasks compared with other environments like JupyterLab, Jupyter Notebook, PyCharm, or VS Code, which offer more specialized features for machine learning model development.

JupyterLab and Jupyter Notebook are both part of the Jupyter ecosystem, offering interactive computing environments for data science and machine learning tasks. The main difference between the two lies in their user interfaces and capabilities.

Jupyter Notebook provides a classic interface where users can create and run code cells, view outputs, and document their workflows using Markdown cells. It offers a straightforward and intuitive environment for prototyping and experimentation. On the other hand, JupyterLab is a more advanced and extensible interface that builds upon the capabilities of Jupyter Notebook. It provides a flexible and integrated environment with features such as multiple panels, tabbed workspaces, and drag-and-drop functionality. JupyterLab offers enhanced support for large-scale projects, allowing users to work with multiple notebooks, text files, terminals, and

CHAPTER 1   OVERVIEW OF SCALABLE DEEP LEARNING PIPELINES ON AWS

custom extensions simultaneously. Additionally, JupyterLab's modular architecture enables users to customize their environment with extensions for specific tasks or workflows.

For quick tests or small code snippets, using a lightweight text editor like Nano to create Python scripts can be sufficient.

In the next chapter, we will demonstrate how to set up some of these tools to create an efficient and productive deep learning development environment.

Summary

In this chapter, we examined a comprehensive deep learning pipeline leveraging the power of Amazon Web Services (AWS) to handle large-scale data processing, model training, and deployment tasks. At the heart of this pipeline lies Amazon S3, serving as the centralized data source for storing raw datasets, preprocessed data, and model artifacts. Data preprocessing is facilitated by PySpark, enabling distributed data processing on large datasets stored in S3. The pipeline utilizes both PyTorch and TensorFlow for building and training deep learning models, harnessing their flexibility and scalability. Amazon EC2 instances provide the necessary computational resources for model training and evaluation, ensuring efficient utilization of cloud resources. Apache Airflow orchestrates the deployment of trained models onto production environments, streamlining the deployment process and ensuring the smooth and logical execution of tasks. We also explored the importance of designing a project directory structure and setting up a virtual environment on EC2, along with the tools used for writing code and developing machine learning models.

In the next chapter, we will guide the reader through the process of setting up an environment on AWS and configuring the necessary tools and services. This sets the stage for subsequent chapters where we will delve deeper into building and deploying scalable pipelines.

CHAPTER 2

Setting Up a Deep Learning Environment on AWS

In the previous chapter, we explored the growing importance of cloud-based deep learning in addressing the challenges posed by the increasing volume of data and the complexity of AI applications. We introduced the deep learning pipeline that integrates powerful frameworks—PySpark, PyTorch, TensorFlow, and Apache Airflow—with Amazon Web Services (AWS) cloud solutions, specifically Amazon S3 and EC2. This pipeline provides a scalable and efficient approach to managing the demands of modern deep learning workflows.

With a solid understanding of the pipeline's components and their roles in deep learning workflows, we now shift our focus to setting up the environment that will enable the reader to apply these concepts in practice. We will walk through the process of configuring the necessary tools and services on AWS to ensure a fully functional setup. This environment will set the stage for building and deploying scalable pipelines in the chapters that follow.

CHAPTER 2   SETTING UP A DEEP LEARNING ENVIRONMENT ON AWS

We will follow the steps below to set up the deep learning environment:

- Creating an AWS Account: This is the first step to access AWS services and resources. By creating an AWS account, you gain access to the AWS Management Console, where you can manage your EC2 instances, S3 buckets, and other AWS services.

- Provisioning EC2: This step creates virtual servers (instances) in the cloud to run applications and workloads. We examine both manual and automated options.

- Setting Up Amazon S3: This step involves creating buckets, configuring access permissions, and uploading or downloading data.

- Creating a Project Directory and Setting Up a Virtual Environment: After provisioning an EC2 instance, you set up a project directory and configure your environment for development or data analysis. This includes tasks like installing necessary software, setting up Python environments, and organizing project files.

- Installing and Configuring Dependencies: This is the step where you will need to install and configure various dependencies to build and deploy scalable deep learning pipelines on AWS effectively. These include installing PySpark, PyTorch, TensorFlow, Airflow, and JupyterLab, among others.

We will explore these steps one by one, starting with creating an AWS account.

CHAPTER 2 SETTING UP A DEEP LEARNING ENVIRONMENT ON AWS

Creating an AWS Account

Before you can dive into setting up a deep learning environment on Amazon Web Services (AWS), you will first need to create an AWS account. This account will grant you access to a wide range of AWS services, providing the foundational infrastructure necessary for building and deploying deep learning pipelines.

Here are the steps to create an AWS account:

- Visit the AWS Website: Navigate to the AWS website at `https://aws.amazon.com/` and click the "Sign In to the Console" button located at the top-right corner of the page.

- Create a New AWS Account: If you're new to AWS, click the "Create a new AWS account" button to begin the signup process. If you already have an AWS account, you can sign in using your existing credentials.

- Provide Account Information: Enter your email address and choose a password for your AWS account. You'll also need to provide some basic information about yourself and your organization.

- Verify Your Identity: AWS will require you to verify your identity using email or phone verification. Follow the prompts to complete this step.

- Choose a Support Plan: You'll be prompted to choose a support plan. For most users getting started with AWS, the free tier offers ample resources for exploration and experimentation.

CHAPTER 2   SETTING UP A DEEP LEARNING ENVIRONMENT ON AWS

- Enter Payment Information: To use AWS services beyond the free tier limits, you'll need to provide payment information. AWS offers pay-as-you-go pricing, meaning you only pay for the services you use, with no long-term contracts or upfront commitments.

- Review and Confirm: Review your account information, service agreement, and pricing details and then confirm your AWS account creation.

Once your AWS account is created, you'll have access to the AWS Management Console, where you can begin provisioning the resources needed for your deep learning environment.

Provisioning Amazon EC2 Instances

Amazon EC2 provides scalable compute capacity in the cloud, allowing you to launch virtual servers, known as instances, to run your applications and workloads. Provisioning EC2 instances is the first step in setting up your deep learning environment, providing the computational power needed for training and inference tasks.

In this section, we will explore both manual provisioning methods and automated provisioning using AWS CloudFormation.

Here are the steps to provision an EC2 instance manually:

- Navigate to the EC2 Dashboard: Sign in to the AWS Management Console, click the "Services" dropdown menu, select "EC2," and then click "EC2 Dashboard" from the EC2 service page.

- Launch Instance: Click the "Launch Instance" button to initiate the instance creation process.

CHAPTER 2   SETTING UP A DEEP LEARNING ENVIRONMENT ON AWS

- Choose an Amazon Machine Image (AMI): Select an AMI that serves as the base for your EC2 instance. You can choose from a variety of pre-configured AMIs, including those optimized for deep learning frameworks like TensorFlow and PyTorch.

- Choose an Instance Type: Select the instance type that best suits your computational needs. For the purpose of this book's project, we have chosen the "t2.xlarge" instance type, which offers 16 GiB of memory and four vCPUs. While this instance does not include a GPU, it is suitable for running demonstrations and experiments described in the book. However, for production-level deep learning tasks, which may require accelerated processing, you would need an instance with a dedicated GPU.

- Configure Instance Details: Customize the instance settings, including the number of instances, network settings, and storage options. For instance, choose a "t2.xlarge" instance with Ubuntu Server 22.04 LTS (which refers to the Ubuntu operating system version 22.04 with Long Term Support), HVM virtualization (Hardware Virtual Machine, providing better performance and compatibility), and SSD Volume Type (Solid State Drive storage, offering faster disk I/O performance).

- Add Storage: Specify the storage volumes attached to your EC2 instance. You can choose from various types of storage, including Amazon Elastic Block Store (EBS) volumes and instance store volumes.

CHAPTER 2   SETTING UP A DEEP LEARNING ENVIRONMENT ON AWS

- Configure Security Groups: Define the security groups for your EC2 instance to control inbound and outbound traffic. Ensure that the necessary ports are open to enable communication with your deep learning environment. Configuring security groups for your EC2 instance involves defining rules that control inbound and outbound traffic to and from the instance. Here's how you can do it:

    - Navigate to the Security Groups Section: After launching your EC2 instance, go to the EC2 Dashboard in the AWS Management Console. Under the "Network & Security" section in the left navigation pane, click "Security Groups."

    - Create a New Security Group: If you don't have an existing security group suitable for your deep learning environment, you can create a new one by clicking the "Create Security Group" button.

    - Define Inbound Rules: In the newly created or selected security group, define inbound rules to control incoming traffic to your EC2 instance. For example, you might want to allow SSH access (port 22) for remote administration, access to Jupyter notebooks or JupyterLab (port 8888), and Apache Airflow's web interface (default port 8080). You can also open additional ports as needed for communication with other services or applications.

    - Define Outbound Rules: Similarly, define outbound rules to control outgoing traffic from your EC2 instance. By default, all outbound traffic is allowed, but you can restrict it based on your specific requirements.

CHAPTER 2   SETTING UP A DEEP LEARNING ENVIRONMENT ON AWS

- Review and Apply Changes: Review the configured security group rules to ensure they meet your security and connectivity needs. Once you're satisfied, click the "Save" or "Apply Changes" button to apply the changes to the security group.

- Associate the Security Group with Your EC2 Instance: Finally, associate the newly created or updated security group with your EC2 instance. You can do this by selecting the instance in the EC2 Dashboard, clicking the "Actions" dropdown menu, and choosing "Security" ➤ "Change Security Groups." Select the desired security group(s) and click "Assign Security Groups."

- Review and Launch: Review the instance configuration details and make any necessary adjustments. Once you're satisfied, click the "Launch" button to launch your EC2 instance.

- Create Key Pair: If you haven't already created an SSH key pair, you'll be prompted to create one. This key pair is used to securely connect to your EC2 instance via SSH.

- Access Your EC2 Instance: Once your EC2 instance is launched, you can access it using SSH or other remote access methods. Start by connecting to your instance and configuring it according to your deep learning requirements.

  - Start and Stop Your EC2 Instance: After your EC2 instance is launched, you may need to start or stop it based on your usage requirements. To do this

CHAPTER 2   SETTING UP A DEEP LEARNING ENVIRONMENT ON AWS

- Start Instance: Navigate to the EC2 Dashboard in the AWS Management Console. Select the instance you want to start, and from the "Actions" dropdown menu, choose "Instance State" ➤ "Start instance."

- Stop Instance: Similarly, to stop an instance, select it from the EC2 Dashboard, and from the "Actions" dropdown menu, choose "Instance State" ➤ "Stop instance."

- Note: If you don't stop the instance after you are done with your work, charges will incur. Ensure to stop the instance when it's no longer needed to avoid unnecessary costs.

By following these steps, you will successfully provision an Amazon EC2 instance.

You can also automate these steps using Amazon CloudFormation—a tool that allows you to define and provision your cloud infrastructure using a simple text file. The script below, e.g., leverages Amazon CloudFormation to automate the creation of an EC2 instance with a t2.xlarge instance type, 40 GB gp3 EBS volume, and a security group allowing SSH access:

```
[In]: AWSTemplateFormatVersion: "2010-09-09"
[In]: Description: Provision a t2.xlarge EC2 instance with a 40 GB gp3
      EBS volume for deep learning tasks

[In]: Resources:
      DeepLearningEC2Instance:
          Type: AWS::EC2::Instance
          Properties:
              InstanceType: t2.xlarge
              KeyName: instance1keypair.pem
```

CHAPTER 2   SETTING UP A DEEP LEARNING ENVIRONMENT ON AWS

```
            SubnetId: subnet-04856bd3f80a406e6
              ImageId: ami-05134c8ef96964280
              BlockDeviceMappings:
              - DeviceName: /dev/sda1
                  Ebs:
                      VolumeSize: 40
                      VolumeType: gp3
              SecurityGroupIds:
              - !Ref InstanceSecurityGroup

    InstanceSecurityGroup:
        Type: AWS::EC2::SecurityGroup
        Properties:
              GroupDescription: Enable SSH access
              VpcId: vpc-09f1a4805e4bd7aa6
              SecurityGroupIngress:
              - IpProtocol: tcp
                  FromPort: 22
                  ToPort: 22
                  CidrIp: 0.0.0.0/0
```

[In]: Outputs:
    InstanceId:
        Description: Instance ID of the newly created EC2
        instance
        Value: !Ref DeepLearningEC2Instance

CHAPTER 2   SETTING UP A DEEP LEARNING ENVIRONMENT ON AWS

This script automates the creation of an EC2 instance with a specific configuration, including an instance type, storage volume, and security settings. Here's a detailed explanation of each part of the script:

- AWSTemplateFormatVersion and Description: The script begins by specifying the version of the CloudFormation template format with AWSTemplateFormatVersion, which is set to "2010-09-09". This indicates the version of the CloudFormation syntax being used. The Description field provides a brief overview of the purpose of the template. In this case, it describes that the script will provision a t2.xlarge EC2 instance with a 40 GB gp3 EBS volume, intended for deep learning tasks.

- Resources Section: The core of the script is the Resources section, where the EC2 instance and its associated configurations are defined. Within this section, there are two key resources: DeepLearningEC2Instance and InstanceSecurityGroup.

    - DeepLearningEC2Instance: This resource represents the EC2 instance itself. The Type is set to AWS::EC2::Instance, indicating that this resource will create an EC2 instance. Several properties are configured under Properties:

        - InstanceType specifies the instance type as t2.xlarge, which is a general-purpose instance suitable for a variety of workloads.

CHAPTER 2   SETTING UP A DEEP LEARNING ENVIRONMENT ON AWS

- KeyName is set to instance1keypair.pem, which is the name of the SSH key pair used to access the instance. This ensures secure access to the instance via SSH.

- SubnetId is set to subnet-04856bd3f80a406e6, indicating the subnet within the specified VPC where the instance will be launched. This allows control over the networking environment in which the instance operates.

- ImageId specifies the AMI (Amazon Machine Image) ID ami-05134c8ef96964280, which defines the operating system and initial software configuration of the instance. This AMI serves as the baseline image for the instance.

- BlockDeviceMappings defines the storage configuration, attaching a 40 GB gp3 EBS (Elastic Block Store) volume to the instance. The DeviceName /dev/sda1 specifies the root volume, and VolumeType is set to gp3, a general-purpose SSD.

- SecurityGroupIds associates the instance with a security group, which is defined later in the script. The security group controls inbound and outbound traffic to the instance.

CHAPTER 2   SETTING UP A DEEP LEARNING ENVIRONMENT ON AWS

- InstanceSecurityGroup: This resource defines the security group associated with the EC2 instance. The Type is set to AWS::EC2::SecurityGroup, indicating that this resource will create a security group. The Properties section specifies the following:

    - GroupDescription provides a brief explanation of the security group's purpose, which in this case is to enable SSH access.

    - VpcId is set to vpc-09f1a4805e4bd7aa6, which links the security group to a specific VPC (Virtual Private Cloud).

    - SecurityGroupIngress defines the inbound traffic rules. The script allows TCP traffic on port 22 (SSH) from any IP address (CidrIp: 0.0.0.0/0). While this is convenient for testing, it is not recommended for production environments due to security concerns.

- Outputs Section: The Outputs section provides information about the resources created by the script. Specifically, it outputs the InstanceId of the newly created EC2 instance. This allows you to easily reference the instance ID after the stack is created, which can be useful for further automation or management tasks.

The above script is in YAML format, but you can also author CloudFormation templates in JSON format. The same CloudFormation features and functions are available in both formats, as demonstrated in the JSON script below:

CHAPTER 2   SETTING UP A DEEP LEARNING ENVIRONMENT ON AWS

```
[In]: {
  "AWSTemplateFormatVersion": "2010-09-09",
  "Description": "Provision a t2.xlarge EC2 instance with a
  40 GB gp3
   EBS volume for deep learning tasks",
  "Resources": {
      "DeepLearningEC2Instance": {
      "Type": "AWS::EC2::Instance",
      "Properties": {
      "InstanceType": "t2.xlarge",
      "KeyName": "instance1keypair.pem",
      "SubnetId": "subnet-04856bd3f80a406e6",
      "ImageId": "ami-05134c8ef96964280",
      "BlockDeviceMappings": [
          {
          "DeviceName": "/dev/sda1",
          "Ebs": {
          "VolumeSize": 40,
          "VolumeType": "gp3"
          }
          }
      ],
      "SecurityGroupIds": [
          {
          "Ref": "InstanceSecurityGroup"
          }
      ]
      }
      },
      "InstanceSecurityGroup": {
      "Type": "AWS::EC2::SecurityGroup",
```

CHAPTER 2   SETTING UP A DEEP LEARNING ENVIRONMENT ON AWS

```
      "Properties": {
      "GroupDescription": "Enable SSH access",
      "VpcId": "vpc-09f1a4805e4bd7aa6",
      "SecurityGroupIngress": [
      {
          "IpProtocol": "tcp",
          "FromPort": 22,
          "ToPort": 22,
          "CidrIp": "0.0.0.0/0"
          }
      ]
      }
      }
  },
  "Outputs": {
      "InstanceId": {
      "Description": "Instance ID of the newly created EC2
      instance",
      "Value": {
      "Ref": "DeepLearningEC2Instance"
      }
      }
  }
}
```

Setting Up Amazon S3

Amazon Simple Storage Service (S3) is a highly scalable storage service offered by Amazon Web Services (AWS). Amazon S3 provides a reliable and cost-effective solution for storing and accessing large volumes of data. It offers high availability, durability, and scalability, making it an ideal

CHAPTER 2   SETTING UP A DEEP LEARNING ENVIRONMENT ON AWS

choice for storing datasets, model checkpoints, training logs, and other resources required for deep learning workflows. By setting up Amazon S3, you can centralize your data storage and easily share resources across multiple EC2 instances and users.

Here are the steps for setting up Amazon S3:

- Navigate to the S3 Dashboard: Sign in to the AWS Management Console, click the "Services" dropdown menu, and select "S3" from the list of available services. This will take you directly to the S3 dashboard, where you can manage your buckets and objects.

- Create a Bucket: Click the "Create bucket" button to create a new S3 bucket. Choose a unique name for your bucket and select the AWS region where you want the bucket to be located. Once the bucket has been created, you will see a message saying, "Successfully created bucket 'your-bucket-name.'"

- Configure Bucket Properties: Configure the properties of your S3 bucket, including the storage class, encryption settings, and versioning options. You can choose the appropriate settings based on your data storage requirements and compliance needs. To configure bucket properties for an S3 bucket, you can follow these steps:

    - Select the Bucket: From the list of buckets displayed on the S3 dashboard, select the bucket for which you want to configure properties.

    - Access Bucket Properties: Once you've selected the bucket, click the "Properties" tab in the bucket details pane. This tab allows you to configure various properties for the selected bucket.

CHAPTER 2   SETTING UP A DEEP LEARNING ENVIRONMENT ON AWS

- Configure Properties: Within the "Properties" tab, you'll find options to configure various settings such as

    - Storage Class: Choose the appropriate storage class for your data, such as Standard, Intelligent-Tiering, Standard-IA (Infrequent Access), One Zone-IA, Glacier, etc. Each storage class has different pricing and access characteristics.

    - Encryption Settings: Enable server-side encryption to encrypt objects stored in the bucket. You can choose between Amazon S3-managed keys (SSE-S3), AWS Key Management Service (SSE-KMS), or customer-provided keys (SSE-C).

    - Versioning: Enable versioning to keep multiple versions of an object in the bucket. This helps protect against accidental deletion or overwrite of objects.

- Save Changes: After configuring the desired properties, make sure to save the changes by clicking the appropriate buttons (e.g., "Save changes" or "Apply").

- Set Bucket Permissions: Define access control settings for your S3 bucket to manage who can access the data stored in the bucket and what actions they can perform. You can set permissions at the bucket and object level. To set bucket permissions and define access control settings for your S3 bucket, follow these steps:

- Select the bucket for which you want to set permissions.

- Once you've selected the bucket, you should see a tab labeled "Permissions."

- Click this tab to access the permissions settings for the selected bucket.

- Under the Permissions tab, you'll find options to set bucket policies, access control lists (ACLs), and block public access settings.

- Click "Bucket Policy" to define a bucket policy using JSON syntax. Bucket policies allow you to specify who can access the bucket and what actions they can perform.

- Alternatively, you can set bucket access control lists (ACLs) by clicking "Access control list (ACL)" under the Permissions tab.

- ACLs enable you to grant specific permissions to predefined groups of users, such as granting read or write access to authenticated users.

- After defining the desired permissions using bucket policies or ACLs, review the settings to ensure they align with your security requirements.

- Once you're satisfied with the permissions configurations, save or apply the changes to enforce the access control settings for your S3 bucket

- If you need to set permissions at the object level (i.e., for individual files within the bucket), you can do so by modifying the object ACLs. This can be done by selecting specific objects within the bucket and adjusting their permissions settings accordingly.

- Upload Data to the Bucket: Once your S3 bucket is set up, you can start uploading data to it. You can upload data directly through the S3 console, use the AWS Command Line Interface (CLI), or integrate S3 with other AWS services for smooth data transfers.

Here are the steps for interacting with S3 from EC2.

To access data stored in your S3 bucket from your EC2 instance, you'll need to configure Identity and Access Management (IAM) roles and policies to grant the necessary permissions. IAM roles allow you to securely authenticate and authorize access to S3 resources without exposing your AWS credentials. Never hardcode AWS credentials in your scripts, as this poses a significant security risk. Instead, consider using the following secure alternatives:

- IAM Roles: Attach an IAM role to your EC2 instance to automatically provide the necessary permissions without embedding credentials in your code.

- Environment Variables: Store sensitive credentials in environment variables that can be accessed by your application without being hardcoded.

- AWS Secrets Manager: Use AWS Secrets Manager to securely store and retrieve credentials, such as database passwords or API keys.

To implement secure access using IAM roles, follow these steps:

1. Sign in to the AWS Management Console and open the IAM console.

2. Navigate to the "Roles" section and click "Create role."

CHAPTER 2   SETTING UP A DEEP LEARNING ENVIRONMENT ON AWS

3. Choose the service that will use this role (in this case, EC2).

4. Attach the necessary policies that grant permissions to access S3 resources (e.g., AmazonS3FullAccess for full access).

5. Review and name your role and then create it.

6. Navigate to the EC2 Dashboard in the AWS Management Console.

7. Select the EC2 instance to which you want to attach the IAM role.

8. Click "Actions" ➤ "Security" ➤ "Modify IAM role."

9. Select the IAM role you created earlier from the dropdown menu and confirm the attachment.

10. Once the IAM role is attached to your EC2 instance, you can access S3 data programmatically using AWS SDKs or command-line tools like the AWS CLI.

Creating a Project Directory

After provisioning an Amazon EC2 instance, creating a project directory is a logical step. This is important for organizing and managing your work efficiently. It provides a structured environment to store all project-related files, including source code, data, documentation, and outputs, facilitating ease of navigation and collaboration. With a designated directory, version control becomes more straightforward, ensuring that changes are tracked accurately and collaboration flows smoothly. Additionally, a well-organized project directory enhances reproducibility by encapsulating all necessary files and dependencies, making it easier to share and replicate results across different environments. It also promotes scalability

CHAPTER 2   SETTING UP A DEEP LEARNING ENVIRONMENT ON AWS

and maintenance, allowing new components to be added without compromising the overall organization. Ultimately, a project directory serves as a foundation for effective project management, documentation, and collaboration, contributing to the success of your project.

In the previous chapter, we outlined a directory structure for a Tesla stock prediction project, but we didn't delve into the code that generated it. The Bash script below, which was executed within JupyterLab in a notebook environment on EC2, establishes this structure with specific subdirectories and files necessary for organizing data, source code, and output files. The tree command at the end displays the structure of the created directories for visual representation:

```
[In]: %%bash

[In]: project_dir="Project Directory Structure"

[In]: mkdir -p "$project_dir"

[In]: mkdir -p "$project_dir/data"
[In]: touch "$project_dir/data/TSLA_stock.csv"

[In]: mkdir -p "$project_dir/src/pyspark_code"
[In]: touch "$project_dir/src/main.py"
[In]: touch "$project_dir/src/pyspark_code/data_preprocessing.py"

[In]: mkdir -p "$project_dir/src/pytorch_code"
[In]: touch "$project_dir/src/pytorch_code/model.py"
[In]: touch "$project_dir/src/pytorch_code/model_training.py"

[In]: mkdir -p "$project_dir/output"
[In]: mkdir -p "$project_dir/output/TSLA_stock.parquet"
[In]: touch "$project_dir/output/TSLA_stock.parquet/_SUCCESS"
[In]: touch "$project_dir/output/TSLA_stock.parquet/part-00000- \
         2c717885-1e92-4091-b73e-c07fb2f42ba8-c000.snappy.
         parquet"
```

CHAPTER 2   SETTING UP A DEEP LEARNING ENVIRONMENT ON AWS

```
[In]: mkdir -p "$project_dir/output/TSLA_stock_train.parquet"
[In]: touch "$project_dir/output/TSLA_stock_train.parquet/_
      SUCCESS"
[In]: touch "$project_dir/output/TSLA_stock_train.parquet/
      part-\ 00000-3931d223-097a-4b7f-a6df-6ddc8abfcb5c-c000.
      snappy.parquet"

[In]: mkdir -p "$project_dir/logs"

[In]: mkdir -p "$project_dir/tests"
[In]: touch "$project_dir/tests/test.py"

[In]: mkdir -p "$project_dir/visualizations"
[In]: touch "$project_dir/visualizations/plot.png"

[In]: touch "$project_dir/README.md"

[In]: touch "$project_dir/config.yaml"

[In]: touch "$project_dir/requirements.txt"

[In]: tree "$project_dir" --dirsfirst
```

Here's what each part of the script does:

- %%bash: This is a cell magic command in Jupyter notebooks that allows you to run Bash commands directly within a code cell. When you include %%bash at the top of a cell in Jupyter, it tells the notebook to interpret the entire cell as Bash commands rather than Python code.

77

CHAPTER 2   SETTING UP A DEEP LEARNING ENVIRONMENT ON AWS

- project_dir="Project Directory Structure": This line defines a variable named project_dir and assigns it the value "Project Directory Structure". This variable will be used to specify the name of the main project directory.

- mkdir -p "$project_dir": This command creates the main project directory specified by the project_dir variable. The -p option ensures that the command creates intermediate directories as needed, and it won't throw an error if the directory already exists.

- mkdir -p "$project_dir/data": Creates a subdirectory named "data" inside the main project directory.

- touch "$project_dir/data/TSLA_stock.csv": Creates an empty file named "TSLA_stock.csv" inside the "data" subdirectory. In a real-world scenario, rather than creating an empty file locally, you would typically upload an existing CSV file to an Amazon S3 bucket. Once the file is stored in S3, you can easily copy it to the local directory on your EC2 instance using commands like aws s3 cp.

- Similar mkdir and touch commands are used to create directories and files within the project directory for different purposes, such as source code (src), output (output), and their respective subdirectories.

- tree "$project_dir" --dirsfirst: This command generates a visual representation of the project directory structure using the tree command. The --dirsfirst option ensures that directories are listed before files in the output.

> **Note** The tree command is not pre-installed on all systems, such as Amazon Linux or Ubuntu on Mac, which we have used for this project. To install the tree command on Ubuntu, we used the following command: `sudo apt-get install tree`; for Amazon Linux: `sudo yum install tree`.

The output of the Bash code, using the tree command, displays the directory structure of project_dir and its subdirectories, showing the created files and folders, as illustrated in Figure 2-1 (saved as an image from JupyterLab notebook).

```
Project Directory Structure
├── data
│   └── TSLA_stock.csv
├── logs
├── output
│   ├── TSLA_stock.parquet
│   │   ├── _SUCCESS
│   │   └── part-00000-2c717885-1e92-4091-b73e-c07fb2f42ba8-c000.snappy.parquet
│   └── TSLA_stock_train.parquet
│       ├── _SUCCESS
│       └── part-00000-3931d223-097a-4b7f-a6df-6ddc8abfcb5c-c000.snappy.parquet
├── src
│   ├── pyspark_code
│   │   └── data_preprocessing.py
│   ├── pytorch_code
│   │   ├── model.py
│   │   └── model_training.py
│   └── main.py
├── tests
│   └── test.py
├── visualizations
│   └── plot.png
├── README.md
├── config.yaml
└── requirements.txt
```

Figure 2-1. Project Directory Structure for Predicting Tesla Stock Price

CHAPTER 2   SETTING UP A DEEP LEARNING ENVIRONMENT ON AWS

Creating a Virtual Environment

In addition to establishing a well-defined project structure, it's advisable to create a virtual environment (e.g., myenv) for managing dependencies and ensuring reproducibility across various environments. This practice isolates the project's dependencies from other Python projects on the system, thereby preventing conflicts and ensuring consistent execution regardless of the system's configuration. Once your EC2 instance is up and running, follow the steps below to create, activate, and deactivate a virtual environment on an Ubuntu operating system:

Step 1: Navigate to your home directory:

[In]: cd /home/ubuntu

Step 2: Create a virtual environment:

[In]: python3 -m venv myenv

This command assumes that Python 3 is installed on your system. Some systems might still default to Python 2.7, which is not compatible with the venv module. You can check your Python version by running

[In]: python3 -version

If Python 3 is not installed, you can install it, e.g., using the following commands on Ubuntu:

[In]: sudo apt-get update
[In]: sudo apt-get install python3

On Amazon Linux:

[In]: sudo yum update
[In]: sudo yum install python3

CHAPTER 2   SETTING UP A DEEP LEARNING ENVIRONMENT ON AWS

The python3 -m venv myenv command creates a virtual environment named "myenv." A virtual environment is an isolated Python environment for your project. You can access this environment from any directory on your EC2 instance. Once activated, you'll be able to use the Python interpreter and any installed packages from within the "myenv" virtual environment regardless of your current working directory. This flexibility allows you to work on your Python projects from any location on your EC2 instance, providing convenience and ease of use.

Let's check what's inside the virtual environment, myenv:

```
[In]: ls myenv
[Out]: bin include lib lib64 pyvenv.cfg
```

These are the standard directories and files created within a Python virtual.

Let's break down what each of these items is:

- bin: This directory contains the executable files for the virtual environment. When you activate the virtual environment, these binaries are added to your PATH so that the Python interpreter, pip, and other tools within the virtual environment take precedence over the global ones.
- include: This directory contains C header files that are needed for compiling certain Python packages that include C extensions.
- lib: This directory contains the Python standard library and other shared libraries needed for running Python scripts within the virtual environment.

Let's look inside the directory:

```
[In]: ls myenv/lib
[Out]: python3.9
```

CHAPTER 2   SETTING UP A DEEP LEARNING ENVIRONMENT ON AWS

- lib64: This directory is a symbolic link to the lib directory. This is done for compatibility with systems that use a lib64 directory for 64-bit libraries.
- pyvenv.cfg: This is a configuration file for the virtual environment. It stores settings and metadata related to the virtual environment.

These directories and files collectively make up the structure of a Python virtual environment. When you activate the virtual environment, the bin directory is added to the beginning of your PATH, ensuring that the Python interpreter and other tools from the virtual environment take precedence.

Step 3: Activate the virtual environment:

[In]: `source myenv/bin/activate`

This command activates the virtual environment. When activated, your shell prompt changes to indicate that you are now working within the "myenv" virtual environment.

Step 4: Deactivate the virtual environment:

[In]: `deactivate`

You use the deactivate command when you want to exit or deactivate the currently activated virtual environment and return to the global Python environment. This is useful when you've finished working on a project or when you want to switch to a different virtual environment or the global Python environment.

> **Note** While virtual environments are effective for managing dependencies within a project, Docker offers an alternative approach. It allows you to containerize your application along with its dependencies, ensuring that it runs consistently across different environments. We will explore Docker in more detail when we demonstrate how to set up a Docker container for Airflow.

Installing and Configuring Dependencies

Once the virtual environment is set up, it is necessary to install and configure various dependencies to build and deploy scalable deep learning pipelines on AWS effectively. This section provides guidance through the process of installing and configuring the most common machine learning libraries and tools within the virtual environment. These include PySpark, PyTorch, TensorFlow, Airflow, and JupyterLab. Additionally, instructions are provided for creating a Databricks account and setting up a workspace in case this framework is preferred for preprocessing input data before storing it on S3 as input for deep learning projects.

Installing PySpark

Follow the steps below to install PySpark:

Step 1: Activate the virtual environment.

If you haven't already activated your virtual environment, navigate to the home directory, e.g., "/home/ubuntu," and activate the "myenv" virtual environment:

```
[In]: cd /home/ubuntu
[In]: source myenv/bin/activate
```

CHAPTER 2   SETTING UP A DEEP LEARNING ENVIRONMENT ON AWS

Step 2: Install PySpark.

Once the virtual environment is activated, install PySpark using pip:

[In]: pip install pyspark
[Out]: Successfully installed py4j-0.10.9.7 pyspark-3.5.1

The line "Successfully installed py4j-0.10.9.7 pyspark-3.5.1" is the output from the pip install pyspark command. It indicates that the PySpark package, along with its dependency Py4J, has been successfully downloaded and installed into your virtual environment ("myenv").

- py4j-0.10.9.7: This is the version of the Py4J library, which is a bridge between Python and Java. PySpark uses Py4J to interact with the Java-based Spark framework.

- pyspark-3.5.1: This is the version of the PySpark library itself that has been installed.

This output confirms that both PySpark and Py4J have been installed correctly and are ready to use within your virtual environment. You can also verify installation is complete by running a Python script that imports PySpark:

[In]: python -c "import pyspark; print(pyspark.__version__)"
[Out]: 3.5.1

This command prints the installed version of PySpark, which, in this case, is 3.5.1.

Installing PyTorch

With the virtual environment activated, install PyTorch using pip. You can choose the appropriate command based on your system configuration and requirements. For example, to install PyTorch with Compute Unified Device Architecture (CUDA) support for GPU acceleration, you can use the following command:

```
[In]: pip install torch torchvision torchaudio
```

After the installation is complete, you can verify it by running a Python script that imports PyTorch:

```
[In]: python -c "import torch; print(torch.__version__)"
[Out]: 2.2.2+cu121
```

This command prints the installed version of PyTorch, which in this case is 2.2.2. This includes CUDA support for GPU acceleration, as indicated by the "+cu121" suffix.

Note CUDA (Compute Unified Device Architecture) is a parallel computing platform and application programming interface (API) model created by NVIDIA. It allows software developers to harness the computational power of NVIDIA GPUs (Graphics Processing Units) for general-purpose processing tasks, including deep learning and scientific simulations.

Installing TensorFlow

Install TensorFlow and Keras using pip as follows:

```
[In]: pip install tensorflow keras
```

CHAPTER 2   SETTING UP A DEEP LEARNING ENVIRONMENT ON AWS

This command installs TensorFlow and Keras, which is now part of TensorFlow as its high-level API for building neural networks.

After the installation is complete, you can verify it by running a Python script that imports TensorFlow and Keras:

```
[In]: python -c "import tensorflow as tf; print(tf.__version__)"
[In]: python -c "import keras; print(keras.__version__)"
[Out]: 2.16.1
[Out]: 3.2.1
```

The output indicates the following versions:

TensorFlow Version: 2.16.1

Keras Version: 3.2.1

Installing Boto3

Boto3 is the Amazon Web Services (AWS) SDK for Python. It allows developers to interact with AWS services using Python code. With Boto3, you can programmatically access and manage various AWS resources such as EC2 instances, S3 buckets, DynamoDB tables, and more.

You can install Boto3 using the pip command below:

```
[In]: pip install boto3
```

After the installation is complete, you can verify it by running a Python script that imports Boto3:

```
[In]: python -c "import boto3; print(boto3.__version__)"
[Out]: 1.34.84
```

The output shows that version 1.34.84 has been successfully installed.

Installing Airflow

There are three primary options for using Apache Airflow on EC2: standalone, Docker-based, and fully managed through Amazon Managed Workflows for Apache Airflow (MWAA). In this section, we provide instructions for setting up the standalone and Docker-based versions of Airflow, while guidance on using MWAA will be provided in Chapter 10.

Standalone Installation

In this subsection, detailed instructions are provided for setting up Airflow on an EC2 instance using a standalone installation. This setup allows for direct control over the Airflow environment without additional layers of abstraction.

Follow these steps to perform this installation:

Step 1: Activate your environment:

```
[In]: source /home/ubuntu/myenv/bin/activate
```

Step 2: Update and install dependencies:

```
[In]: sudo apt update
[In]: sudo apt install python3-pip sqlite3 libpq-dev
```

- sudo apt update: This command updates the package lists for upgrades.

- sudo apt install python3-pip sqlite3 libpq-dev: Installs necessary dependencies including Python 3 pip, SQLite, and PostgreSQL development files.

CHAPTER 2   SETTING UP A DEEP LEARNING ENVIRONMENT ON AWS

Step 3: Install Apache Airflow with PostgreSQL support:

```
[In]: pip install "apache-airflow[postgres]==2.5.0" \
    --constraint \
    "https://raw.githubusercontent.com/apache/airflow/
    constraints- \
    2.5.0/constraints-3.7.txt"
```

- This line installs Apache Airflow with PostgreSQL support. It specifies version 2.5.0 and installs additional dependencies from a constraints file.

Step 4: Initialize Airflow database:

```
[In]: airflow db init
```

This line initializes the Airflow metadata database.

Step 5: Install and configure PostgreSQL:

```
[In]: sudo apt-get install postgresql postgresql-contrib
[In]: sudo -i -u postgres
[In]: psql
```

Code explanation:

- Line 1: Installs PostgreSQL and additional contrib package.
- Line 2: Switches to the PostgreSQL user.
- Line 3: Enters the PostgreSQL command-line interface.

Step 6: In the PostgreSQL shell:

```
[In]: CREATE DATABASE airflow;
[In]: CREATE USER your-username WITH PASSWORD 'your-password';
[In]: GRANT ALL PRIVILEGES ON DATABASE airflow TO your-username;
```

```
[In]: \q
[In]: exit
```

- CREATE DATABASE airflow;: Creates a database named "airflow"
- CREATE USER your-username WITH PASSWORD 'your-password';: Creates a user with a specified username and password
- GRANT ALL PRIVILEGES ON DATABASE airflow TO your-username;: Grants all privileges on the "airflow" database to the created user
- \q: Exits the PostgreSQL shell
- exit: Exits the PostgreSQL user session

These instructions create a PostgreSQL database for Airflow to store its metadata. Additionally, the code creates a dedicated user with a password for accessing this database. It's crucial to replace "your-username" with a username of your choice and "your-password" with a strong password. This user will have full privileges on the "airflow" database, ensuring that Airflow can manage its metadata effectively. After executing these commands, you'll exit the PostgreSQL shell and return to the command line.

By creating a separate database and user for Airflow, you ensure better security and organization of your PostgreSQL instance, isolating Airflow's data from other databases and users. Additionally, using a dedicated user with limited privileges reduces the risk of unauthorized access to the Airflow metadata.

CHAPTER 2   SETTING UP A DEEP LEARNING ENVIRONMENT ON AWS

Step 7: Update Airflow configuration:

In this step, we set up the necessary configurations and settings of Apache Airflow:

```
[In]: cd airflow
[In]: sed -i \
      's#sqlite:////home/ubuntu/airflow/airflow.db \
      #postgresql+psycopg2://your-username: \
      your- password@localhost/airflow#g' airflow.cfg
[In]: sed -i 's#SequentialExecutor#LocalExecutor#g' airflow.cfg
```

Code explanation:

- Line 1: Changes directory to the Airflow configuration directory.

- Line 2: Replaces the SQLite connection string in the Airflow configuration file with a PostgreSQL connection string. You need to replace "your-username" and "your-password" with the actual username and password you have set up for the PostgreSQL database in Step 6. This ensures that Airflow connects to the correct database with the appropriate credentials.

- Line 3: Replaces the SequentialExecutor with the LocalExecutor in the Airflow configuration file.

The purpose of switching from SQLite to PostgreSQL and from SequentialExecutor to LocalExecutor in Apache Airflow configuration is to enhance scalability and robustness, as explained below:

Switching from SQLite to PostgreSQL:

- SQLite is suitable for development and testing purposes but may not be suitable for production environments, especially when dealing with multiple Airflow users, high concurrency, and large volumes of data.

90

CHAPTER 2   SETTING UP A DEEP LEARNING ENVIRONMENT ON AWS

- PostgreSQL is a powerful, open source relational database management system known for its scalability, reliability, and feature-rich capabilities. By switching to PostgreSQL, you can better handle concurrent access, manage larger datasets, and ensure data integrity in production environments.

Switching from SequentialExecutor to LocalExecutor:

- The SequentialExecutor in Airflow runs tasks sequentially in a single process, which may become a performance bottleneck as the number of tasks and their dependencies grow.

- The LocalExecutor, on the other hand, allows Airflow to parallelize task execution across multiple worker processes, improving performance and scalability. Each task can be executed independently, making better use of available computing resources.

Step 8: Create an admin user:

```
[In]: airflow users create -u your-username -f your-first-
      name -l \
      your-last-name -r Admin -e your@mail.com
```

Creating an admin user in Apache Airflow is essential for managing and administrating the Airflow instance securely. By creating an admin user, you establish a user with elevated privileges, allowing them to perform administrative tasks such as managing connections, variables, pools, and users, as well as accessing sensitive information like DAG runs and logs. It's crucial to choose strong passwords and limit access to necessary personnel only when creating admin users. This helps prevent unauthorized access and protects sensitive information within Airflow. Assigning administrative roles to specific users also helps enforce security

CHAPTER 2   SETTING UP A DEEP LEARNING ENVIRONMENT ON AWS

policies within the Airflow instance. Admin users have the authority to configure access controls, set permissions, and monitor user activity, ensuring the integrity and security of the Airflow environment.

Step 9: Start Airflow web server and scheduler:

```
[In]: airflow webserver &
[In]: airflow scheduler
```

- airflow webserver &: This command starts the Airflow web server in the background. The web server provides a user interface (UI) for managing and monitoring Airflow workflows. By running the web server in the background with the & symbol, you allow it to continue running even after you log out of the terminal session. This ensures that the web UI remains accessible even when you are not actively logged into the server.

- airflow scheduler: This command starts the Airflow scheduler. The scheduler is responsible for triggering task instances based on the defined schedule (e.g., cron expressions) and dependencies between tasks (defined in Directed Acyclic Graphs or DAGs). It continuously monitors the DAGs for tasks that need to be executed and triggers them according to their schedule and dependencies.

It's recommended to run these commands in separate terminal sessions. Running them separately allows for better monitoring and troubleshooting of each component. Tools like tmux or screen provide session management features, enabling you to detach and reattach to terminal sessions, ensuring that the web server and scheduler continue running even after you log out of the terminal session.

CHAPTER 2   SETTING UP A DEEP LEARNING ENVIRONMENT ON AWS

Step 10: Access Airflow UI:

- Open a web browser and navigate to your EC2 instance's public IP address or domain name, followed by the Airflow port (default is 8080):

    [In]: http://<your_ec2_public_ip>:8080

- You will be prompted to enter your username and password.

- Enter the credentials you created earlier, and you should be logged into the Airflow UI.

To set up port 8080 for accessing the Airflow UI, you typically need to ensure that your EC2 instance's security group allows inbound traffic on port 8080 and that Airflow is configured to listen on port 8080.

Here's how you can do it:

EC2 Security Group Configuration:

- Go to the AWS Management Console and navigate to the EC2 Dashboard.

- Select the instance running Airflow.

- In the instance details pane at the bottom, find the Security Groups section and click the associated security group.

- In the security group settings, click the "Inbound rules" tab.

- Click "Edit inbound rules" and add a rule allowing traffic on port 8080 from your IP address or from anywhere (not recommended for production).

- Save the changes.

CHAPTER 2   SETTING UP A DEEP LEARNING ENVIRONMENT ON AWS

Configure Airflow to Listen on Port 8080:

- Connect to your EC2 instance via SSH.
- Navigate to the Airflow configuration directory using cd command:

  [In]: cd airflow

- Edit the airflow.cfg file using a text editor, e.g., Nano:

  [In]: nano airflow.cfg

- Find the web_server_port parameter in the configuration file and set it to 8080:

  [In]: web_server_port = 8080

- Save the changes and exit the text editor.
- After saving the changes to the airflow.cfg file, you'll need to restart the Airflow web server for the changes to take effect. This can typically be done by stopping and restarting the Airflow web server process.
- You can restart the Airflow web server by killing the current session and then starting a new one, as follows:
  - Find the process ID (PID) of the Airflow web server:

    [In]: ps aux | grep 'airflow webserver'

  - Identify the PID associated with the Airflow web server process.
  - Kill the Airflow web server process using its PID:

    [In]: kill <PID>

- Start the Airflow web server again:

```
[In]: airflow webserver &
```

Docker-Based Setup

Running Apache Airflow in Docker offers significant advantages over a traditional standalone installation, providing a streamlined, consistent environment that simplifies setup, reduces configuration errors, and enhances portability. Docker encapsulates Airflow and its dependencies within containers, ensuring reliable workflow execution across different systems, whether in development, testing, or production.

To set up Apache Airflow in Docker, you need to install Docker and Docker Compose. Docker packages applications into containers, including all necessary code, libraries, and dependencies, ensuring consistency across various environments. Docker Compose further simplifies this by allowing you to define and manage multi-container applications, which is particularly useful for complex setups like Apache Airflow, where services such as the web server, scheduler, and database must work together.

The Apache Airflow project provides an official `docker-compose.yml` file that defines all the required services, including the Airflow web server for workflow management, the scheduler for task execution, Postgres for metadata storage, Redis for task queue management, and workers and triggerers for executing tasks and handling deferred operations.

The easiest way to install Docker on a Linux operating system, such as Ubuntu, for testing purposes is to use the convenience script provided by Docker at `https://get.docker.com/`. After installing Docker on your EC2 instance, you set up and run Airflow using the official Docker Compose configuration.

CHAPTER 2 SETTING UP A DEEP LEARNING ENVIRONMENT ON AWS

Follow these steps:

Step 1: Create the script file.

- Once you're connected to your EC2 instance via SSH, start by creating a new shell script file:

  [In]: nano install-docker.sh

- Copy and paste the entire script content provided by Docker into this file. This script is available at https://get.docker.com/ and automates the Docker installation process.

Step 2: Make the script executable.

- After pasting the script, save and close the file by pressing Ctrl + X, then Y, and Enter.

- Set the script as executable by running the following command:

  [In]: chmod +x install-docker.sh

Step 3: Run the script with root privileges.

- Execute the script to install Docker with administrator privileges:

  [In]: sudo ./install-docker.sh

- The script will detect your Linux distribution and version and then proceed with the Docker installation accordingly.

Step 4: Verify Docker installation.

- After the script completes, verify that Docker is installed correctly by running

  [In]: docker -version

- You should see the version of Docker installed on your instance, confirming that the installation was successful.

Step 5: Run a test Docker container.

- To ensure Docker is working correctly, run a test Docker container:

  ```
  [In]: sudo docker run hello-world
  ```

- This command will download a test image and run it, displaying a confirmation message if successful.

Step 6: Install Docker Compose.

- Use the following to download and install Docker Compose:

  ```
  [In]: sudo curl -L \
        "https://github.com/docker/compose/releases/latest/download/" \
        "docker-compose-$(uname -s | tr '[:upper:]' '[:lower:]')" \
        "-$(uname -m)" -o /usr/bin/docker-compose

  [In]: sudo chmod 755 /usr/bin/docker-compose
  ```

  The first command downloads the latest Docker Compose binary from the official GitHub repository, selecting the correct version for your system and saving it to /usr/bin/docker-compose. The second command makes the Docker Compose binary executable by setting the appropriate permissions.

CHAPTER 2   SETTING UP A DEEP LEARNING ENVIRONMENT ON AWS

- Confirm the installation by checking the version of Docker Compose:

  [In]: docker-compose –version

Step 7: Download the Airflow Docker Compose configuration.

- Apache Airflow provides an official Docker Compose setup to make running Airflow easy. Download the docker-compose.yaml file, which contains the configuration for running Airflow and its associated services (like PostgreSQL and Redis) in Docker containers:

  [In]: curl -LfO \

    'https://airflow.apache.org/docs/apache-airflow/' \

    '2.7.2/docker-compose.yaml'

Step 8: Prepare the environment.

- Create the directories that Airflow will use for DAGs, logs, and plugins:

  [In]: mkdir -p ./dags ./logs ./plugins

- Create a .env file to set the necessary environment variables, including your user ID and group ID to avoid permission issues:

  [In]: echo -e "AIRFLOW_UID=$(id -u)\nAIRFLOW_GID=0" > .env

CHAPTER 2 SETTING UP A DEEP LEARNING ENVIRONMENT ON AWS

Step 9: Initialize the Airflow database.

- Initialize the Airflow database by running the following command:

    [In]: docker compose up airflow-init

Step 10: Start Airflow services.

- After the database is initialized, start the Airflow web server, scheduler, and other necessary services:

    [In]: docker compose up -d

- Ensure that all the Airflow containers are running by checking their status:

    [In]: docker compose ps

Step 11: Access the Airflow web interface.

- Obtain the Public IP of Your EC2 Instance:
    - You can find the public IP in the AWS Management Console under the details of your EC2 instance.
- Open the Required Port in the Security Group:
    - By default, Airflow's web server runs on port 8080.
    - Ensure that port 8080 is open in the security group associated with your EC2 instance. You can add an inbound rule in your security group to allow traffic on port 8080 from your IP address.

        Example inbound rule:
        - Type: Custom TCP Rule
        - Protocol: TCP

CHAPTER 2   SETTING UP A DEEP LEARNING ENVIRONMENT ON AWS

- Port Range: 8080

- Source: Your IP (or 0.0.0.0/0 for open access)

- Access Airflow in Your Browser:

- Open a web browser and navigate to

  [In]: http://<your-ec2-public-ip>:8080

  - Replace <your-ec2-public-ip> with the actual public IP of your EC2 instance.

  - You should see the Airflow login page/Airflow dashboard.

Installing JupyterLab

JupyterLab is an open source web-based interactive development environment (IDE) for Jupyter notebooks, code, and data. It provides a flexible and extensible architecture for working with code, data, and visualization in a variety of programming languages, including Python.

Below are step-by-step instructions to install and configure JupyterLab on EC2 instance:

Step 1: Install JupyterLab:

- Connect to your EC2 instance and open command prompt.

- Activate your virtual environment:

  [In]: source myenv/bin/activate

- Run the following command to download and install JupyterLab and its dependencies using pip:

  [In]: pip install jupyterlab

CHAPTER 2   SETTING UP A DEEP LEARNING ENVIRONMENT ON AWS

Step 2: Generate Jupyter Notebook configuration:

- Run the code below to generate the Jupyter Notebook configuration file:

    [In]: jupyter notebook --generate-config

    Generating the Jupyter Notebook configuration file allows you to customize various aspects of your Jupyter Notebook environment, such as specifying default settings, configuring security options, and setting up authentication mechanisms.

Step 3: Set password for JupyterLab:

[In]: jupyter notebook password

Step 4: Start JupyterLab:

- Run the following command to start JupyterLab on port 8888 and listen on all available IP addresses:

    [In]: jupyter lab --ip=0.0.0.0 --port=8888 --no-browser

Step 5: Configure security group and access JupyterLab:

- Update Security Group: Ensure that your EC2 instance's security group allows inbound traffic on port 8888.
  To do this

  - Go to the EC2 Dashboard in the AWS Management Console.

  - In the left-hand menu, under Network & Security, select Security Groups.

  - Choose the security group associated with your EC2 instance.

CHAPTER 2   SETTING UP A DEEP LEARNING ENVIRONMENT ON AWS

- Click the Inbound rules tab and then Edit inbound rules.

- Add a rule that allows Custom TCP traffic on port 8888 from your IP or from any IP (0.0.0.0/0) for testing purposes (use caution with this option for security reasons).

- Access JupyterLab: Open a web browser and navigate to http://<your_ec2_instance_ip>:8888.

You should now be able to access Jupyter your EC2 instance using the password you generated.

Setting Up a Databricks Account and Workspace

If you prefer to use Databricks for preprocessing your data, the steps below provide an overview of creating a Databricks account, setting up a workspace, and connecting it with your AWS account:

Step 1: Sign up for Databricks.

- Go to the Databricks website at https://www.databricks.com/.

- Click the "Try Databricks" button at the top-right corner.

- Fill in the required information to create your account, including your first and last name, email address, company, title, and country, and pick your cloud provider (in this case, AWS; the others being Microsoft Azure and Google Cloud Platform).

CHAPTER 2　SETTING UP A DEEP LEARNING ENVIRONMENT ON AWS

Step 2: Confirm your email.

- After signing up, check your email for a verification message from Databricks.
- Click the verification link in the email to confirm your email address.

Step 3: Log into Databricks.

- Once your email is confirmed, log into the Databricks platform using your credentials.

Step 4: Create a workspace.

- After logging in, you will be prompted to create a Databricks workspace.
- Choose a name for your workspace.
- Follow the on-screen instructions to complete the workspace creation process.

Step 5: Explore Databricks.

- Once your workspace is created, you can explore the Databricks environment.
- Familiarize yourself with the workspace interface, notebooks, and other features.

Step 6: Access S3 from Databricks.

- When working with Databricks, it's possible to use different methods to access an S3 bucket. For testing and quick setups, you can run the code below in Databricks:

```
[In]: ACCESS_KEY = "your-access-key"
[In]: SECRET_KEY = "your-secret-key"
```

CHAPTER 2   SETTING UP A DEEP LEARNING ENVIRONMENT ON AWS

```
[In]: AWS_BUCKET_NAME = "your-bucket-name"
[In]: MOUNT_NAME = "/mnt/datalake-central"

[In]: dbutils.fs.mount(
          source=f"s3a://{ACCESS_KEY}:{SECRET_KEY}
          @{AWS_BUCKET_NAME}/",
              mount_point=MOUNT_NAME,
              extra_configs={"fs.s3a.access.key":
              ACCESS_KEY,
                        "fs.s3a.secret.key":
                        SECRET_KEY}
      )
```

This code has two main parts—variable assignment and mounting the S3 bucket:

Variable Assignment:

- ACCESS_KEY: Retrieves the AWS access key from environment variables.

- SECRET_KEY: Retrieves the AWS secret key from environment variables.

- AWS_BUCKET_NAME: Stores the name of the S3 bucket you want to mount.

- MOUNT_NAME: Specifies the mount point for the S3 bucket. In this case, the mount name is "datalake-central," but it can be any name. Including /mnt/ before the mount name is a common convention in Linux environments, especially when mounting external storage devices or network shares.

CHAPTER 2   SETTING UP A DEEP LEARNING ENVIRONMENT ON AWS

Mounting the S3 Bucket:

- dbutils.fs.mount: This function is used to mount the S3 bucket. It takes several parameters:
  - source: The URL of the S3 bucket, including the access key and secret key. This format allows Databricks to authenticate with the S3 bucket using the provided credentials.
  - mount_point: The local mount point where the S3 bucket will be mounted.
  - extra_configs: Additional configurations, in this case, the access key and secret key for the S3 bucket. The extra_configs parameter provides additional configurations for the mount operation. Here, it includes the access key and secret key again, which are necessary for accessing the S3 bucket.

By executing this code, you will mount the specified S3 bucket to the specified mount point in your Databricks environment, allowing you to interact with the S3 bucket as if it were a local file system.

You can obtain the access key and secret key required for the code from the AWS Management Console as follows:

- Go to the AWS Management Console (https://aws.amazon.com/console/) and sign in with your AWS account credentials.

- Once you're logged in, navigate to the IAM (Identity and Access Management) service by selecting it from the list of services in the AWS Management Console.

- In the IAM console, select "Users" from the navigation pane on the left.

105

CHAPTER 2   SETTING UP A DEEP LEARNING ENVIRONMENT ON AWS

- Choose the IAM user for which you want to create an access key and secret key pair.

- In the "Security credentials" tab, scroll down to the "Access keys" section.

- Click the "Create access key" button.

- Select "Application running on an AWS compute service" under "Use case" and confirm the message "I understand the above recommendation and want to proceed to create an access key" at the bottom of the screen.

- Once the access key is created, you'll see the access key ID and the secret access key. Take note of these credentials as they will be displayed only once. If you lose them, you'll need to create a new access key.

- Ensure that the IAM user or role associated with the access key pair has the necessary permissions for S3 actions. This can be achieved by attaching the appropriate policies such as AmazonS3FullAccess.

One issue with this method is the security risk arising from hardcoding credentials in your code, which can lead to accidental exposure. Another issue relates to maintenance: updating credentials requires code changes, which can be cumbersome.

A better solution is to use IAM roles, which AWS manages securely without needing to specify the credentials required to access S3 in your code. This approach is more secure and scalable, especially in production environments.

An IAM role is an AWS identity that you can create and assign specific permissions to, allowing trusted entities (such as EC2 instances running Databricks) to assume the role and perform actions in AWS. When an IAM role is attached to the EC2 instances hosting your Databricks cluster,

AWS automatically provides temporary credentials to the instances. These credentials are rotated and managed by AWS, eliminating the need to embed access keys in your code.

Follow these steps to set up and use IAM roles in Databricks:

Step 1: Create an IAM role with S3 access.

- Sign in to the AWS Management Console and go to the IAM service.
- Create a new role:
  - Click Roles in the left-hand menu and then click Create role.
  - Select EC2 as the trusted entity because the role will be assumed by the EC2 instances running Databricks.
- Attach policies:
  - Attach the AmazonS3FullAccess policy to the role, or create a custom policy with specific S3 permissions tailored to your needs.
- Complete the role creation:
  - Name the role (e.g., DatabricksS3AccessRole) and finish the creation process.

Step 2: Attach the IAM role to the Databricks cluster.

- In the AWS Management Console, navigate to the EC2 Dashboard.
- Identify the EC2 instances that make up your Databricks cluster.
- Attach the IAM role:

CHAPTER 2 SETTING UP A DEEP LEARNING ENVIRONMENT ON AWS

- Select each instance, go to Actions ➤ Security ➤ Modify IAM role, and attach the DatabricksS3AccessRole you created.

Step 3: Access S3 from Databricks without hardcoding credentials.

- With the IAM role successfully created and attached to your Databricks cluster, you're now ready to securely access your S3 bucket without the need to hardcode any credentials. The following code snippet demonstrates how to mount the S3 bucket using the IAM role:

```
AWS_BUCKET_NAME = "your-bucket-name"
MOUNT_NAME = "/mnt/datalake-central"
dbutils.fs.mount(
    source=f"s3a://{AWS_BUCKET_NAME}/",
    mount_point=MOUNT_NAME
)
```

You can test if the mount has succeeded by running the code below:

```
[In]: MOUNT_NAME = "/mnt/datalake-central"
[In]: mounts = dbutils.fs.mounts()
[In]: mount_point_exists = any(
        mount.mountPoint == MOUNT_NAME for mount in mounts
    )
[In]: if mount_point_exists:
        print("S3 bucket was successfully mounted.")
[In]: else:
        print("S3 bucket mount failed.")

[Out]: S3 bucket was successfully mounted.
```

CHAPTER 2   SETTING UP A DEEP LEARNING ENVIRONMENT ON AWS

Here's a breakdown of each line of code:

- MOUNT_NAME = "/mnt/datalake-central": This line defines a variable named MOUNT_NAME and assigns it the value "/mnt/datalake-central". This variable stores the path to the mount point in the Databricks file system where the S3 bucket is mounted.

- mounts = dbutils.fs.mounts(): This line retrieves a list of all currently mounted file systems in the Databricks environment using the mounts() function from the dbutils.fs module. Each item in the list represents a mounted file system.

- mount_point_exists = any(mount.mountPoint == MOUNT_NAME for mount in mounts): This line checks if the specified mount point exists among the mounted file systems. It uses a generator expression with the any() function to iterate over each mounted file system (mount) and checks if its mountPoint attribute matches the value stored in the MOUNT_NAME variable.

- if mount_point_exists:: This line starts an if statement, which checks if the mount_point_exists variable is True. If the specified mount point exists among the mounted file systems, the condition evaluates to True, and the code block inside the if statement will be executed.

- print("S3 bucket was successfully mounted."): This line prints a message indicating that the S3 bucket was successfully mounted. It is executed if the specified mount point exists among the mounted file systems.

CHAPTER 2   SETTING UP A DEEP LEARNING ENVIRONMENT ON AWS

- else:: This line starts the else block, which executes if the condition in the if statement (checking if the mount point exists) evaluates to False. In other words, if the specified mount point does not exist among the mounted file systems, the code block inside the else statement will be executed.

- print("S3 bucket mount failed."): This line prints a message indicating that the mount operation for the S3 bucket failed. It is executed if the specified mount point does not exist among the mounted file systems.

Now that you have successfully mounted the S3 bucket onto Databricks using the dbutils.fs.mount function, you can start writing to and reading from S3 inside Databricks. The code below creates a taco DataFrame, saves it in the mounted S3, and then reads it again:

```
[In]: data_to_write = [("Chicken Taco", 5.99),
                       ("Beef Taco", 6.49),
                       ("Vegetarian Taco", 4.99)]
[In]: df = spark.createDataFrame(data_to_write, ["Taco", "Price"])
[In]: df = df.repartition(1)
[In]: MOUNT_NAME = "/mnt/datalake-central"
[In]: df.write.mode("overwrite").parquet(
          MOUNT_NAME + "/taco_data_parquet"
      )
[In]: df_read = spark.read.parquet(MOUNT_NAME + "/taco_data_parquet")
[In]: df_read.show()
```

CHAPTER 2   SETTING UP A DEEP LEARNING ENVIRONMENT ON AWS

Let's go through the different sections of the code:

1. Creating DataFrame:

   ```
   [In]: data_to_write = [("Chicken Taco", 5.99),
                          ("Beef Taco", 6.49),
                          ("Vegetarian Taco", 4.99)]
   [In]: df = spark.createDataFrame(data_to_write,
         ["Taco", "Price"])
   ```

   This section creates a DataFrame named df with taco data. It uses the spark.createDataFrame() method to create a DataFrame from a list of tuples data_to_write, where each tuple contains a taco name and its price. The column names "Taco" and "Price" are specified in the second argument.

2. Saving DataFrame:

   ```
   [In]: df = df.repartition(1)
   [In]: MOUNT_NAME = "/mnt/datalake-central"
   [In]: df.write.mode("overwrite").parquet(
             MOUNT_NAME + "/taco_data_parquet"
         )
   ```

   This section repartitions the DataFrame to a single partition using df.repartition(1). A partition in Spark is a logical division of the dataset that allows for parallel processing. Then, it defines the mount point /mnt/datalake-central. Finally, it saves the DataFrame to the mounted S3 bucket as a Parquet file named "taco_data_parquet," overwriting existing files if they exist, using df.write.mode("overwrite"). parquet().

CHAPTER 2 SETTING UP A DEEP LEARNING ENVIRONMENT ON AWS

The file will be saved on S3 on the EC2 instance in a folder named "taco_data_parquet/" as

```
part-00000-tid-6689481035778156686-b47d6778-
9a4c-48b3-950c-3ac770289ab1- \
453-1- c000.snappy.parquet
```

along with other files such as the "_SUCCESS" file, which indicates that the write operation completed successfully. The snappy file contains the actual data, while the other files can be ignored as they are part of Spark's internal processes.

3. Reading DataFrame:

```
[In]: df_read = spark.read.parquet
        (MOUNT_NAME + "/taco_data_parquet")
```

This section reads the Parquet file "taco_data_parquet" from the mounted S3 bucket back into a DataFrame named df_read using spark.read.parquet().

4. Displaying DataFrame:

```
[In]: df_read.show()
```

This final section displays the contents of the DataFrame df_read using the show() method. The output would show the contents of the DataFrame with columns "Taco" and "Price" as follows:

Taco	Price
Chicken Taco	5.99
Beef Taco	6.49
Vegetarian Taco	4.99

Summary

In this chapter, we guided the reader through the process of setting up an environment for deep learning on Amazon Web Services (AWS) and configuring the necessary tools and services. More specifically, we provided detailed instructions on how to create an AWS account, provision an EC2 instance, set up Amazon S3, create a project directory, set up a virtual environment, and install and configure dependencies, specifically PySpark, PyTorch, TensorFlow, Airflow, Boto3, and JupyterLab. We also provided instructions on how to set up a Databricks account and workspace.

In the next chapter, we will cover the process of preparing data for deep learning tasks using PySpark. We will delve into crucial aspects of data exploration, data preprocessing, feature engineering, and formatting data for PyTorch and TensorFlow.

CHAPTER 3

Data Preparation with PySpark for Deep Learning

This chapter covers the process of preparing data for deep learning tasks using PySpark. We delve into the crucial aspects of data preprocessing, feature engineering, and formatting data for PyTorch and TensorFlow.

Data preprocessing includes a series of steps aimed at cleaning and transforming raw data to make it suitable for modeling purposes. These steps include addressing missing values, scaling features, and encoding categorical variables. To illustrate these concepts, the chapter provides code examples including the assembly of feature vectors using VectorAssembler and their subsequent scaling using StandardScaler.

Feature engineering plays an important role in enhancing the quality and predictive power of input features. It involves the creation of new features or the transformation of existing ones to encapsulate additional information. Techniques may include generating polynomial features, deriving interaction terms, and handling datetime features such as calculating the difference between two timestamps or the time elapsed since a particular event.

CHAPTER 3   DATA PREPARATION WITH PYSPARK FOR DEEP LEARNING

Furthermore, the chapter addresses the crucial step of formatting data for deep learning tasks. After preprocessing and feature engineering, the processed data needs to be converted into tensor formats compatible with PyTorch and TensorFlow for efficient processing by these frameworks. We provide code examples to illustrate the conversion of data into PyTorch and TensorFlow tensors, along with the organization of data into batches using DataLoader objects created by data loader functions.

Before diving into these topics, it's important to first consider how PySpark's parallel processing capabilities can enhance the efficiency of data handling, particularly with large datasets. Techniques such as repartitioning, broadcasting joins, caching, using efficient file formats such as Parquet, optimizing shuffle partitions, and enabling Adaptive Query Execution (AQE) will play a key role in optimizing the performance of these data preparation tasks.

To illustrate these concepts of PySpark's parallel processing and data preparation for large datasets and deep learning, we will use a historical dataset containing Tesla stock prices. This dataset will serve as a demonstration for deep learning tasks in PyTorch and TensorFlow, specifically in regression models covered in Chapters 4 and 5, respectively.

The Dataset

We downloaded the Tesla stock price dataset in CSV format from Yahoo! Finance website and stored it in an AWS S3 bucket using the following link:

```
https://finance.yahoo.com/quote/TSLA/history/
```

Let's first explore this dataset before using it to illustrate PySpark's parallel processing and data preparation for deep learning with PyTorch and TensorFlow. Data exploration serves as a foundation for understanding the characteristics and patterns within the dataset, guiding the subsequent preprocessing and feature engineering steps. Our exploration will include printing the first ten rows of the dataset,

CHAPTER 3 DATA PREPARATION WITH PYSPARK FOR DEEP LEARNING

performing descriptive statistics, checking for missing values, and visualizing the data. Through this exploration using PySpark, we gain insights into the data structure and distribution, laying the groundwork for effective data preparation for deep learning tasks.

We begin by importing the necessary libraries for the script:

```
[In]: import boto3
[In]: import matplotlib.pyplot as plt
[In]: from pyspark.sql import SparkSession
[In]: from pyspark.sql.functions import col
[In]: import pyspark.sql.functions as F
[In]: import numpy as np
```

Following are the explanations for the libraries:

- boto3: This library is used for interacting with AWS services. It will be utilized to download the Tesla stock CSV file from an S3 bucket in the copy_file_from_s3 function.

- matplotlib.pyplot: This library will be employed to create visualizations based on DataFrame in the visualize_data method.

- SparkSession from pyspark.sql: This is the entry point to programming Spark with the DataFrame API. It will be initialized to create a SparkSession object for data processing operations.

- col from pyspark.sql.functions: This function will be utilized to reference DataFrame columns in the check_for_null_values method.

CHAPTER 3   DATA PREPARATION WITH PYSPARK FOR DEEP LEARNING

- F from pyspark.sql.functions: This is an alias for pyspark.sql.functions. It's imported to abbreviate function calls to PySpark functions, used in various methods.

- numpy as np: This library will be utilized for numerical operations and to adjust tick placement in visualizations.

We then define a class named DataProcessor, which is intended to handle various data processing tasks within the context of a SparkSession environment. Inside the class, we define several methods for loading the data, performing descriptive statistics, checking for null values, and visualizing the data:

```
[In]: class DataProcessor:
          def __init__(self, spark_session):
              self.spark = spark_session
```

- Within the class definition, there is an __init__ method serving as the constructor.

- The __init__ method takes one parameter, spark_session, representing a SparkSession object.

- Inside the __init__ method, the self.spark attribute is assigned the spark_session object passed as an argument. This assignment ensures that each instance of the DataProcessor class has access to a SparkSession object.

Inside the class, we start by defining the load_data method, which is designed to load Tesla stock price data from a CSV file using SparkSession:

CHAPTER 3　DATA PREPARATION WITH PYSPARK FOR DEEP LEARNING

```
[In]: def load_data(self, file_path: str):
          """
          Load stock price data from a CSV file using
          SparkSession.
          """
          try:
              df = self.spark.read.csv(
                  file_path,
                  header=True,
                  inferSchema=True
              )
              return df
          except Exception as e:
              print(f"Error loading data: {str(e)}")
              return None
```

- The method takes two parameters: self, a reference to the instance of the class, and file_path, a string representing the path to the CSV file containing Tesla stock price data.

- A documentation string (docstring) enclosed in triple quotes provides a brief explanation of the method's purpose, which is to load stock price data from a CSV file using SparkSession.

- The method begins with a try block, indicating that the code inside will be executed, and any exceptions that occur will be caught and handled.

- Inside the try block, the method uses the read.csv method of the SparkSession object (self.spark) to read the CSV file specified by the file_path parameter. The

119

CHAPTER 3   DATA PREPARATION WITH PYSPARK FOR DEEP LEARNING

arguments header=True and inferSchema=True are provided to specify that the first row of the CSV file contains column headers and to infer the schema of the DataFrame based on the data types of the columns.

- If the data is successfully loaded, the method returns the DataFrame containing the stock price data. In case an exception occurs during the data loading process (e.g., file not found, invalid file format), the method catches the exception and executes the code within the except block.

- Within the except block, an error message is printed indicating the nature of the error, and None is returned to signify that the data loading process was unsuccessful.

Subsequently, the print_first_n_rows method inside the class prints the first n rows of the DataFrame:

```
[In]: def print_first_n_rows(self, df, n=10):
        """Print the first n rows of the DataFrame."""
        print(f"First {n} rows of the DataFrame:")
        df.show(n)
```

- The method takes three parameters: self, a reference to the instance of the class; df, the DataFrame whose first n rows are to be printed; and n, an integer representing the number of rows to print (default is 10).

- A docstring provides a brief explanation of the method's purpose, which is to print the first n rows of the DataFrame.

CHAPTER 3 DATA PREPARATION WITH PYSPARK FOR DEEP LEARNING

- The method begins by printing a message indicating the number of rows that will be printed, using f-string formatting to include the value of n.

- The show method of the DataFrame (df) is called with the argument n, which specifies the number of rows to display.

Moving forward, the calculate_descriptive_statistics method calculates descriptive statistics of the DataFrame:

```
[In]: def calculate_descriptive_statistics(self, df):
        """Calculate descriptive statistics of the
        DataFrame."""
        print("Descriptive Statistics:")
        df.summary().show()
```

- The method takes two parameters: self, a reference to the instance of the class, and df, the DataFrame for which descriptive statistics are to be calculated.

- A docstring provides a brief explanation of the method's purpose, which is to calculate descriptive statistics of the DataFrame.

- The method begins by printing a message indicating that descriptive statistics are being calculated.

- The summary method of the DataFrame (df) is called. This method computes summary statistics including count, mean, standard deviation, and minimum and maximum values.

Continuing with the data exploration methods inside the DataProcessor class, the visualize_data method below creates visualizations based on DataFrame:

121

CHAPTER 3   DATA PREPARATION WITH PYSPARK FOR DEEP LEARNING

```
[In]: def visualize_data(self, df):
          """Create visualizations based on DataFrame."""
          print("Data Visualization:")
          try:
              df_pd = df.toPandas()
              plt.figure(figsize=(10, 6))
              plt.plot(df_pd['Date'], df_pd['Close'])
              plt.xlabel('Date')
              plt.ylabel('Closing Price')
              plt.title('Tesla Stock Closing Prices Over Time')
              plt.xticks(rotation=45, ha='right')
              plt.gca().invert_xaxis()
              plt.xticks(
                  np.arange(
                      0,
                      len(df_pd['Date']),
                      step=max(len(df_pd['Date']) // 10, 1)
                  )
              )
              plt.show()
          except Exception as e:
              print(f"Error visualizing data: {str(e)}")
```

- The method takes two parameters: self, a reference to the instance of the class, and df, the DataFrame from which visualizations are to be created.

- A docstring provides a brief explanation of the method's purpose, which is to create visualizations based on the DataFrame.

- The method begins by printing a message indicating that data visualization is being performed.

CHAPTER 3   DATA PREPARATION WITH PYSPARK FOR DEEP LEARNING

- Inside a try block, the method converts the Spark DataFrame (df) into a Pandas DataFrame (df_pd) using the toPandas method. This conversion is necessary for plotting with Matplotlib.

- Matplotlib is then used to create a line plot of the Tesla stock closing prices over time. The plt.plot function is called with df_pd['Date'] as the x-axis data and df_pd['Close'] as the y-axis data.

- The method sets various properties of the plot, such as labels for the x- and y-axes, title, and rotation of x-axis ticks for better readability.

- The x-axis is inverted using plt.gca().invert_xaxis() to ensure that the dates are displayed in chronological order.

- The tick positions on the x-axis are adjusted using plt.xticks to improve readability. This line ensures that there are approximately ten ticks on the x-axis.

- Finally, the plot is displayed using plt.show().

- If an exception occurs during the data visualization process (e.g., invalid data for plotting), the method catches the exception and executes the code within the except block. It prints an error message indicating the nature of the error.

Transitioning to the next step, the check_for_null_values method checks for missing values in the DataFrame:

```
[In]: def check_for_null_values(self, df):
          """Check for null values in the DataFrame."""
          print("Null Value Check:")
```

CHAPTER 3   DATA PREPARATION WITH PYSPARK FOR DEEP LEARNING

```
        null_counts = df.select([col(c).isNull()\
            .cast("int").alias(c) for c in df.columns])\
            .agg(*[F.sum(c).alias(c) for c in df.columns])\
            .toPandas()
        print(null_counts)
```

- The method takes two parameters: self, a reference to the instance of the class, and df, the DataFrame in which null values are to be checked.

- A docstring provides a brief explanation of the method's purpose, which is to check for null values in the DataFrame.

- The method begins by printing a message indicating that a null value check is being performed.

- Inside a single line, the method uses a combination of Spark DataFrame operations to calculate the number of null values for each column:

    - df.select(...): Selects the columns of the DataFrame.

    - [col(c).isNull().cast("int").alias(c) for c in df.columns]: Constructs a list comprehension to iterate over each column (c) in the DataFrame's columns. For each column, it creates an expression to cast the null values to integer (1 for null, 0 for non-null) and alias the result with the column name.

    - .agg(*[F.sum(c).alias(c) for c in df.columns]): Aggregates the counts of null values for each column using the sum function. The * before the list comprehension is used to unpack the list comprehension into separate arguments for the agg function.

CHAPTER 3   DATA PREPARATION WITH PYSPARK FOR DEEP LEARNING

- .toPandas(): This method converts the result DataFrame to a Pandas DataFrame to make it easier to display and analyze the results. Converting the DataFrame to a Pandas format allows for smooth integration with the Pandas ecosystem, enabling users to leverage its rich set of functionalities for further analysis and visualization.
- The result DataFrame, containing the counts of null values for each column, is stored in the null_counts variable.
- The null_counts DataFrame is then printed to display the counts of null values for each column.

Turning now to the functions outside the DataProcessor class, we have the copy_file_from_s3 function, which copies the CSV file from the S3 bucket to the local file path on EC2 instance:

```
[In]: def copy_file_from_s3(
        bucket_name: str,
        file_key: str,
        local_file_path: str
    ):
        """
        Copy file from S3 bucket to local file path.
        """
        try:
            s3 = boto3.client('s3')
            s3.download_file(bucket_name, file_key, local_file_path)
            print(f"File downloaded from S3 bucket {bucket_name}"
```

125

CHAPTER 3 DATA PREPARATION WITH PYSPARK FOR DEEP LEARNING

```
            f"to local file path: {local_file_path}"
        )
    except Exception as e:
        print(f"Error copying file from S3: {str(e)}")
```

- Inside the function, a try–except block is used to handle any potential errors that may occur during the file copying process.

- Within the try block

  - An S3 client object (s3) is created using the boto3.client('s3') method. This client provides access to the Amazon S3 service.

  - The s3.download_file() method is then called to download the file from the specified S3 bucket (bucket_name) and key (file_key) to the local file path (local_file_path).

  - After the file is successfully downloaded, a confirmation message is printed indicating the source S3 bucket and the destination local file path.

- If an exception occurs during the file copying process (e.g., due to network issues or incorrect permissions), it is caught by the except block.

- Inside the except block

  - An error message is printed, indicating the nature of the error that occurred during the file copying process. This helps in diagnosing and troubleshooting any issues that arise.

CHAPTER 3   DATA PREPARATION WITH PYSPARK FOR DEEP LEARNING

Moving on to the main function, this is the main entry point of the script. It initializes SparkSession, copies the file from S3, initializes DataProcessor, loads data, and performs data processing tasks:

```
[In]: def main(s3_bucket_name: str, s3_file_key: str, local_
      file_path:
          str):
          """
          Main function for stock price prediction.

          This function initializes a SparkSession, copies
          a file from an S3 bucket to a local file path,
          initializes a DataProcessor object, loads data
          from the local file, and performs various data
          processing tasks.

          :param s3_bucket_name:
              Name of the S3 bucket containing the file
          :param s3_file_key:
              Key of the file within the S3 bucket
          :param local_file_path:
              Local file path to save the downloaded file
          """
          spark = (SparkSession.builder
                  .appName("StockPricePrediction")
                  .getOrCreate())

          copy_file_from_s3(
              s3_bucket_name,
              s3_file_key,
              local_file_path
          )
```

CHAPTER 3   DATA PREPARATION WITH PYSPARK FOR DEEP LEARNING

```
    data_processor = DataProcessor(spark)

    df = data_processor.load_data(local_file_path)
    if df is not None:
        data_processor.print_first_n_rows(df)
        data_processor.calculate_descriptive_
        statistics(df)
        data_processor.check_for_null_values(df)
        data_processor.visualize_data(df)
    else:
        print("Error loading data. Exiting.")
```

- The function takes three parameters: s3_bucket_name, s3_file_key, and local_file_path, which represent the name of the S3 bucket containing the file, the key of the file within the S3 bucket, and the local file path to save the downloaded file, respectively.

- Inside the function

  - A SparkSession object is created using the SparkSession.builder method, with an application name of "StockPricePrediction". If a SparkSession already exists, it will be reused; otherwise, a new one will be created.

  - The copy_file_from_s3 function is called to copy the file from the specified S3 bucket (s3_bucket_name) and key (s3_file_key) to the local file path (local_file_path).

  - A DataProcessor object is initialized with the SparkSession created earlier.

CHAPTER 3 DATA PREPARATION WITH PYSPARK FOR DEEP LEARNING

- Data is loaded from the local file using the DataProcessor's load_data method. If the DataFrame is successfully loaded (df is not None), various data processing tasks are performed using methods of the DataProcessor object:

    - Printing the first n rows of the DataFrame (print_first_n_rows)

    - Calculating descriptive statistics of the DataFrame (calculate_descriptive_statistics)

    - Checking for null values in the DataFrame (check_for_null_values)

    - Creating visualizations based on DataFrame (visualize_data)

- If an error occurs during data loading, an error message is printed, and the program exits with the message "Error loading data. Exiting."

Proceeding to the final step, the conditional block checks if the script is being run directly as the main program. It sets up parameters for S3 bucket name, file key, and local file path and then calls the main function with these parameters:

```
[In]: if __name__ == "__main__":
         s3_bucket_name = 'instance1bucket'
         s3_file_key = 'TSLA_stock.csv'
         local_file_path = '/home/ubuntu/airflow/dags/TSLA_stock.csv'
         main(s3_bucket_name, s3_file_key, local_file_path)
```

CHAPTER 3   DATA PREPARATION WITH PYSPARK FOR DEEP LEARNING

- Inside the block, three variables are defined:
  - s3_bucket_name: A string variable representing the name of the S3 bucket where the file is located
  - s3_file_key: A string variable representing the key of the file within the S3 bucket
  - local_file_path: A string variable representing the local file path where the file will be downloaded
- The main function is called with these variables as arguments. This function orchestrates the entire process of data processing, including initializing SparkSession, copying the file from the S3 bucket, loading the data, and performing various data processing tasks.

By encapsulating the code within the main function and executing it conditionally, the script can be imported as a module in other scripts without executing the main functionality. This enhances code reusability and modularity. However, when running the script in Airflow as a Directed Acyclic Graph (DAG) in Chapter 8, the main function is not needed because Airflow orchestrates the execution of tasks within the DAG. Each task in the DAG can utilize specific functions or classes from the script without executing the main functionality.

To run the script, we first save it as tesla_stock_exploration.py, activate the virtual environment on the EC2 instance, and then run the Python script using the following commands:

```
[In]: cd /home/ubuntu
[In]: source myenv/bin/activate
[In]: python3 tesla_stock_exploration.py
```

CHAPTER 3   DATA PREPARATION WITH PYSPARK FOR DEEP LEARNING

Let's now examine the output from the code.

The first output is a message that will be produced by the copy_file_from_s3 function, indicating that the CSV file was successfully downloaded from an S3 bucket to the local file path:

```
[Out]: File downloaded from S3 bucket instance1bucket to
       local file
       path: /home/ubuntu/airflow/dags/TSLA_stock.csv
```

The second output is a table with the first ten rows of the DataFrame, generated by the print_first_n_rows method. This displays the Date, Open, High, Low, Close, and Volume columns of the Tesla stock. Printing the first ten observations provides a simple and effective way to obtain a quick overview of the dataset and verify that everything is in order before proceeding with more detailed analysis or processing:

Date	Open	High	Low	Close	Volume
2/23/24	195.31	197.57	191.50	191.97	78,670,300
2/22/24	194.00	198.32	191.36	197.41	92,739,500
2/21/24	193.36	199.44	191.95	194.77	103,844,000
2/20/24	196.13	198.60	189.13	193.76	104,545,800
2/16/24	202.06	203.17	197.40	199.95	111,173,600
2/15/24	189.16	200.88	188.86	200.45	120,831,800
2/14/24	185.30	188.89	183.35	188.71	81,203,000
2/13/24	183.99	187.26	182.11	184.02	86,759,500
2/12/24	192.11	194.73	187.28	188.13	95,498,600
2/09/24	190.18	194.12	189.48	193.57	84,476,300

CHAPTER 3   DATA PREPARATION WITH PYSPARK FOR DEEP LEARNING

The third output is a summary statistic for each numerical column in the dataset, produced by the calculate_descriptive_statistics method within the DataProcessor class:

Summary	Open	High	Low	Close	Volume
count	1258	1258	1258	1258	1,258
mean	176.37	180.31	172.10	176.31	133,933,068
stddev	105.44	107.68	102.90	105.28	85,052,921
min	12.07	12.45	11.80	11.93	29,401,800
25%	57.20	59.10	55.59	57.63	81,203,000
50%	202.59	208.00	198.50	203.33	109,536,700
75%	251.45	256.59	246.35	251.92	157,577,100
max	411.47	414.50	405.67	409.97	914,082,000

The output provides a comprehensive summary of the numerical data in the dataset, including measures of central tendency, dispersion, and distribution. It helps in understanding the overall characteristics and variability of the data.

Below is an interpretation of this data:

Count:

- There are 1,258 observations for each numerical column.

Mean:

- The average values for each numerical column are as follows:

Open: 176.4
High: 180.3
Low: 172.1

CHAPTER 3   DATA PREPARATION WITH PYSPARK FOR DEEP LEARNING

Close: 176.3
Volume: 133,933,068
Standard Deviation:

- The standard deviation values for each numerical column are as follows:

Open: 105.4
High: 107.7
Low: 102.9
Close: 105.3
Volume: 85,052,921

These standard deviation values indicate the amount of deviation or spread from the mean value for each column. A higher standard deviation implies more dispersion in the data, while a lower standard deviation suggests less dispersion.

Minimum:

- The minimum values for each numerical column are as follows:

Open: 12.1
High: 12.5
Low: 11.8
Close: 11.9
Volume: 29,401,800

25th Percentile (Q1):

- The 25th percentile values for each numerical column are as follows:

Open: 57.2
High: 59.1
Low: 55.6
Close: 57.6
Volume: 81,203,000

CHAPTER 3　DATA PREPARATION WITH PYSPARK FOR DEEP LEARNING

These percentile values indicate the data distribution. For instance, the 25th percentile for the Open column is 57.2, indicating that 25% of the data points in this column fall below this value.

50th Percentile (Median or Q2):

- The median values for each numerical column are as follows:

Open: 202.6
High: 208.0
Low: 198.5
Close: 203.3
Volume: 109,536,700

These median values indicate the central tendency of the data within each column. In other words, they represent the value that divides the dataset into two equal halves. The median is often referred to as the 50th percentile because it divides the dataset into two equal parts, with 50% of the data points falling below it and 50% above it.

75th Percentile (Q3):

- The 75th percentile values for each numerical column are as follows:

Open: 251.5
High: 256.6
Low: 246.4
Close: 251.9
Volume: 157,577,100

These percentile values indicate the distribution of data within each column. Specifically, the 75th percentile value represents the value below which 75% of the data points in the column fall. The 75th percentile is often used in statistical analysis to understand the spread or variability of the data, particularly in relation to the median and other quartiles.

CHAPTER 3 DATA PREPARATION WITH PYSPARK FOR DEEP LEARNING

Maximum:

- The maximum values for each numerical column are as follows:

Open: 411.5

High: 414.5

Low: 405.7

Close: 409.9

Volume: 914,082,000

The next output, produced by the check_for_null_values method, indicates that for each column in the dataset, there are no null values present:

Date	Open	High	Low	Close	Volume
0	0	0	0	0	0

Handling null values is a critical aspect of machine learning because missing data can significantly impact the performance and accuracy of models. Imputation techniques are commonly used to address missing values. One simple method is mean/median/mode imputation, where missing values are replaced with the mean, median, or mode of the respective column.

Predictive imputation involves using machine learning models to predict and fill in missing values based on other features in the dataset. This method can provide more accurate imputations but requires additional modeling.

Deletion strategies involve removing data with missing values. Row deletion is used when the proportion of missing values is small; rows with missing data are removed, which is simple but may lead to the loss of valuable information if not done carefully. Column deletion may be employed if a column has a high percentage of missing values and is not crucial for analysis. This approach simplifies the dataset but might discard potentially useful information.

CHAPTER 3　DATA PREPARATION WITH PYSPARK FOR DEEP LEARNING

Feature engineering can also play a role in handling missing data. Creating a binary indicator variable to flag the presence of missing values helps the model learn patterns associated with missing data. This technique allows the model to account for the fact that data is missing and potentially use this information to improve predictions.

The fact that there are no missing values across any of the columns in the Tesla stock price dataset is a positive indicator, suggesting that the dataset is complete in terms of missing data. With no null values to handle, there is no immediate need for imputation or deletion of rows or columns containing missing values. The absence of null values simplifies the data preprocessing phase and instills confidence in the reliability of subsequent analyses or modeling tasks.

The final output is a graph of the stock closing price over time produced by the visualize_data method (Figure 3-1).

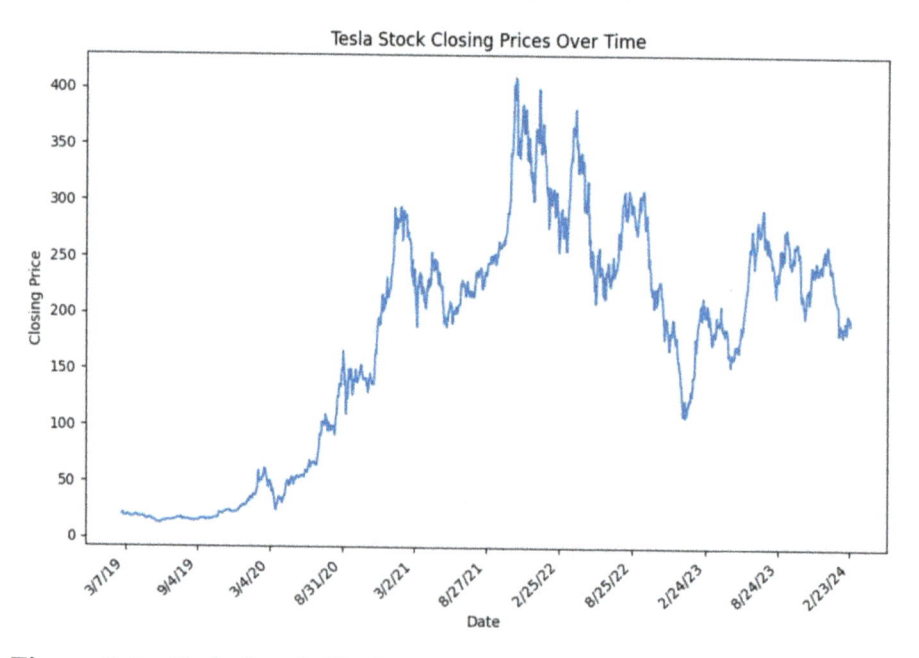

Figure 3-1. Tesla Stock Closing Price

CHAPTER 3   DATA PREPARATION WITH PYSPARK FOR DEEP LEARNING

Plotting the target variable over time as part of data exploration provides a visual summary of the data, aids in identifying patterns and anomalies, and guides subsequent modeling decisions, ultimately leading to more informed and effective predictive modeling. In our case, plotting the actual closing prices serves as a valuable reference point for evaluating the performance of our predictive models, as we will see in Chapters 4 and 5. When we generate predictions using our deep learning model, we can overlay the predicted closing prices on the same plot as the actual closing prices. This allows us to visually compare how well our model's predictions align with the actual values over time.

PySpark's Parallel Processing

In PySpark, parallel processing enhances the efficiency and speed of computations by breaking tasks into independent subtasks that run concurrently. This is achieved by distributing the processing across multiple data partitions. The ability to distribute computations across multiple cores and nodes in a cluster enables the processing of large datasets quickly.

To illustrate how parallelism can be optimized, we first focus on the technique of repartitioning. To do this, let's revise the Tesla stock price exploration code by removing functions that are not essential for demonstrating the impact of this technique, such as calculating descriptive statistics, visualizing data, and checking for null values. We will also add functionality to print the number of partitions before and after repartitioning to highlight its effect on data distribution and include a step to repartition the DataFrame into a specified number of partitions (e.g., 10), demonstrating how this operation can potentially enhance data processing performance.

CHAPTER 3   DATA PREPARATION WITH PYSPARK FOR DEEP LEARNING

To start with, the code below adds functionality to print the number of partitions before and after repartitioning:

```
[In]: def print_partition_info(self, df):
          """Print the number of partitions in the
          DataFrame."""
          num_partitions = df.rdd.getNumPartitions()
          print(f"number of partitions: {num_partitions}")
```

This code retrieves and displays the number of partitions in the Tesla stock price DataFrame using the getNumPartitions() method on the DataFrame's underlying RDD (Resilient Distributed Dataset). This helps us understand how data is distributed across partitions before and after repartitioning.

To repartition the DataFrame into a specified number of partitions (e.g., 10), we update the main function in the Tesla stock price exploration code as follows:

```
[In]: def main(
          s3_bucket_name: str,
          s3_file_key: str,
          local_file_path: str
      ):
          ...
          # Print initial partition info
          data_processor.print_partition_info(df)

          # Repartition the DataFrame
          repartitioned_df = df.repartition(10)
          print("After repartitioning:")
          data_processor.print_partition_info(repartitioned_df)

          # Print the first few rows of the repartitioned
          DataFrame
          data_processor.print_first_n_rows(repartitioned_df)
```

CHAPTER 3   DATA PREPARATION WITH PYSPARK FOR DEEP LEARNING

The main function now includes steps to print partition information before and after repartitioning. The data_processor.print_partition_info(df) method displays the number of partitions before repartitioning. The DataFrame is then repartitioned into ten partitions using df.repartition(10), redistributing the data to potentially improve performance or better manage data distribution. Finally, the updated partition information is printed to show the effects of repartitioning.

Below is the complete code for repartitioning using the Tesla stock price dataset:

```
[In]: import boto3
[In]: from pyspark.sql import SparkSession

[In]: class DataProcessor:
          def __init__(self, spark_session):
              self.spark = spark_session

          def load_data(self, file_path: str):
              """
              Load stock price data from a CSV file using
              SparkSession.
              """
              try:
                  df = self.spark.read.csv(
                      file_path, header=True, inferSchema=True
                  )
                  return df
              except Exception as e:
                  print(f"Error loading data: {str(e)}")
                  return None

          def print_first_n_rows(self, df, n=10):
              """Print the first n rows of the DataFrame."""
```

CHAPTER 3 DATA PREPARATION WITH PYSPARK FOR DEEP LEARNING

```
            print(f"First {n} rows of the DataFrame:")
            df.show(n)

        def print_partition_info(self, df):
            """Print the number of partitions in the
            DataFrame."""
            num_partitions = df.rdd.getNumPartitions()
            print(f"Initial number of partitions: {num_
            partitions}")
```

```
[In]: def copy_file_from_s3(
        bucket_name: str,
        file_key: str,
        local_file_path: str
    ):
        """
        Copy file from S3 bucket to local file path.
        """
        try:
            s3 = boto3.client('s3')
            s3.download_file(bucket_name, file_key, local_
            file_path)
            print(
                f"File downloaded from S3 bucket {bucket_
                name} to "
                f"local file path: {local_file_path}"
            )
         except Exception as e:
            print(f"Error copying file from S3: {str(e)}")
```

```
[In]: def main(
        s3_bucket_name: str,
        s3_file_key: str,
        local_file_path: str
```

140

):
    """
    Main function to demonstrate repartitioning of stock
    price data.

    This function initializes a SparkSession, copies
    a file from an S3 bucket to a local file path,
    initializes a DataProcessor object, loads data from
    the local file, prints partition info, repartitions
    the DataFrame, and prints partition info again.

    :param s3_bucket_name:
        Name of the S3 bucket containing the file
    :param s3_file_key:
        Key of the file within the S3 bucket
    :param local_file_path:
        Local file path to save the downloaded file
    """
    # Initialize SparkSession
    spark = SparkSession.builder \
        .appName("StockPriceRepartitioning") \
        .getOrCreate()

    # Copy file from S3 bucket
    copy_file_from_s3(
        s3_bucket_name,
        s3_file_key,
        local_file_path
)

    # Initialize DataProcessor
    data_processor = DataProcessor(spark)

```
        # Load data
        df = data_processor.load_data(local_file_path)
        if df is not None:
            # Print initial partition info
            data_processor.print_partition_info(df)

            # Repartition the DataFrame
            repartitioned_df = df.repartition(10)
            print("After repartitioning:")
            data_processor.print_partition_
            info(repartitioned_df)

            # Print the first few rows of the
            repartitioned DataFrame
            data_processor.print_first_n_
            rows(repartitioned_df)
        else:
            print("Error loading data. Exiting.")
```
```
[In]: if __name__ == "__main__":
        s3_bucket_name = 'instance1bucket'
        s3_file_key = 'TSLA_stock.csv'
        local_file_path = '/home/ubuntu/airflow/dags/TSLA_
        stock.csv'
        main(s3_bucket_name, s3_file_key, local_file_path)
```

[Out]:

Initial number of partitions: 1

After repartitioning:
number of partitions: 10

CHAPTER 3   DATA PREPARATION WITH PYSPARK FOR DEEP LEARNING

First 10 rows of the DataFrame:

Date	Open	High	Low	Close	Volume
9/26/19	15.38	16.22	15.16	16.17	178267500
8/25/22	302.36	302.96	291.60	296.07	53230000
9/13/23	270.07	274.98	268.10	271.30	111673700
11/26/19	22.35	22.37	21.81	21.93	119211000
5/17/21	191.85	196.58	187.07	192.28	97171200
6/4/19	12.07	12.93	11.97	12.91	207112500
4/3/19	19.15	19.74	19.14	19.45	118791000
6/23/21	210.67	219.07	210.01	218.86	93297600
11/20/20	166.00	167.50	163.02	163.20	98735700
12/22/20	216.00	216.63	204.74	213.45	155148000

In this example, the initial DataFrame had one partition, which was then repartitioned into ten partitions. With only one partition, parallel processing is limited because Spark can only process one chunk of data at a time. If the cluster has more available cores, they may remain underutilized since only one task is executed. By increasing the number of partitions from one to ten using the repartition(10) function, the Tesla stock price data is redistributed across ten partitions, making each partition smaller. Spark can now execute up to ten tasks in parallel. This means more data is processed simultaneously, reducing the overall time required to complete the analysis.

CHAPTER 3   DATA PREPARATION WITH PYSPARK FOR DEEP LEARNING

The number of partitions can be optimized based on the workload and data characteristics using Coalesce. For example, the code below reduces the number of partitions from ten to five in the previously repartitioned DataFrame:

```
[In]: coalesced_df = repartitioned_df.coalesce(5)
[In]: data_processor.print_partition_info(coalesced_df)
```

In the process of repartition, PySpark shuffles data between partitions, which is typically a costly operation in terms of time and resources. The number of these shuffles is defined by the spark.sql.shuffle.partitions configuration. By default, this is set to 200 in Spark 3.x, but it can be adjusted based on the size of the data and the resources available in the cluster. The code below shows how to set the number of shuffle partitions:

```
[In]: spark.conf.set("spark.sql.shuffle.partitions", "20")
```

There are other techniques to improve performance and parallelism in PySpark. One of these is saving the repartitioned DataFrame in Parquet format:

```
[In]: repartitioned_df.write.parquet(
        "/home/ubuntu/airflow/dags/tesla_stock.parquet")
```

Caching the repartitioned DataFrame to keep it in memory for faster access during subsequent operations is another technique to help improve performance:

```
[In]: repartitioned_df.cache()
```

This is especially useful when performing multiple actions on the same DataFrame, as it avoids recomputing the DataFrame each time it is used.

Another technique to improve performance when processing large datasets in PySpark is broadcasting. This involves efficiently joining a large DataFrame (e.g., repartitioned_df) with a smaller DataFrame (e.g., small_df):

CHAPTER 3   DATA PREPARATION WITH PYSPARK FOR DEEP LEARNING

```
[In]: from pyspark.sql.functions import broadcast
[In]: small_df = spark.createDataFrame([
        ("2023-08-25", 5.5),
        ("2023-08-26", 6.0), ],
       ["Date", "AdditionalData"])
# Broadcast join
[In]: joined_df = repartitioned_df.join(
        broadcast(small_df),
        on="Date",
        how="left"
      )
```

Broadcasting the smaller DataFrame allows Spark to distribute it to all worker nodes, thus avoiding the shuffle operation that would be required for a regular join. This can significantly improve performance for joins involving large and small datasets.

Saving large datasets in Parquet format is another method to improve performance when processing large datasets in PySpark as Parquet supports compression and optimized read operations. Saving the DataFrame in Parquet format helps in reducing storage space and improving the speed of read operations due to its optimized layout.

The code below saves the repartitioned Tesla stock price dataset as Parquet:

```
[In]: repartitioned_df.write.parquet(
        "/home/ubuntu/airflow/dags/tesla_stock.parquet")
```

Finally, enabling Adaptive Query Execution (AQE) optimizes query execution based on runtime statistics. It can adjust query plans to improve performance by making decisions like optimizing join strategies or adjusting the number of shuffle partitions based on the actual data processed:

145

CHAPTER 3   DATA PREPARATION WITH PYSPARK FOR DEEP LEARNING

```
[In]: spark.conf.set("spark.sql.adaptive.enabled", "true")
```

> **Note**   Adaptive Query Execution (AQE) is primarily associated with SQL operations in Apache Spark, particularly when using the DataFrame and Spark SQL APIs. However, since the DataFrame API in PySpark is an abstraction built on top of Spark SQL, many operations you perform with DataFrames (like groupBy, join, agg, etc.) are internally translated into SQL queries. Therefore, AQE can optimize these operations as well.

Below is an updated version of the Tesla Stock price repartition script incorporating the above techniques:

```
[In]: import boto3
[In]: from pyspark.sql import SparkSession
[In]: from pyspark.sql.functions import broadcast

[In]: class DataProcessor:
          def __init__(self, spark_session):
              self.spark = spark_session

          def load_data(self, file_path: str):
              try:
                  df = self.spark.read.csv(
                      file_path,
                      header=True,
                      inferSchema=True
                  )
                  return df
              except Exception as e:
                  print(f"Error loading data: {str(e)}")
                  return None
```

CHAPTER 3   DATA PREPARATION WITH PYSPARK FOR DEEP LEARNING

```
    def print_first_n_rows(self, df, n=10):
        print(f"First {n} rows of the DataFrame:")
        df.show(n)

    def print_partition_info(self, df):
        num_partitions = df.rdd.getNumPartitions()
        print(f"Number of partitions: {num_partitions}")
```

[In]:
```
def copy_file_from_s3(
    bucket_name: str,
    file_key: str,
    local_file_path: str
):
    try:
        s3 = boto3.client('s3')
        s3.download_file(bucket_name, file_key, local_
        file_path)
        print(f"File downloaded from S3 bucket
        {bucket_name} "
              f"to local file path: {local_file_path}"
        )
    except Exception as e:
        print(f"Error copying file from S3: {str(e)}")
```

[In]:
```
def main(
    s3_bucket_name: str,
    s3_file_key: str,
    local_file_path: str
):
    spark = SparkSession.builder \
        .appName("StockPriceOptimization") \
        .config("spark.sql.adaptive.enabled", "true") \
        .config("spark.sql.shuffle.partitions", "20") \
        .getOrCreate()
```

147

CHAPTER 3   DATA PREPARATION WITH PYSPARK FOR DEEP LEARNING

```
copy_file_from_s3(
    s3_bucket_name,
    s3_file_key,
    local_file_path
)
data_processor = DataProcessor(spark)
df = data_processor.load_data(local_file_path)

if df is not None:
    data_processor.print_partition_info(df)

    # Repartition the DataFrame
    repartitioned_df = df.repartition(10)
    data_processor.print_partition_
    info(repartitioned_df)

    # Coalesce partitions
    coalesced_df = repartitioned_df.coalesce(5)
    data_processor.print_partition_info(coalesced_df)

    # Example broadcast join
    small_df = spark.createDataFrame([
        ("2023-08-25", 5.5),
        ("2023-08-26", 6.0),
    ], ["Date", "AdditionalData"])
    joined_df = repartitioned_df.join(
        broadcast(small_df), on="Date", how="left")

    # Save as Parquet
    repartitioned_df.write.parquet(
        "/home/ubuntu/airflow/dags/tesla_stock.
        parquet"
    )
```

CHAPTER 3 DATA PREPARATION WITH PYSPARK FOR DEEP LEARNING

```
        # Print the first few rows
        data_processor.print_first_n_
        rows(repartitioned_df)
    else:
        print("Error loading data. Exiting.")
```
```
[In]: if __name__ == "__main__":
        s3_bucket_name = 'instance1bucket'
        s3_file_key = 'TSLA_stock.csv'
        local_file_path = '/home/ubuntu/airflow/dags/TSLA_
        stock.csv'
        main(s3_bucket_name, s3_file_key, local_file_path)
```

Data Preparation with PySpark for PyTorch

After exploring the Tesla stock dataset and demonstrating how to improve PySpark's parallel processing, we are now in a position to examine the topic of data preparation for deep learning frameworks, first focusing on PyTorch. Specifically, we examine how to perform data preprocessing, feature engineering, and data formatting using PySpark. We begin by importing necessary modules and then proceed to demonstrate each step of the data preparation process:

```
[In]: import logging
[In]: from pyspark.sql import SparkSession
[In]: from pyspark.ml.feature import VectorAssembler,
      StandardScaler
[In]: from pyspark.sql import DataFrame
[In]: import numpy as np
[In]: import torch
[In]: from torch.utils.data import DataLoader, TensorDataset
[In]: import boto3
```

CHAPTER 3   DATA PREPARATION WITH PYSPARK FOR DEEP LEARNING

This code imports necessary modules and libraries for performing data preparation tasks and interacting with S3. More specifically

- logging: The logging module will be used throughout the code to log messages, warnings, and errors during program execution.

- SparkSession from pyspark.sql: SparkSession will be used in the load_data() function to create a SparkSession object for reading CSV data and creating DataFrame objects.

- VectorAssembler and StandardScaler from pyspark. ml.feature: These will be used in the preprocess_data() function for feature engineering. VectorAssembler will assemble raw feature columns into a single feature vector column, while StandardScaler will scale input features to have zero mean and unit standard deviation.

- DataFrame from pyspark.sql: DataFrame will be used extensively throughout the code to represent and manipulate data in tabular format.

- numpy as np: NumPy will be used in the process of converting Spark DataFrame columns to NumPy arrays for further processing.

- torch and DataLoader, TensorDataset from torch.utils. data: These will be used in the create_data_loader() function to create a PyTorch DataLoader for batch processing. DataLoader, TensorDataset, and torch will be used to handle tensors and create data loaders for efficient data loading during model training.

CHAPTER 3   DATA PREPARATION WITH PYSPARK FOR DEEP LEARNING

- boto3: Boto3 will be used in the copy_file_from_s3() function to interact with AWS S3 and download a file from the specified S3 bucket to a local file path.

In the next step, we set up logging for the Python script:

```
[In]: logging.basicConfig(level=logging.INFO)
[In]: logger = logging.getLogger(__name__)
```

- logging.basicConfig(level=logging.INFO): This line configures the logging system. It sets the logging level to INFO, which means that only messages with a severity level of INFO and above will be logged. Other possible levels include DEBUG, WARNING, ERROR, and CRITICAL.

- logger = logging.getLogger(name): This line creates a logger object named logger. The name variable represents the current module's name. When Python executes a script directly (as the main program), name is set to "main." However, when a script is imported as a module in another script, name is set to the name of the module. This allows modules to identify themselves when logging messages. By default, this logger is set to the root logger of the logging hierarchy. It allows for logging messages from different parts of the codebase and controlling their behavior independently. The logger can be used to log messages using methods like logger.info(), logger.warning(), and logger.error().

Moving forward, we define a function named load_data, which loads Tesla stock price data from a CSV file using SparkSession. The function takes a file path as input and returns a DataFrame:

CHAPTER 3   DATA PREPARATION WITH PYSPARK FOR DEEP LEARNING

```
[In]: def load_data(file_path: str) -> DataFrame:
          """
          Load stock price data from a CSV file using
          SparkSession.
          """
          spark = (SparkSession.builder
                   .appName("StockPricePrediction")
                   .getOrCreate())
          df = spark.read.csv(
                   file_path,
                   header=True,
                   inferSchema=True
          )
          return df
```

Here's a breakdown of what the function does:

- def load_data(file_path: str) -> DataFrame:: This line defines the function load_data with one parameter file_path of type str. The return annotation -> DataFrame indicates that the function returns a DataFrame object.

- """Load stock price data from a CSV file using SparkSession.""": This is a docstring that provides a brief description of what the function does.

- spark = SparkSession.builder: This line creates a SparkSession object named spark using the builder method. SparkSession is the entry point to Spark functionality in PySpark.

- .appName("StockPricePrediction") : This sets the application name to "StockPricePrediction". This name will be visible in the Spark UI.

CHAPTER 3   DATA PREPARATION WITH PYSPARK FOR DEEP LEARNING

- .getOrCreate(): This method retrieves an existing SparkSession or creates a new one if it doesn't exist.
- df = spark.read.csv(file_path, header=True, inferSchema=True): This line reads the CSV file located at the specified file_path using the spark.read.csv method. It sets header=True to indicate that the first row of the CSV file contains the column names and inferSchema=True to automatically infer the data types of the columns.
- return df: Finally, the function returns the DataFrame df containing the data loaded from the CSV file.

Transitioning to the next step, we define a function named preprocess_data that takes a DataFrame df as input and returns a DataFrame. The function preprocesses the data by assembling feature vectors using VectorAssembler, scaling them using StandardScaler, and performing feature engineering:

```
[In]: def preprocess_data(df: DataFrame) -> DataFrame:
          """
          Preprocess the data by assembling feature vectors
          using VectorAssembler, scaling them using
          StandardScaler, and performing feature engineering.
          """
          assembler = VectorAssembler(
              inputCols=['Open', 'High', 'Low', 'Volume'],
              outputCol='features'
          )
          df = assembler.transform(df)
          logger.info(
```

153

CHAPTER 3   DATA PREPARATION WITH PYSPARK FOR DEEP LEARNING

```
        "First 5 observations after assembling features:"
    )
    df.show(5, truncate=False)
    df = perform_feature_engineering(df)
    logger.info(
        "First 5 observations after feature engineering:"
    )
    df.show(5, truncate=False)
    scaler = StandardScaler(
        inputCol="features",
        outputCol="scaled_features",
        withStd=True, withMean=True
    )
    scaler_model = scaler.fit(df)
    df = scaler_model.transform(df)
    df = df.select('scaled_features', 'Close')
    return df
```

- def preprocess_data(df: DataFrame) -> DataFrame:: This line defines the function preprocess_data with one parameter df of type DataFrame. The return annotation -> DataFrame indicates that the function returns a DataFrame object.

- """"Preprocess the data by assembling feature vectors using VectorAssembler, scaling them using StandardScaler, and performing feature engineering."""": This is a docstring that provides a brief description of what the function does.

CHAPTER 3   DATA PREPARATION WITH PYSPARK FOR DEEP LEARNING

- assembler = VectorAssembler(inputCols=['Open', 'High', 'Low', 'Volume'], outputCol='features'): This line creates a VectorAssembler object named assembler with specified input columns ('Open', 'High', 'Low', 'Volume') and output column name 'features'. The VectorAssembler combines multiple columns into a single vector column.

- df = assembler.transform(df): This line applies the VectorAssembler to the DataFrame df, creating a new column named 'features' that contains the assembled feature vectors.

- logger.info("First 5 observations after assembling features:"): This logs an informational message indicating that the first five observations after assembling features will be shown.

- df.show(5, truncate=False): This line displays the first five observations of the DataFrame df after assembling features, without truncating the column values.

- df = perform_feature_engineering(df): This line applies additional feature engineering steps to the DataFrame df by calling the perform_feature_engineering function.

- logger.info("First 5 observations after feature engineering:"): This logs an informational message indicating that the first five observations after feature engineering will be shown.

- df.show(5, truncate=False): This line displays the first five observations of the DataFrame df after feature engineering, without truncating the column values.

CHAPTER 3   DATA PREPARATION WITH PYSPARK FOR DEEP LEARNING

- scaler = StandardScaler(inputCol="features", outputCol="scaled_features", withStd=True, withMean=True): This line creates a StandardScaler object named scaler with specified input column ("features"), output column ("scaled_features"), and options for scaling (withStd=True, withMean=True).

- scaler_model = scaler.fit(df): This fits the StandardScaler to the DataFrame df, creating a scaler model.

- df = scaler_model.transform(df): This line transforms the DataFrame df using the scaler model, scaling the 'features' column and storing the scaled values in a new column named 'scaled_features'.

- df = df.select('scaled_features', 'Close'): This line selects the 'scaled_features' and 'Close' columns from the DataFrame df.

- return df: Finally, the function returns the preprocessed DataFrame df.

Progressing to the next step, we define a function named perform_feature_engineering that takes a DataFrame df as input and returns a DataFrame. The function performs feature engineering operations to enhance the data's predictive power. It adds new columns to the DataFrame df to capture relevant information from existing features, including

- Price Range Interaction: Represents the difference between High and Low prices

- Price Change Interaction: Represents the difference between Close and Open prices

- Volume–Price Interaction: Represents the product of Volume and Close prices

CHAPTER 3 DATA PREPARATION WITH PYSPARK FOR DEEP LEARNING

```
[In]: def perform_feature_engineering(df: DataFrame) ->
      DataFrame:
          """
          Perform feature engineering operations to enhance the
          data's predictive power.

          This function adds new columns to the DataFrame df to
          capture relevant information from existing features,
          including:

          - Price Range Interaction:
              Represents the difference between High and
              Low prices.
          - Price Change Interaction:
              Represents the difference between Close and
              Open prices.
          - Volume-Price Interaction:
              Represents the product of Volume and
              Close prices.

          Parameters:
              df (DataFrame):
                  The input DataFrame containing the raw data.

          Returns:
              DataFrame:
                  The DataFrame df with additional engineered
                  features.
          """
          df = df.withColumn('price_range', df['High'] -
          df['Low'])
          df = df.withColumn('price_change', df['Close'] -
          df['Open'])
```

157

CHAPTER 3 DATA PREPARATION WITH PYSPARK FOR DEEP LEARNING

```
    df = df.withColumn(
            'volume_price_interaction',
            df['Volume'] * df['Close']
    )

    return df
```

- def perform_feature_engineering(df: DataFrame) -> DataFrame:: This line defines the function perform_feature_engineering with one parameter df of type DataFrame. The return annotation -> DataFrame indicates that the function returns a DataFrame object.

- """ Perform feature engineering operations to enhance the data's predictive power...""": This is a docstring that provides a brief description of what the function does.

- df = df.withColumn('price_range', df['High'] - df['Low']): This line adds a new column named 'price_range' to the DataFrame df, which represents the difference between the 'High' and 'Low' prices. It uses PySpark's withColumn function to achieve that.

- df = df.withColumn('price_change', df['Close'] - df['Open']): This line adds another new column named 'price_change' to the DataFrame df, which represents the difference between the 'Close' and 'Open' prices.

- df = df.withColumn('volume_price_interaction', df['Volume'] * df['Close']): This line adds a third new column named 'volume_price_interaction' to the DataFrame df, which represents the product of the 'Volume' and 'Close' prices.

- return df: Finally, the function returns the DataFrame df after performing the feature engineering operations.

CHAPTER 3   DATA PREPARATION WITH PYSPARK FOR DEEP LEARNING

To proceed, we define the print_first_5_observations function, which prints the first five rows of a DataFrame along with a custom heading:

```
[In]: def print_first_5_observations(df: DataFrame,
        heading: str):
        """
        Print the first 5 observations of the
        DataFrame with a
        custom heading.
        """
        logger.info(heading)
        df.show(5, truncate=False)
```

- df: DataFrame: The DataFrame containing the data to be printed.

- heading: str: The custom heading to be displayed before printing the DataFrame.

- logger.info(heading): Logs the custom heading using the logger.info() function. This heading provides context for the printed DataFrame.

- df.show(5, truncate=False): Calls the show() method on the DataFrame df with an argument of 5 and truncate=False. This method displays the first five rows of the DataFrame without truncating any column values.

CHAPTER 3   DATA PREPARATION WITH PYSPARK FOR DEEP LEARNING

Subsequently, we define the convert_to_tensor function, which is responsible for converting features and labels into PyTorch tensors:

```
[In]: def convert_to_tensor(features, labels):
          """
          Convert features and labels to PyTorch tensors.
          """
          tensor_features = torch.tensor(
              features,
              dtype=torch.float32
          )
          tensor_labels = torch.tensor(labels, dtype=torch.float32)
          return tensor_features, tensor_labels
```

- features: The features data to be converted into PyTorch tensors.

- labels: The labels data to be converted into PyTorch tensors.

- The convert_to_tensor function uses the torch.tensor() function from the PyTorch library to convert the features and labels into PyTorch tensors.

- The dtype=torch.float32 argument specifies that the data type of the resulting tensors should be 32-bit floating-point numbers.

- The converted tensors are assigned to the variables tensor_features and tensor_labels, respectively.

- Finally, the function returns a tuple (tensor_features, tensor_labels), where tensor_features is a PyTorch tensor containing the converted features and tensor_labels is a PyTorch tensor containing the converted labels.

CHAPTER 3   DATA PREPARATION WITH PYSPARK FOR DEEP LEARNING

Continuing on, we define the create_data_loader function, which is responsible for creating a PyTorch DataLoader object for batch processing of data:

```
[In]: def create_data_loader(features, labels, batch_size):
          """
          Create a PyTorch DataLoader for batch processing.
          """
          tensor_features, tensor_labels = convert_to_tensor(
              features,
              labels
          )
          dataset = TensorDataset(tensor_features, tensor_labels)
          data_loader = DataLoader(
              dataset,
              batch_size=batch_size,
              shuffle=True
          )
          return data_loader
```

The create_data_loader function has the following parameters:

- features: The features data to be included in the DataLoader
- labels: The corresponding labels data to be included in the DataLoader
- batch_size: The desired batch size for processing the data

CHAPTER 3 DATA PREPARATION WITH PYSPARK FOR DEEP LEARNING

In terms of functionality

- The function first calls the convert_to_tensor() function to convert the features and labels into PyTorch tensors.

- Using the converted tensors, it creates a PyTorch TensorDataset, which is a dataset wrapper that holds features and labels together.

- The TensorDataset allows iteration over batches of data during training.

- Finally, the function creates a PyTorch DataLoader object, which helps in iterating over batches of data from the dataset.

- The batch_size parameter specifies the number of samples per batch, and the shuffle=True parameter ensures that the data is shuffled before each epoch to prevent the model from learning sequence patterns.

The create_data_loader function returns a PyTorch DataLoader object ready for use in training or validation processes. This function is crucial for efficiently handling large datasets during the training of deep learning models. It enables batch processing, which helps in speeding up the training process and optimizing memory usage.

In the next step, we define the copy_file_from_s3, which is responsible for copying a file from an Amazon S3 bucket to a local file path:

```
[In]: def copy_file_from_s3(
          bucket_name: str,
          file_key: str,
          local_file_path: str
    ):
        """
```

CHAPTER 3   DATA PREPARATION WITH PYSPARK FOR DEEP LEARNING

```
    Copy file from S3 bucket to local file path.
    """
    s3 = boto3.client('s3')
    s3.download_file(bucket_name, file_key, local_
    file_path)
    logger.info(
        f"File downloaded from S3 bucket {bucket_name}"
        f"to local file path: {local_file_path}"
    )
```

Here's an explanation of the function in terms of parameters and functionality:

Parameters:

- bucket_name: The name of the S3 bucket from which the file will be copied

- file_key: The key (or path) of the file within the S3 bucket

- local_file_path: The local file path where the copied file will be saved

Functionality:

- The function first creates a connection to the Amazon S3 service using the boto3 library.

- Then, it uses the download_file method of the S3 client to download the specified file from the given S3 bucket (bucket_name) and file key (file_key) to the specified local file path (local_file_path).

- After the file is successfully downloaded, it logs an informational message using the logger object to indicate that the file has been downloaded from the S3 bucket to the local file path.

163

CHAPTER 3   DATA PREPARATION WITH PYSPARK FOR DEEP LEARNING

The function doesn't return any value explicitly, but it performs the file copy operation as described above. This function is useful for retrieving files stored in Amazon S3 buckets and bringing them into the local environment for further processing or analysis. It can be used in various scenarios, such as downloading datasets for machine learning tasks, fetching configuration files, or transferring data between different storage systems.

Continuing with the process of data preparation, we define the main function, which serves as the entry point for the script or application:

```
[In]: def main():
          s3_bucket_name = 'instance1bucket'
          s3_file_key = 'TSLA_stock.csv'
          local_file_path = '/home/ubuntu/airflow/dags/TSLA_stock.csv'
          copy_file_from_s3(
              s3_bucket_name,
              s3_file_key,
              local_file_path
          )

          df = load_data(local_file_path)
          print_first_5_observations(df, "First 5 observations:")
          df = preprocess_data(df)
          print_first_5_observations(
              df,
              "First 5 observations after scaling:"
          )

          df_features = np.array(
              df.select('scaled_features')
```

164

CHAPTER 3   DATA PREPARATION WITH PYSPARK FOR DEEP LEARNING

```
        .rdd.map(lambda x: x.scaled_features.toArray())
        .collect()
    )
    df_labels = np.array(
        df.select('Close')
        .rdd.map(lambda x: x.Close)
        .collect()
    )
    batch_size = 32
    train_loader = create_data_loader(
        df_features,
        df_labels,
        batch_size
    )
    logger.info(
        "First 5 observations after scaling"
        "and converting to tensors:"
    )
    print(train_loader.dataset[:5])
```

This main function orchestrates the entire data preparation process, from copying the data file from an S3 bucket to local storage to preprocessing the data, converting it into suitable formats for model training, and finally creating a DataLoader for efficient batch processing.

Let's examine these steps one by one:

Copying File from S3 Bucket:

- The function specifies the name of the S3 bucket (s3_bucket_name), the key of the file within the bucket (s3_file_key), and the local file path where the file will be copied (local_file_path).

CHAPTER 3   DATA PREPARATION WITH PYSPARK FOR DEEP LEARNING

- Then, it calls the copy_file_from_s3 function to copy the file from the specified S3 bucket to the local file path.

Loading Data:

- The function loads the data from the local file path (local_file_path) using the load_data function and assigns it to the DataFrame variable df.

Printing First Five Observations:

- The function prints the first five observations of the DataFrame (df) using the print_first_5_observations function with a custom heading "First 5 observations:".

Preprocessing Data:

- The function preprocesses the loaded data by calling the preprocess_data function and assigns the preprocessed DataFrame back to df.

- After preprocessing, it again prints the first five observations of the DataFrame with a custom heading "First 5 observations after scaling:".

Converting Data to NumPy Arrays:

- The function converts the Spark DataFrame columns containing scaled features and the 'Close' values to NumPy arrays (df_features and df_labels, respectively).

- These arrays will be used as input features and labels for training the model.

CHAPTER 3   DATA PREPARATION WITH PYSPARK FOR DEEP LEARNING

Creating DataLoader:

- The function creates a PyTorch DataLoader (train_loader) for batch processing using the create_data_loader function. This DataLoader will be used to iterate over batches of training data during model training.

Printing First Five Observations as Tensors:

- The function logs an informational message indicating that the first five observations after scaling and converting to tensors will be printed.
- Then, it prints the first five observations of the DataLoader's dataset.

In the final process of data preparation, we have the conditional statement, which checks whether the script is being run as the main program or if it is being imported as a module into another script:

```
[In]: if __name__ == "__main__":
          main()
```

If the script is being executed directly (i.e., as the main program), the condition __name__ == "__main__" evaluates to True, and the code block inside the if statement is executed. If the script is imported as a module into another script, the condition __name__ == "__main__" evaluates to False, and the code block inside the if statement is not executed. In this context, main() is called when the script is run directly, meaning that the main function will be executed, initiating the data preparation process. This structure allows the script to be both runnable on its own and importable as a module into other scripts without causing unintended side effects.

CHAPTER 3   DATA PREPARATION WITH PYSPARK FOR DEEP LEARNING

To run the script, we first save it as pyspark_pytorch_preparation.py, activate the virtual environment on the EC2 instance, and then run it using the following commands:

[In]: cd /home/ubuntu
[In]: source myenv/bin/activate
[In]: python3 pyspark_pytorch_preparation.py

Let's now examine the output from the code.

The first output, produced by the print_first_5_observations function, shows the first five observations of the DataFrame:

Date	Open	High	Low	Close	Volume
2/23/24	195.31	197.57	191.50	191.97	78,670,300
2/22/24	194.00	198.32	191.36	197.41	92,739,500
2/21/24	193.36	199.44	191.95	194.77	103,844,000
2/20/24	196.13	198.60	189.13	193.76	104,545,800
2/16/24	202.06	203.17	197.40	199.95	111,173,600

The table consists of six columns: Date, Open, High, Low, Close, and Volume. Each row represents a specific date, and each column represents different attributes of the Tesla stock on that date. This output provides a glimpse of the raw data loaded from the CSV file before any preprocessing or feature engineering is applied. Understanding the initial structure and content of the data is crucial for identifying any initial trends or patterns and deciding on the appropriate preprocessing steps.

The second output, produced by the preprocess_data function, displays the first five observations after assembling the features:

CHAPTER 3   DATA PREPARATION WITH PYSPARK FOR DEEP LEARNING

Date	Open	High	Low	Close	Volume	Features
2/23/24	195.31	197.57	191.50	191.97	78,670,300	[195.31, 197.57, 191.50, 78670300]
2/22/24	194.00	198.32	191.36	197.41	92,739,500	[194.00, 198.32, 191.36, 92739500]
2/21/24	193.36	199.44	191.95	194.77	103,844,000	[193.36, 199.44, 191.95, 103844000]
2/20/24	196.13	198.60	189.13	193.76	104,545,800	[196.13, 198.60, 189.13, 104545800]
2/16/24	202.06	203.17	197.40	199.95	111,173,600	[202.06, 203.17, 197.40, 111173600]

Here's an explanation of each column in the table:

- Date: The date of the stock price observation.
- Open: The opening price of the stock on that date.
- High: The highest price of the stock during the trading session on that date.
- Low: The lowest price of the stock during the trading session on that date.
- Close: The closing price of the stock on that date.
- Volume: The trading volume (number of shares traded) of the stock on that date.
- Features: This column represents the assembled feature vector for each observation. The features are extracted from the 'Open', 'High', 'Low', and 'Volume' columns and combined into a single vector. Each element in the vector corresponds to one of these

CHAPTER 3   DATA PREPARATION WITH PYSPARK FOR DEEP LEARNING

features. For example, the first row indicates that on February 23, 2024, the stock opened at $195.31, reached a high of $197.57 and a low of $191.50, closed at $191.97, and had a trading volume of 78,670,300 shares. The feature vector corresponding to this observation is [195.31, 197.57, 191.50, 78,670,300]. Similarly, the subsequent rows represent the stock price observations for the following dates, along with their respective feature vectors.

Assembling feature vectors using VectorAssembler in PySpark simplifies the workflow by consolidating multiple input columns into a single vector column. This consolidation streamlines the feature engineering process, facilitating easier management and manipulation of features within the dataset. Furthermore, leveraging Spark's distributed computing capabilities, VectorAssembler enables efficient data processing, particularly advantageous for handling large-scale datasets. By assembling feature vectors, Spark can parallelize computations across multiple nodes in a cluster, enhancing processing speed and scalability.

The third output shows the first five observations after feature engineering. Since the Open, High, Low, Close, Volume, and Features columns are the same as in the previous table, we have omitted them for clarity. Instead, this table focuses on the newly added columns:

Date	...	Price Range	Price Change	Volume-Price Interaction
2/23/24	...	6.07	-3.34	15,102,337,569.07
2/22/24	...	6.96	3.41	18,307,705,065.98
2/21/24	...	7.49	1.41	20,225,696,295.38
2/20/24	...	9.47	-2.37	20,256,793,685.27
2/16/24	...	5.77	-2.11	22,229,160,986.48

CHAPTER 3   DATA PREPARATION WITH PYSPARK FOR DEEP LEARNING

Below is an explanation of each new column in the table:

- price_range: Represents the difference between High and Low prices
- price_change: Represents the difference between Close and Open prices
- volume_price_interaction: Represents the product of Volume and Close prices

Let's break down how each of these new columns was generated, using data for 2/23/24:

Price Range:

- 197.57−191.50=6.07

Price Change:

- 191.97−195.31=−3.34

Volume-Price Interaction:

- 78,670,300×191.97=15,102,337,569.07

Each of these engineered features can provide valuable information that may enhance the predictive performance of a model for forecasting Tesla stock prices:

- The price range can provide insights into the volatility of the stock. High price ranges may indicate periods of market uncertainty or significant price fluctuations, while low price ranges may suggest relative stability. Incorporating this feature allows the model to capture the magnitude of price movements, which can be valuable for predicting future price changes.

CHAPTER 3   DATA PREPARATION WITH PYSPARK FOR DEEP LEARNING

- The price change provides information about the direction of price movements during a trading session. Positive price changes indicate that the stock closed higher than it opened, while negative price changes indicate the opposite. By including this feature, the model can capture intraday price dynamics and potential trends in investor sentiment or market momentum.

- The interaction between volume and closing prices reflects the level of trading activity relative to price movements. High trading volumes combined with significant price changes may indicate strong market participation and investor interest in the stock. Conversely, low trading volumes accompanied by price changes could signal weak market sentiment or lack of investor conviction. By incorporating this feature, the model can account for the interplay between trading activity and price dynamics, which can provide additional insights into market behavior and potential price movements.

The next output shows the first five observations after scaling:

scaled_features	Close
[0.1796, 0.1602, 0.1885, -0.6497]	191.97
[0.1672, 0.1672, 0.1872, -0.4843]	197.41
[0.1612, 0.1776, 0.1929, -0.3538]	194.77
[0.1874, 0.1698, 0.1655, -0.3455]	193.76
[0.2437, 0.2123, 0.2459, -0.2676]	199.95

CHAPTER 3   DATA PREPARATION WITH PYSPARK FOR DEEP LEARNING

This table shows the scaled features and the corresponding Close prices for the first five observations after scaling. The scaled_features column contains the scaled values of the features, and the Close column contains the corresponding closing prices.

Scaling is an important step in data preparation as it ensures that the features are on a similar scale, making them more suitable for model training. The scaled features are represented as values between 0 and 1, which facilitates more consistent and effective modeling. Ultimately, scaling contributes to better model performance, stability, and interpretability, leading to more accurate predictions and insights from the data.

The final output displays the first five observations after scaling and converting to tensors. The Features column contains the scaled feature values, and the Close column contains the corresponding closing prices:

Features	Close
[0.1797, 0.1602, 0.1885, -0.6497]	191.97
[0.1672, 0.1672, 0.1872, -0.4843]	197.41
[0.1612, 0.1776, 0.1929, -0.3538]	194.77
[0.1874, 0.1698, 0.1655, -0.3455]	193.76
[0.2437, 0.2123, 0.2459, -0.2676]	199.95

This is the same output as in the previous table since all we have done here is convert the data into tensors. The output remains the same because converting the data into tensors doesn't alter its values; it simply changes the data structure to a format suitable for processing with PyTorch.

Converting the data to tensors allows us to prepare the input features and target variables in a format suitable for training deep learning models using PyTorch. While the values of the data remain unchanged

CHAPTER 3   DATA PREPARATION WITH PYSPARK FOR DEEP LEARNING

after conversion, the tensor representation enables efficient processing and training within the PyTorch framework, ultimately facilitating the development of accurate and effective predictive models.

Data Preparation with PySpark for TensorFlow

The code so far was designed to work specifically with PyTorch. However, with minor modifications, such as replacing PyTorch-specific functions with TensorFlow equivalents, it can be adapted to fit TensorFlow for deep learning tasks.

The code below demonstrates this modification to perform data preprocessing, feature engineering, and data formatting for TensorFlow using PySpark. We start by importing necessary modules and then proceed to demonstrate each step of the data preparation process:

```
[In]: import logging
[In]: from pyspark.sql import SparkSession
[In]: from pyspark.ml.feature import VectorAssembler, StandardScaler
[In]: from pyspark.sql import DataFrame
[In]: import numpy as np
[In]: import tensorflow as tf
[In]: from tensorflow.keras.utils import to_categorical
[In]: import boto3
```

- logging: The logging module will be used throughout the code to log messages, warnings, and errors during program execution.

- SparkSession from pyspark.sql: SparkSession will be used in the load_data() function to create a SparkSession object for reading CSV data and creating DataFrame objects.

174

CHAPTER 3   DATA PREPARATION WITH PYSPARK FOR DEEP LEARNING

- VectorAssembler and StandardScaler from pyspark. ml.feature: These will be used in the preprocess_data() function for feature engineering. VectorAssembler will assemble raw feature columns into a single feature vector column, while StandardScaler will scale input features to have zero mean and unit standard deviation.

- DataFrame from pyspark.sql: DataFrame will be used extensively throughout the code to represent and manipulate data in tabular format.

- numpy as np: NumPy will be used in the process of converting Spark DataFrame columns to NumPy arrays for further processing.

- tensorflow as tf: TensorFlow will be used in the create_ data_loader() function to create a TensorFlow Dataset for batch processing.

- boto3: Boto3 will be used in the copy_file_from_s3() function to interact with AWS S3 and download a file from the specified S3 bucket to a local file path.

In the next step, we set up logging:

```
[In]: logging.basicConfig(level=logging.INFO)
[In]: logger = logging.getLogger(__name__)
```

- These lines configure the logging system to output log messages with a level of INFO or higher. A logger object named logger is then created for the current script.

CHAPTER 3   DATA PREPARATION WITH PYSPARK FOR DEEP LEARNING

Moving forward, we create the load_data function, which facilitates the loading of Tesla stock price data from a CSV file using SparkSession:

```
[In]: def load_data(file_path: str) -> DataFrame:
          """
          Load stock price data from a CSV file using
          SparkSession.
          """
          spark = (SparkSession.builder
                  .appName("StockPricePrediction")
                  .getOrCreate())
          df = spark.read.csv(
                  file_path,
                  header=True,
                  inferSchema=True
          )
          return df
```

- The function takes a single argument file_path of type string and returns a DataFrame object.

- A docstring enclosed in triple quotes provides a brief description of the function's purpose, which is to load stock price data from a CSV file using SparkSession.

- Inside the function, a SparkSession object named spark is created using the builder pattern. The builder method is called to initialize the SparkSession configuration. The appName method sets the name of the application to "StockPricePrediction". Finally, getOrCreate() ensures that an existing SparkSession is reused or a new one is created as needed.

CHAPTER 3   DATA PREPARATION WITH PYSPARK FOR DEEP LEARNING

- The function reads data from a CSV file located at the specified file_path using the SparkSession spark. It utilizes the read.csv method to read the CSV file, with header=True indicating that the first row contains column headers and inferSchema=True instructing Spark to automatically infer the data types of columns.

- After loading the data into a DataFrame named df, the function returns this DataFrame. The returned DataFrame contains the Tesla stock price data loaded from the CSV file and can be used for further processing or analysis within the program.

Following this, we define the preprocess_data function to preprocess the input DataFrame by assembling feature vectors, scaling features, and performing feature engineering. The function uses PySpark's VectorAssembler and StandardScaler for these tasks:

```
[In]: def preprocess_data(df: DataFrame) -> DataFrame:
        """
        Preprocess the data by assembling feature vectors
        using VectorAssembler, scaling them using
        StandardScaler, and performing feature engineering.
        """
        assembler = VectorAssembler(
            inputCols=['Open', 'High', 'Low', 'Volume'],
            outputCol='features'
        )
        df = assembler.transform(df)
        logger.info(
            "First 5 observations after assembling features:"
```

177

CHAPTER 3 DATA PREPARATION WITH PYSPARK FOR DEEP LEARNING

```
)
df.show(5, truncate=False)

df = perform_feature_engineering(df)
logger.info(
    "First 5 observations after feature engineering:"
)
df.show(5, truncate=False)

scaler = StandardScaler(
    inputCol="features",
    outputCol="scaled_features",
    withStd=True, withMean=True
)
scaler_model = scaler.fit(df)
df = scaler_model.transform(df)
df = df.select('scaled_features', 'Close')
return df
```

- The function takes a DataFrame df as input and returns a DataFrame after preprocessing.

- A docstring provides a brief description of the function's purpose, which is to preprocess the data by assembling feature vectors, scaling them using StandardScaler, and performing feature engineering.

- Inside the function, a VectorAssembler named assembler is created to assemble feature vectors using the input columns 'Open', 'High', 'Low', and 'Volume', with the output column named 'features'. The transform method is then applied to transform the input DataFrame df using the assembler.

CHAPTER 3　DATA PREPARATION WITH PYSPARK FOR DEEP LEARNING

- Logging information is provided using the logger.info function to indicate the progress of the preprocessing steps. The first five observations of the DataFrame after assembling features are displayed using the df.show method.

- The perform_feature_engineering function is called to perform additional feature engineering operations on the DataFrame df.

- Another preprocessing step involves scaling the features using StandardScaler. A StandardScaler object scaler is created with input and output columns specified, along with options for standardization. The fit method is then applied to the DataFrame to obtain a scaler model, which is used to transform the DataFrame.

- Finally, the DataFrame is modified to select only the 'scaled_features' and 'Close' columns before returning it from the function. This DataFrame contains the preprocessed data ready for further analysis or modeling.

In the next step, we create the perform_feature_engineering function, which applies various feature engineering operations to the DataFrame, such as creating interaction terms and extracting useful information from existing features:

```
[In]: def perform_feature_engineering(df: DataFrame) ->
       DataFrame:
           """
           Perform feature engineering operations such as
           creating interaction terms, generating polynomial
```

CHAPTER 3   DATA PREPARATION WITH PYSPARK FOR DEEP LEARNING

features, or extracting useful information from existing features.
"""

```
df = df.withColumn('price_range', df['High'] - df['Low'])
df = df.withColumn('price_change', df['Close'] - df['Open'])

df = df.withColumn(
        'volume_price_interaction',
        df['Volume'] * df['Close']
)

return df
```

- The function takes a DataFrame df as input and returns a DataFrame after performing feature engineering.

- A docstring provides a brief description of the function's purpose, which is to perform feature engineering operations such as creating interaction terms.

- Inside the function, feature engineering operations are performed on the DataFrame df:

  - A new column 'price_range' is created by subtracting the 'Low' price from the 'High' price for each observation. This captures the range of prices for each stock.

  - Another new column 'price_change' is created by subtracting the 'Open' price from the 'Close' price. This represents the change in price over a given period.

180

CHAPTER 3   DATA PREPARATION WITH PYSPARK FOR DEEP LEARNING

- Additionally, a column 'volume_price_interaction' is generated by multiplying the 'Volume' column with the 'Close' price. This interaction term captures the relationship between trading volume and stock price.

- After applying these feature engineering operations, the modified DataFrame df is returned from the function, containing the newly created columns along with the original features. These engineered features can be valuable for improving the performance of machine learning models or gaining insights from the data.

Moving on, we create the print_first_5_observations function, which prints the first five observations of a DataFrame with a custom heading:

```
[In]: def print_first_5_observations(df: DataFrame,
      heading: str):
          """
          Print the first 5 observations of the DataFrame with
          a custom heading.
          """
          logger.info(heading)
          df.show(5, truncate=False)
```

- The function takes two arguments: a DataFrame df and a string heading. It does not return any value.

- A docstring provides a brief description of the function's purpose, which is to print the first five observations of the DataFrame with a custom heading.

- Inside the function, the logger.info function is used to log the custom heading passed to the function.

181

CHAPTER 3   DATA PREPARATION WITH PYSPARK FOR DEEP LEARNING

- The df.show method is then employed to display the first five observations of the DataFrame df. The truncate=False argument ensures that the displayed columns are not truncated, allowing the full content of each column to be shown.

In the subsequent step, we define the create_data_loader function, which creates a TensorFlow Dataset for batch processing:

```
[In]: def create_data_loader(features, labels, batch_size):
      """
      Create a TensorFlow Dataset for batch processing.
      """
      dataset = tf.data.Dataset.from_tensor_slices(
          (features, labels)
      )
      dataset = dataset.shuffle(buffer_size=10000)
      dataset = batch(batch_size)
      return dataset
```

- The function takes three arguments: features, labels, and batch_size. It does not return any value directly but returns a TensorFlow Dataset object.

- A docstring provides a brief description of the function's purpose, which is to create a TensorFlow Dataset for batch processing.

- Inside the function, a TensorFlow Dataset is created using the tf.data.Dataset.from_tensor_slices method. This method slices the input tensors features and labels along the first dimension to create individual elements of the dataset. Each element consists of a pair of corresponding features and labels.

- The created dataset is then shuffled using the shuffle method with a buffer size of 10000. Shuffling helps in randomizing the order of elements in the dataset, which can be beneficial during training to prevent model overfitting.

- Subsequently, the batch method is applied to the shuffled dataset to group elements into batches of size batch_size. Batching allows for efficient processing of data in batches during training, reducing memory usage and improving computational efficiency.

- Finally, the function returns the created dataset, which is now ready for use in training machine learning models with TensorFlow. This dataset can be iterated over in batches to train models effectively.

Carrying forward, we create the copy_file_from_s3 function, which copies a file from an Amazon S3 bucket to a local file path:

```
[In]: def copy_file_from_s3(bucket_name: str, file_key: str,
          local_file_path: str):
        """
        Copy file from S3 bucket to local file path.
        """
        s3 = boto3.client('s3')
        s3.download_file(bucket_name, file_key, local_
        file_path)
        logger.info(
            f"File downloaded from S3 bucket {bucket_name}"
            f"to local file path: {local_file_path}"
        )
```

CHAPTER 3   DATA PREPARATION WITH PYSPARK FOR DEEP LEARNING

- The function takes three arguments: bucket_name, file_key, and local_file_path. It does not return any value.

- A docstring provides a brief description of the function's purpose, which is to copy a file from an S3 bucket to a local file path.

- Inside the function, an instance of the boto3.client for Amazon S3 is created using s3 = boto3.client('s3'). This client allows interaction with the S3 service.

- The s3.download_file method is then invoked to download the file from the specified S3 bucket (bucket_name) and file key (file_key) to the local file path (local_file_path). This method handles the actual file transfer from S3 to the local system.

- After the file has been successfully downloaded, logging information is provided using the logger.info function to indicate that the file has been downloaded from the S3 bucket to the local file path. The message includes details such as the S3 bucket name and the local file path where the file has been saved.

In the next step, we define the main function, which orchestrates the entire process by calling the necessary functions in a sequential manner. It copies a file from an S3 bucket, loads the data, preprocesses it, converts it to NumPy arrays, and creates a TensorFlow Dataset for batch processing:

```
[In]: def main():
          s3_bucket_name = 'instance1bucket'
          s3_file_key = 'TSLA_stock.csv'
          local_file_path = '/home/ubuntu/airflow/dags/TSLA_stock.csv'
          copy_file_from_s3(
```

```python
    s3_bucket_name,
    s3_file_key,
    local_file_path
)
df = load_data(local_file_path)
print_first_5_observations(df, "First 5
observations:")
df = preprocess_data(df)
print_first_5_observations(
    df,
    "First 5 observations after scaling:"
)

df_features = np.array(
    df.select('scaled_features')
    .rdd.map(lambda x: x.scaled_features.toArray())
    .collect()
)
df_labels = np.array(
    df.select('Close')
    .rdd.map(lambda x: x.Close)
    .collect()
)

batch_size = 32
train_dataset = create_data_loader(
    df_features,
    df_labels,
    batch_size
)
```

```
    for features, labels in train_dataset.take(1):
        logger.info(
            "First 5 observations after scaling"
            "and converting to TensorFlow Dataset:"
        )
        logger.info(features[:5])
        logger.info("First 5 labels with no scaling:")
        logger.info(labels[:5])
```

- The function begins by initializing variables s3_bucket_name, s3_file_key, and local_file_path with values representing the S3 bucket name, file key (path within the bucket), and the local file path where the S3 file will be downloaded.

- It then calls the copy_file_from_s3 function, passing the S3 bucket name, file key, and local file path as arguments. This function is responsible for downloading the specified file from the S3 bucket to the local file path.

- After downloading the file, the function proceeds to load the data from the local file path into a DataFrame df using the load_data function.

- The print_first_5_observations function is then called twice to display the first five observations of the DataFrame df with custom headings. The first call prints the initial observations before any preprocessing, while the second call prints the observations after scaling.

CHAPTER 3   DATA PREPARATION WITH PYSPARK FOR DEEP LEARNING

- Next, the feature engineering process is applied to the DataFrame df by calling the preprocess_data function. This function assembles feature vectors, scales them using StandardScaler, and performs additional feature engineering operations.

- The preprocess_data function returns the modified DataFrame df, which now contains scaled features and additional engineered features.

- Subsequently, the features and labels are extracted from the DataFrame df and converted into NumPy arrays df_features and df_labels, respectively. These arrays are required for creating a TensorFlow Dataset for training the machine learning model.

- A TensorFlow Dataset train_dataset is created using the create_data_loader function, which batches the features and labels into batches of a specified size (batch_size).

- Finally, a loop iterates over the first batch of the train_dataset, logging the first five observations of the features after scaling and the first five labels.

Finally, the conditional block below ensures that the main function is executed only when the script is run directly, not when it is imported as a module in another script:

```
[In]: if __name__ == "__main__":
        main()
```

- The __name__ variable is a special built-in variable in Python that represents the name of the current module. When the Python interpreter runs a script directly, it sets the value of __name__ to "__main__".

CHAPTER 3 DATA PREPARATION WITH PYSPARK FOR DEEP LEARNING

- The conditional statement checks if the value of __name__ is equal to "__main__". If it is, it means that the script is being executed as the main program, and the code block beneath the conditional statement will be executed.

- In this case, the code block consists of a single function call: main(). This call executes the main function defined earlier in the script, which orchestrates the execution of various tasks such as downloading data, preprocessing, and preparing the data for training.

By wrapping the main() function call inside the if __name__ = = "__main__": conditional statement, the script can be imported as a module into other scripts without automatically executing the main() function. This allows the functions and classes defined in the script to be used as reusable components in other Python programs.

To run the script, we first save it as pyspark_tensorflow_preparation.py, activate the virtual environment on the EC2 instance, and then run the Python script using the following commands:

```
[In]: cd /home/ubuntu
[In]: source myenv/bin/activate
[In]: python3 pyspark_tensorflow_preparation.py
```

Let's now examine the output from the code.

The first output, produced by the print_first_5_observations function, displays the first five rows of the DataFrame before any processing. It includes columns such as Date, Open, High, Low, Close, and Volume. This output provides a glimpse into the data that will be used for training the TensorFlow model:

CHAPTER 3   DATA PREPARATION WITH PYSPARK FOR DEEP LEARNING

Date	Open	High	Low	Close	Volume
2/23/24	195.31	197.57	191.50	191.97	78,670,300
2/22/24	194.00	198.32	191.36	197.41	92,739,500
2/21/24	193.36	199.44	191.95	194.77	103,844,000
2/20/24	196.13	198.60	189.13	193.76	104,545,800
2/16/24	202.06	203.17	197.4	199.95	111,173,600

The second output, produced by the preprocess_data(df) function, shows the first five rows of the DataFrame after assembling feature vectors using VectorAssembler. It adds a new column named "features," which contains the assembled features:

Date	Open	High	Low	Close	Volume	Features
2/23/24	195.31	197.57	191.50	191.97	78,670,300	[195.31, 197.57, 191.5, 78,670,300]
2/22/24	194.00	198.32	191.36	197.41	92,739,500	[194.0, 198.32, 191.36, 92,739,500]
2/21/24	193.36	199.44	191.95	194.77	103,844,000	[193.36, 199.44, 191.95, 103,844,000]
2/20/24	196.13	198.60	189.13	193.76	104,545,800	[196.13, 198.6, 189.13, 104,545,800]
2/16/24	202.06	203.17	197.40	199.95	111,173,600	[202.06, 203.17, 197.4, 111,173,600]

The third output, produced by the perform_feature_engineering(df) function, displays the first five rows of the DataFrame after performing feature engineering operations. New columns such as "price_range,"

CHAPTER 3   DATA PREPARATION WITH PYSPARK FOR DEEP LEARNING

"price_change," and "volume_price_interaction" are added based on the original columns. Since the Open, High, Low, Close, Volume, and Features columns are the same as in the previous table, we have omitted them from the below table so that to focus on the new columns:

Date	Price Range	Price Change	Volume-Price Interaction
2/23/24	6.07	-3.34	15,102,337,569.07
2/22/24	6.96	3.41	18,307,705,065.96
2/21/24	7.49	1.41	20,225,696,295.38
2/20/24	9.47	-2.37	20,256,793,685.27
2/16/24	5.77	-2.11	22,229,160,986.48

The fourth output, produced by the preprocess_data(df) function, presents the first five rows of the DataFrame after scaling the features using StandardScaler. It includes the "scaled_features" column along with the "Close" column:

Scaled Features	Close
[0.18, 0.16, 0.19, -0.65]	191.97
[0.17, 0.17, 0.19, -0.48]	197.41
[0.16, 0.18, 0.19, -0.35]	194.77
[0.19, 0.17, 0.17, -0.35]	193.76
[0.24, 0.21, 0.25, -0.27]	199.95

The final output displays the first five observations after scaling and converting to tensors. In theory, the output should be the same as in the previous table since all we have done here is convert the data into tensors.

CHAPTER 3   DATA PREPARATION WITH PYSPARK FOR DEEP LEARNING

However, the reader may observe different results due to the shuffling of the data, which can lead to variations in the order of the samples within each batch.

Bringing It All Together

In this section, we bring together the individual code snippets presented throughout the exploration, parallel processing, and preparation process. This demonstrates how each step integrates to form a comprehensive data exploration and preparation pipeline. By combining the various techniques and functionalities explored earlier, we showcase the end-to-end workflow for exploring and preparing data with PySpark for deep learning tasks using PyTorch and TensorFlow.

Data Exploration with PySpark

In this subsection, we combine the code snippets for exploring the Tesla stock dataset, including printing the first ten rows of the dataset, performing descriptive statistics, checking for missing values, and visualizing the Close price over time in a line chart:

```
[In]: import boto3
[In]: import matplotlib.pyplot as plt
[In]: from pyspark.sql import SparkSession
[In]: from pyspark.sql.functions import col
[In]: import pyspark.sql.functions as F
[In]: import numpy as np

[In]: class DataProcessor:
          def __init__(self, spark_session):
              self.spark = spark_session
```

CHAPTER 3   DATA PREPARATION WITH PYSPARK FOR DEEP LEARNING

```
[In]: def load_data(self, file_path: str):
          """
          Load stock price data from a CSV file using
          SparkSession.
          """
          try:
              df = self.spark.read.csv(
                  file_path,
                  header=True,
                  inferSchema=True
              )
              return df
          except Exception as e:
              print(f"Error loading data: {str(e)}")
              return None

[In]: def print_first_n_rows(self, df, n=10):
          """Print the first n rows of the DataFrame."""
          print(f"First {n} rows of the DataFrame:")
          df.show(n)

[In]: def calculate_descriptive_statistics(self, df):
          """Calculate descriptive statistics of the
          DataFrame."""
          print("Descriptive Statistics:")
          df.summary().show()

[In]: def visualize_data(self, df):
          """Create visualizations based on DataFrame."""
          print("Data Visualization:")
          try:
              df_pd = df.toPandas()
              plt.figure(figsize=(10, 6))
```

CHAPTER 3   DATA PREPARATION WITH PYSPARK FOR DEEP LEARNING

```
        plt.plot(df_pd['Date'], df_pd['Close'])
        plt.xlabel('Date')
        plt.ylabel('Closing Price')
        plt.title('Tesla Stock Closing Prices Over Time')
        plt.xticks(rotation=45, ha='right')
        plt.gca().invert_xaxis()
        plt.xticks(np.arange(0, len(df_pd['Date']),
            step=max(len(df_pd['Date']) // 10, 1)))
        plt.show()
    except Exception as e:
        print(f"Error visualizing data: {str(e)}")
```

[In]:
```
def check_for_null_values(self, df):
    """Check for null values in the DataFrame."""
    print("Null Value Check:")
    null_counts = df.select([col(c).isNull()\
        .cast("int").alias(c) for c in df.columns])\
        .agg(*[F.sum(c).alias(c) for c in df.columns])\
        .toPandas()
    print(null_counts)
```

[In]:
```
def copy_file_from_s3(bucket_name: str, file_key: str,
    local_file_path: str):
    """
    Copy file from S3 bucket to local file path.
    """
    try:
        s3 = boto3.client('s3')
        s3.download_file(bucket_name, file_key, local_
        file_path)
        print(
```

193

```
                f"File downloaded from S3 bucket
                {bucket_name}"
                f"to local file path: {local_file_path}"
            )
        except Exception as e:
            print(f"Error copying file from S3: {str(e)}")

[In]: def main(
          s3_bucket_name: str,
          s3_file_key: str,
          local_file_path: str
      ):
          """
          Main function for stock price prediction.

          This function initializes a SparkSession, copies
          a file from an S3 bucket to a local file path,
          initializes a DataProcessor object, loads data
          from the local file, and performs various data
          processing tasks.

          :param s3_bucket_name:
              Name of the S3 bucket containing the file
          :param s3_file_key:
              Key of the file within the S3 bucket
          :param local_file_path:
              Local file path to save the downloaded file
          """

          spark = (SparkSession.builder
                   .appName("StockPricePrediction")
                   .getOrCreate())
```

CHAPTER 3   DATA PREPARATION WITH PYSPARK FOR DEEP LEARNING

```
        copy_file_from_s3(
            s3_bucket_name,
            s3_file_key,
            local_file_path
        )
        data_processor = DataProcessor(spark)

        df = data_processor.load_data(local_file_path)
        if df is not None:
            data_processor.print_first_n_rows(df)
            data_processor.calculate_descriptive_
            statistics(df)
            data_processor.check_for_null_values(df)
            data_processor.visualize_data(df)
        else:
            print("Error loading data. Exiting.")
```

```
[In]: if __name__ == "__main__":
        s3_bucket_name = 'instance1bucket'
        s3_file_key = 'TSLA_stock.csv'
        local_file_path = '/home/ubuntu/airflow/dags/TSLA_
        stock.csv'
        main(s3_bucket_name, s3_file_key, local_file_path)
```

Parallel Processing in PySpark

In this subsection, we present the code that has been used to demonstrate the various techniques to speed up processing of large datasets in PySpark:

```
[In]: import boto3
[In]: from pyspark.sql import SparkSession
[In]: from pyspark.sql.functions import broadcast
```

CHAPTER 3   DATA PREPARATION WITH PYSPARK FOR DEEP LEARNING

```
[In]: class DataProcessor:
          def __init__(self, spark_session):
              self.spark = spark_session

          def load_data(self, file_path: str):
              try:
                  df = self.spark.read.csv(
                      file_path,
                      header=True,
                      inferSchema=True
                  )
                  return df
              except Exception as e:
                  print(f"Error loading data: {str(e)}")
                  return None

          def print_first_n_rows(self, df, n=10):
              print(f"First {n} rows of the DataFrame:")
              df.show(n)

          def print_partition_info(self, df):
              num_partitions = df.rdd.getNumPartitions()
              print(f"Number of partitions: {num_partitions}")

[In]: def copy_file_from_s3(
          bucket_name: str,
          file_key: str,
          local_file_path: str
      ):
          try:
              s3 = boto3.client('s3')
              s3.download_file(bucket_name, file_key, local_
              file_path)
```

```
            print(f"File downloaded from S3 bucket 
            {bucket_name} "
                f"to local file path: {local_file_path}"
            )
        except Exception as e:
            print(f"Error copying file from S3: {str(e)}")
[In]: def main(
        s3_bucket_name: str,
        s3_file_key: str,
        local_file_path: str
    ):
        spark = SparkSession.builder \
            .appName("StockPriceOptimization") \
            .config("spark.sql.adaptive.enabled", "true") \
            .config("spark.sql.shuffle.partitions", "20") \
            .getOrCreate()

        copy_file_from_s3(
            s3_bucket_name,
            s3_file_key,
            local_file_path
        )
        data_processor = DataProcessor(spark)
        df = data_processor.load_data(local_file_path)

        if df is not None:
            data_processor.print_partition_info(df)

            # Repartition the DataFrame
            repartitioned_df = df.repartition(10)
            data_processor.print_partition_
            info(repartitioned_df)
```

```
# Coalesce partitions
coalesced_df = repartitioned_df.coalesce(5)
data_processor.print_partition_info(coalesced_df)

# Example broadcast join
small_df = spark.createDataFrame([
    ("2023-08-25", 5.5),
    ("2023-08-26", 6.0),
], ["Date", "AdditionalData"])
joined_df = repartitioned_df.join(
    broadcast(small_df), on="Date", how="left")

# Save as Parquet
repartitioned_df.write.parquet(
    "/home/ubuntu/airflow/dags/tesla_stock.
    parquet"
)

# Print the first few rows
data_processor.print_first_n_
rows(repartitioned_df)
        else:
            print("Error loading data. Exiting.")
```

[In]:
```
if __name__ == "__main__":
    s3_bucket_name = 'instance1bucket'
    s3_file_key = 'TSLA_stock.csv'
    local_file_path = '/home/ubuntu/airflow/dags/TSLA_stock.csv'
    main(s3_bucket_name, s3_file_key, local_file_path)
```

Data Preparation with PySpark for PyTorch

In this subsection, we combine the individual code snippets we have used to prepare the Tesla stock dataset for a PyTorch deep learning model:

```
[In]: import logging
[In]: from pyspark.sql import SparkSession
[In]: from pyspark.ml.feature import VectorAssembler,
      StandardScaler
[In]: from pyspark.sql import DataFrame
[In]: import numpy as np
[In]: import torch
[In]: from torch.utils.data import DataLoader, TensorDataset
[In]: import boto3

[In]: logging.basicConfig(level=logging.INFO)
[In]: logger = logging.getLogger(__name__)

[In]: def load_data(file_path: str) -> DataFrame:
        """
        Load stock price data from a CSV file using
        SparkSession.
        """
        spark = (SparkSession.builder
                .appName("StockPricePrediction")
                .getOrCreate())
        df = spark.read.csv(
                file_path,
                header=True,
                inferSchema=True
        )
        return df
```

CHAPTER 3   DATA PREPARATION WITH PYSPARK FOR DEEP LEARNING

```
[In]: def preprocess_data(df: DataFrame) -> DataFrame:
          """
          Preprocess the data by assembling feature vectors
          using VectorAssembler, scaling them using
          StandardScaler, and performing feature engineering.
          """
          assembler = VectorAssembler(
              inputCols=['Open', 'High', 'Low', 'Volume'],
              outputCol='features'
          )
          df = assembler.transform(df)
          logger.info(
              "First 5 observations after assembling features:"
          )
          df.show(5, truncate=False)
          df = perform_feature_engineering(df)
          logger.info(
              "First 5 observations after feature engineering:"
          )
          df.show(5, truncate=False)
          scaler = StandardScaler(
              inputCol="features",
              outputCol="scaled_features",
              withStd=True, withMean=True
          )
          scaler_model = scaler.fit(df)
          df = scaler_model.transform(df)
          df = df.select('scaled_features', 'Close')
          return df
```

CHAPTER 3 DATA PREPARATION WITH PYSPARK FOR DEEP LEARNING

```
[In]: def perform_feature_engineering(df: DataFrame) ->
      DataFrame:
          """
          Perform feature engineering operations to enhance the
          data's predictive power.

          This function adds new columns to the DataFrame
          `df` to capture relevant information from existing
          features, including:

          - Price Range Interaction:
              Represents the difference between High and
              Low prices.
          - Price Change Interaction:
              Represents the difference between Close and
              Open prices.
          - Volume-Price Interaction:
              Represents the product of Volume and
              Close prices.

          Parameters:
              df (DataFrame):
                  The input DataFrame containing the raw data.

          Returns:
              DataFrame:
                  The DataFrame `df` with additional engineered
                  features.
          """
          df = df.withColumn('price_range', df['High'] -
          df['Low'])
```

CHAPTER 3   DATA PREPARATION WITH PYSPARK FOR DEEP LEARNING

```
        df = df.withColumn('price_change', df['Close'] -
        df['Open'])

        df = df.withColumn(
            'volume_price_interaction',
            df['Volume'] * df['Close']
        )

        return df

[In]: def print_first_5_observations(df: DataFrame,
      heading: str):
        """
        Print the first 5 observations of the DataFrame with
        a custom heading.
        """
        logger.info(heading)
        df.show(5, truncate=False)

[In]: def convert_to_tensor(features, labels):
        """
        Convert features and labels to PyTorch tensors.
        """
        tensor_features = torch.tensor(
            features,
            dtype=torch.float32
        )
        tensor_labels = torch.tensor(labels, dtype=torch.
        float32)
        return tensor_features, tensor_labels
```

```
[In]: def create_data_loader(features, labels, batch_size):
          """
          Create a PyTorch DataLoader for batch processing.
          """
          tensor_features, tensor_labels = convert_to_tensor(
              features,
              labels
          )
          dataset = TensorDataset(tensor_features,
          tensor_labels)
          data_loader = DataLoader(
              dataset,
              batch_size=batch_size,
              shuffle=True
          )
          return data_loader
[In]: def copy_file_from_s3(
          bucket_name: str,
          file_key: str,
          local_file_path: str
      ):
          """
          Copy file from S3 bucket to local file path.
          """
          s3 = boto3.client('s3')
          s3.download_file(bucket_name, file_key, local_
          file_path)
          logger.info(
              f"File downloaded from S3 bucket {bucket_name}"
              f"to local file path: {local_file_path}"
          )
```

CHAPTER 3   DATA PREPARATION WITH PYSPARK FOR DEEP LEARNING

```
[In]: def main():
          s3_bucket_name = 'instance1bucket'
          s3_file_key = 'TSLA_stock.csv'
          local_file_path = '/home/ubuntu/airflow/dags/TSLA_
          stock.csv'
          copy_file_from_s3(
              s3_bucket_name,
              s3_file_key,
              local_file_path
          )

          df = load_data(local_file_path)
          print_first_5_observations(df, "First 5
          observations:")
          df = preprocess_data(df)
          print_first_5_observations(
              df,
              "First 5 observations after scaling:"
          )

          df_features = np.array(
              df.select('scaled_features')
              .rdd.map(lambda x: x.scaled_features.toArray())
              .collect()
          )
          df_labels = np.array(
              df.select('Close')
              .rdd.map(lambda x: x.Close).collect()
          )

          batch_size = 32
          train_loader = create_data_loader(
```

CHAPTER 3   DATA PREPARATION WITH PYSPARK FOR DEEP LEARNING

```
        df_features, df_labels,
        batch_size
    )
    logger.info(
        "First 5 observations after scaling"
        "and converting to tensors:"
    )
    print(train_loader.dataset[:5])
[In]: if __name__ == "__main__":
        main()
```

Data Preparation with PySpark for Tensorflow

In this subsection, we consolidate the individual code snippets previously employed in preparing the Tesla stock dataset for utilization in a TensorFlow deep learning model:

```
[In]: import logging
[In]: from pyspark.sql import SparkSession
[In]: from pyspark.ml.feature import VectorAssembler,
      StandardScaler
[In]: from pyspark.sql import DataFrame
[In]: import numpy as np
[In]: import tensorflow as tf
[In]: from tensorflow.keras.utils import to_categorical
[In]: import boto3

[In]: logging.basicConfig(level=logging.INFO)
[In]: logger = logging.getLogger(__name__)

[In]: def load_data(file_path: str) -> DataFrame:
```

CHAPTER 3   DATA PREPARATION WITH PYSPARK FOR DEEP LEARNING

```
    """
    Load stock price data from a CSV file using
    SparkSession.
    """
    spark = (SparkSession.builder
            .appName("StockPricePrediction")
            .getOrCreate())
    df = spark.read.csv(
            file_path,
            header=True,
            inferSchema=True
    )
    return df
```

[In]:
```
def preprocess_data(df: DataFrame) -> DataFrame:
    """
    Preprocess the data by assembling feature vectors
    using VectorAssembler, scaling them using
    StandardScaler, and performing feature engineering.
    """
    assembler = VectorAssembler(
        inputCols=['Open', 'High', 'Low', 'Volume'],
        outputCol='features'
    )
    df = assembler.transform(df)
    logger.info(
        "First 5 observations after assembling features:"
    )
    df.show(5, truncate=False)

    df = perform_feature_engineering(df)
```

CHAPTER 3   DATA PREPARATION WITH PYSPARK FOR DEEP LEARNING

```
        logger.info(
            "First 5 observations after feature engineering:"
        )
        df.show(5, truncate=False)

        scaler = StandardScaler(
            inputCol="features",
            outputCol="scaled_features",
            withStd=True, withMean=True
        )
        scaler_model = scaler.fit(df)
        df = scaler_model.transform(df)
        df = df.select('scaled_features', 'Close')
        return df

[In]: def perform_feature_engineering(df: DataFrame) ->
      DataFrame:
        """
        Perform feature engineering operations such as
        creating interaction terms, generating polynomial
        features, or extracting useful information from
        existing features.
        """
        df = df.withColumn('price_range', df['High'] -
        df['Low'])
        df = df.withColumn('price_change', df['Close'] -
        df['Open'])

        df = df.withColumn(
            'volume_price_interaction',
            df['Volume'] * df['Close']
        )

        return df
```

CHAPTER 3   DATA PREPARATION WITH PYSPARK FOR DEEP LEARNING

```
[In]: def print_first_5_observations(df: DataFrame,
      heading: str):
          """
          Print the first 5 observations of the DataFrame with
          a custom heading.
          """
          logger.info(heading)
          df.show(5, truncate=False)
[In]: def create_data_loader(features, labels, batch_size):
          """
          Create a TensorFlow Dataset for batch processing.
          """
          dataset = tf.data.Dataset.from_tensor_slices(
              (features, labels)
          )
          dataset = dataset.shuffle(buffer_size=10000)
          dataset = batch(batch_size)
          return dataset
[In]: def copy_file_from_s3(bucket_name: str, file_key: str,
      local_file_path: str):
          """
          Copy file from S3 bucket to local file path.
          """
          s3 = boto3.client('s3')
          s3.download_file(bucket_name, file_key, local_
          file_path)
          logger.info(
              f"File downloaded from S3 bucket {bucket_name}"
              f"to local file path: {local_file_path}"
          )
```

CHAPTER 3   DATA PREPARATION WITH PYSPARK FOR DEEP LEARNING

```
[In]: def main():
          s3_bucket_name = 'instance1bucket'
          s3_file_key = 'TSLA_stock.csv'
          local_file_path = '/home/ubuntu/airflow/dags/TSLA_
          stock.csv'
          copy_file_from_s3(
              s3_bucket_name,
              s3_file_key,
              local_file_path
          )
          df = load_data(local_file_path)
          print_first_5_observations(df, "First 5
          observations:")
          df = preprocess_data(df)
          print_first_5_observations(
              df,
              "First 5 observations after scaling:"
          )
          df_features = np.array(
              df.select('scaled_features')
              .rdd.map(lambda x: x.scaled_features.toArray())
              .collect()
          )
          df_labels = np.array(
              df.select('Close').rdd.map(lambda x: x.Close)
              .collect()
          )
          batch_size = 32
          train_dataset = create_data_loader(
              df_features,
```

```
                df_labels,
                batch_size
            )

            for features, labels in train_dataset.take(1):
                logger.info(
                    "First 5 observations after scaling and"
                    "converting to TensorFlow Dataset:"
                )
                logger.info(features[:5])
                logger.info("First 5 labels with no scaling:")
                logger.info(labels[:5])

[In]: if __name__ == "__main__":
        main()
```

Summary

In this chapter, we explored how PySpark's parallel processing capabilities can enhance the efficiency of data handling, particularly with large datasets. We examined techniques such as repartitioning, broadcasting joins, caching, saving data as Parquet, optimizing shuffle partitions, and enabling Adaptive Query Execution (AQE) to improve the performance of these data preparation tasks.

We also covered the process of preparing data for deep learning tasks using PySpark. We explored crucial aspects of this process, including data exploration (loading the data, printing the first ten rows of the dataset, performing descriptive statistics, checking for null values, and visualizing the data), data preparation (preprocessing and feature engineering), and data formatting for PyTorch and TensorFlow.

Exploring data serves as the foundation for understanding its characteristics and patterns. Preprocessing involves cleaning and transforming raw data to make it suitable for modeling purposes, while feature engineering plays an important role in enhancing the quality and predictive power of input features. Formatting ensures that the processed data is converted into tensor formats compatible with PyTorch and TensorFlow for efficient processing.

In the next chapter, we will embark on our journey of building, training, and evaluating deep learning algorithms, starting with PyTorch regression.

CHAPTER 4

Deep Learning with PyTorch for Regression

In this chapter, we begin our exploration of deep learning models using PyTorch. This framework, along with TensorFlow (which we will explore in Chapters 5 and 7), represents the industry standard for building and training neural networks. Here, we focus on regression, a fundamental problem in machine learning where the goal is to predict continuous numerical values.

Deep learning approaches to regression tasks offer several advantages over traditional algorithms like linear regression. Deep learning models can automatically learn intricate patterns and relationships from complex datasets, capturing nonlinearities and interactions that may be challenging for linear models to handle effectively. Additionally, deep learning models have the ability to scale with large datasets and can potentially provide better predictive performance when dealing with datasets that include multiple features, such as those related to stock prices, trading volumes, and other factors influencing the financial markets.

However, it's essential to acknowledge that deep learning models often require significantly more computational resources and data compared with traditional algorithms, rendering them computationally intensive and potentially impractical for applications with limited resources. This is where AWS comes in. Amazon Web Services offers scalable and flexible cloud computing solutions, providing the necessary infrastructure for deploying and running deep learning models efficiently. Furthermore, deep learning models are often considered black boxes, making them more difficult to interpret than traditional algorithms like linear regression. They also require extensive hyperparameter tuning, an issue we will revisit in Chapter 9. Nonetheless, deep learning models continue to demonstrate remarkable performance across a wide range of tasks and remain an area of active research and development.

To demonstrate how to use PyTorch for regression tasks, we will build, train, and evaluate a neural network model and use it to predict Tesla stock prices. This type of prediction is crucial in finance, where accurately forecasting stock prices can inform investment decisions and risk management strategies. It's important to note, however, that the purpose of this model in this chapter is for demonstration purposes, and we are not aiming to develop a model for commercial use.

The Dataset

We will utilize a historical daily dataset of Tesla stock obtained from Yahoo! Finance. This dataset was downloaded in CSV format and subsequently stored on AWS S3 within a bucket named "instance1bucket" from the following link:

https://finance.yahoo.com/quote/TSLA/history/

We explored this dataset in detail in the previous chapter. In this section, we will load the CSV file, copy it from the S3 bucket to a local directory on the EC2 instance, and print the first five rows to refresh our memory of what the features and target variable look like.

CHAPTER 4   DEEP LEARNING WITH PYTORCH FOR REGRESSION

We start by importing the necessary modules:

```
[In]: import subprocess
[In]: from pyspark.sql import SparkSession
[In]: import logging
```

- import subprocess: This line imports the subprocess module, which is commonly used to execute shell commands or interact with the system's shell. In our code, subprocess will be used to execute the aws s3 cp command for copying the CSV file from the S3 bucket to the local directory.

- from pyspark.sql import SparkSession: This line imports the SparkSession class from the pyspark.sql module. SparkSession serves as the starting point for utilizing Spark's Dataset and DataFrame API, providing an interface for programming tasks within the Spark framework. It allows us to create DataFrame and Dataset objects, interact with Spark SQL, and configure Spark properties. In our upcoming code, SparkSession will be used to create a Spark application and load the CSV file into a DataFrame.

- import logging: This line imports the logging module, which provides a framework for generating log messages from Python programs. It is commonly used for debugging, monitoring, and troubleshooting Python applications and scripts. In our code, logging is used to log informational and error messages during the data loading and copying processes, providing feedback and aiding in debugging.

CHAPTER 4   DEEP LEARNING WITH PYTORCH FOR REGRESSION

Next, we define the load_data function, which loads the Tesla stock price data from a CSV file using SparkSession:

```
[In]: def load_data(file_path: str):
          """
          Load stock price data from a CSV file using
          SparkSession.
          """
          spark = (SparkSession.builder
                  .appName("StockPricePrediction")
                  .getOrCreate())
          df = spark.read.csv(
                  file_path,
                  header=True,
                  inferSchema=True
          )
          return df
```

Let's break down the function:

- def load_data(file_path: str):: This line defines the function load_data with one parameter, file_path, which is a string representing the path to the CSV file.

- """ Load stock price data from a CSV file using SparkSession. """: This is a docstring, which provides documentation for the function. It describes the purpose of the function, which is to load stock price data from a CSV file using SparkSession.

- spark = SparkSession.builder: This line instantiates a SparkSession object using the builder method, enabling the configuration of Spark properties and settings.

CHAPTER 4   DEEP LEARNING WITH PYTORCH FOR REGRESSION

- .appName("StockPricePrediction"): This sets the application name to "StockPricePrediction". This is optional but can be useful for identifying the application in Spark's monitoring tools. When a Spark job is executed within an Airflow DAG (Directed Acyclic Graph), as we will demonstrate in Chapter 8, setting a descriptive application name can greatly assist in tracking and managing the Spark job within Airflow's interface and logs.

- .getOrCreate(): This method either returns an existing SparkSession if there is one running in the current environment or creates a new one if none exists.

- df = spark.read.csv(file_path, header=True, inferSchema=True): This line uses the read.csv() method of the SparkSession object to read the CSV file located at file_path into a DataFrame (df). The header=True argument indicates that the first row of the CSV file contains the column names, and inferSchema=True instructs Spark to automatically infer the data types of each column.

- return df: Finally, the function returns the DataFrame df containing the loaded data.

The next step is to define the copy_and_print_data function, which copies the CSV file from an S3 bucket on AWS to a local directory on the EC2 instance and prints the first five observations:

```
[In]: def copy_and_print_data():
          """
          Copy a CSV file from an S3 bucket to a local
          directory and
          print the first 5 observations.
```

217

```
"""
    s3_bucket_path = "s3://instance1bucket/TSLA_
    stock.csv"
    local_file_path = "/home/ubuntu/airflow/dags/TSLA_
    stock.csv"
    logging.basicConfig(level=logging.INFO)
    logger = logging.getLogger(__name__)

    try:
        subprocess.run(
            [
                "aws", "s3", "cp",
                s3_bucket_path,
                local_file_path
            ],
            check=True
        )
        df = load_data(local_file_path)
        print("First 5 observations of the DataFrame:")
        df.show(5)

    except FileNotFoundError:
        logger.error(f"Data file not found at {s3_
        bucket_path}")
    except subprocess.CalledProcessError as e:
        logger.error(f"Error copying data from S3: {e}")
```

Let's break down the function:

- def copy_and_print_data():: This line defines the function copy_and_print_data with no parameters. The function copies a CSV file from an S3 bucket on AWS to a local directory on EC2 instance and prints the first five observations.

CHAPTER 4 DEEP LEARNING WITH PYTORCH FOR REGRESSION

- """ Copy a CSV file from an S3 bucket to a local directory and print the first 5 observations. """: This is a docstring, providing a brief description of what the function does.

- s3_bucket_path = "s3://instance1bucket/TSLA_stock.csv": This line defines the path to the CSV file in the AWS S3 bucket.

- local_file_path = "/home/ubuntu/airflow/dags/TSLA_stock.csv": This line defines the path where the CSV file will be copied to locally on the EC2 instance.

- logging.basicConfig(level=logging.INFO): This line configures the logging module to display INFO-level log messages.

- logger = logging.getLogger(__name__): This line initializes a logger object specific to this module.

- try:: This begins a try–except block to handle potential exceptions that may occur during execution.

- subprocess.run(["aws", "s3", "cp", s3_bucket_path, local_file_path], check=True): This line executes a subprocess to copy the CSV file from the S3 bucket to the local directory on the EC2 instance using the AWS CLI command aws s3 cp. The check=True parameter ensures that an exception is raised if the subprocess command does not complete successfully.

- df = load_data(local_file_path): This line calls the load_data function to load the CSV file into a DataFrame (df).

219

CHAPTER 4   DEEP LEARNING WITH PYTORCH FOR REGRESSION

- print("First 5 observations of the DataFrame:"): This line prints a descriptive message indicating that the first five observations of the DataFrame will be printed.

- df.show(5): This line displays the first five rows of the DataFrame df using the PySpark show() method.

- except FileNotFoundError:: This except clause handles a FileNotFoundError exception, which occurs if the specified file is not found.

- except subprocess.CalledProcessError as e:: This except clause handles a CalledProcessError exception, which occurs if there is an error when executing the subprocess command.

- logger.error(f"Data file not found at {s3_bucket_path}"): This line logs an error message if the specified file is not found.

- logger.error(f"Error copying data from S3: {e}"): This line logs an error message if there is an error when executing the subprocess command.

Finally, we define the main function:

```
[In]: if __name__ == "__main__":
         copy_and_print_data()
```

- if __name__ == "__main__": This is a commonly used line of code to execute a Python program only when the script is run directly, as opposed to being imported as a module into another script. It checks if the special variable __name__ is equal to "__main__".

220

This condition evaluates to True if the script is being run directly and False if it is being imported as a module. This is a common best practice in Python programming, as it allows the script to be used both as a standalone program and as a reusable module.

- copy_and_print_data(): This line calls the function copy_and_print_data() if the script is run directly. In other words, when the Python interpreter executes this script directly, it will invoke the copy_and_print_data() function.

We save the script as "spark_data_processing.py" in the "/home/ubuntu/airflow/dags" directory. To execute it, we have the option of navigating to this directory using the following command in the JupyterLab interface terminal or executing the script directly on the EC2 command-line interface:

```
[In]: cd /home/ubuntu/airflow/dags
```

Next, we activate the virtual environment "myenv" with the command

```
[In]: source myenv/bin/activate
```

Finally, with the virtual environment activated, we run the script using the command

```
[In]: python3 spark_data_processing.py
```

This initiates the execution of the script, which involves copying the Tesla stock CSV file from the S3 bucket to the local directory and printing the first five observations of the DataFrame.

After running the code, we obtain the following output:

```
[Out]: First 5 observations of the DataFrame:
```

Date	Open	High	Low	Close	Volume
2/23/24	195.31	197.57	191.50	191.97	78670300
2/22/24	194.00	198.32	191.36	197.41	92739500
2/21/24	193.36	199.44	191.95	194.77	103844000
2/20/24	196.13	198.60	189.13	193.76	104545800
2/16/24	202.06	203.17	197.40	199.95	111173600

The daily dataset ranges from February 26, 2019, to February 23, 2024 (1,258 observations). The output displays the first five observations from this historical dataset. Each trading day captures real-time fluctuations in the Tesla stock price. These observations highlight key indicators—Open, High, Low, Close, and Volume—illustrating price movements over a specific timeframe. Their definitions are as follows:

- Open: The opening price of Tesla stock on the given date
- High: The highest price of Tesla stock reached during the trading session on the given date
- Low: The lowest price of Tesla stock reached during the trading session on the given date
- Close: The closing price of Tesla stock on the given date
- Volume: The trading volume, representing the total number of shares traded on the given date

CHAPTER 4 DEEP LEARNING WITH PYTORCH FOR REGRESSION

Predicting Tesla Stock Price with PyTorch

In this section, we utilize the dataset explored in the previous section to implement a deep learning algorithm for predicting Tesla's stock prices. In this dataset, the features of the model consist of Open, High, Low, and Volume, while the target variable is the Close price. The opening price of the stock can provide valuable information about market sentiment at the beginning of a trading day, reflecting investors' initial reactions to news, events, or overnight developments. Similarly, trading volume is important because high trading volumes often indicate increased investor interest and can signal potential price movements. Volume can also be used as a proxy for liquidity, which is essential for accurate price discovery. As for the highest and lowest prices reached during a trading session, they provide insights into price volatility and trading range. High prices indicate the highest level of buying interest, while low prices represent the lowest level of selling interest. Analyzing the range between high and low prices can help identify trends, support, and resistance levels.

By incorporating these features into our predictive model, we can capture both fundamental market dynamics (such as opening sentiment and trading activity) and technical aspects (such as price volatility and trading range), which are essential for predicting stock prices accurately. However, it's worth noting that the purpose of our model is illustration rather than providing a complete account. While these features are valuable, incorporating additional features such as technical indicators, fundamental data, or market sentiment could further enhance the predictive power of our model. This, however, is outside the scope of this chapter.

We leverage the capabilities of PySpark for data processing and PyTorch for building, training, and evaluating the neural network model. The Python code is designed to handle an end-to-end workflow, from data loading and preprocessing to model training, evaluation, and prediction. In Chapter 8, we will demonstrate how to deploy this code using Apache Airflow.

CHAPTER 4   DEEP LEARNING WITH PYTORCH FOR REGRESSION

The entire modeling process consists of the following ten steps:

    Step 1: Imports

    Step 2: Logging Setup

    Step 3: Data Loading

    Step 4: Data Preprocessing

    Step 5: DataLoader Creation

    Step 6: Model Training

    Step 7: Model Evaluation

    Step 8: Main

    Step 9: Helper

    Step 10: Execution

Let's dive deep into the code step by step:

Step 1: Imports

In this initial step, we introduce the libraries and modules imported in the code. These imports enable data processing, modeling, and logging:

```
[In]: import logging
[In]: import subprocess
[In]: from typing import Tuple
[In]: import numpy as np
[In]: import torch
[In]: import torch.nn as nn
[In]: import torch.optim as optim
[In]: from torch.utils.data import DataLoader, TensorDataset
[In]: from pyspark.sql import SparkSession
[In]: from pyspark.ml.feature import VectorAssembler,
       StandardScaler
[In]: from pyspark.sql import DataFrame
```

CHAPTER 4 DEEP LEARNING WITH PYTORCH FOR REGRESSION

Here's a breakdown of the imports and their significance:

- logging: This library is utilized for generating log messages during the execution of the code. It enables tracking the program's progress, debugging issues, and monitoring activities. In our code, logging will be used to log informational and error messages during various stages such as data loading, preprocessing, model training, and evaluation.

- subprocess: This module enables the code to execute system commands and interact with other programs. It's used in our code for copying data from an AWS S3 bucket to a local directory on EC2 before data processing.

- typing: This module allows specifying the expected types of function arguments and return values. It will be employed in our code in function annotations for specifying argument and return types.

- numpy: This is a fundamental package for scientific computing with Python. It provides support for large, multidimensional arrays and matrices, along with a collection of mathematical functions to operate on these arrays. In our code, NumPy is used for converting Spark DataFrame columns to NumPy arrays.

- torch, torch.nn, torch.optim: These are components of PyTorch, a machine learning library primarily used for deep learning applications. They are used in our code for defining and training the neural network model.

- DataLoader, TensorDataset: These are classes from PyTorch's torch.utils.data module. DataLoader is used to load datasets in batches during training and

CHAPTER 4 DEEP LEARNING WITH PYTORCH FOR REGRESSION

- inference, while TensorDataset is a dataset wrapper that allows convenient access to tensors. They are used in the code for creating data loaders for training and testing the neural network model.

- pyspark.sql, pyspark.ml.feature, DataFrame: These are components of PySpark and will be used in our code for loading, preprocessing, and manipulating structured data in Spark DataFrames before converting them into PyTorch tensors for training the neural network model. Specifically, pyspark.sql is used for Spark DataFrame operations and pyspark.ml.feature for data preprocessing tasks such as feature assembling and scaling, and DataFrame represents the DataFrame class used in PySpark for data manipulation and analysis.

Step 2: Logging Setup

In this step, we configure the logging system to ensure effective monitoring of the code's execution and to facilitate debugging and tracking of program progress:

```
[In]: logging.basicConfig(level=logging.INFO)
[In]: logger = logging.getLogger(__name__)
```

Here is what the two lines mean:

- logging.basicConfig(level=logging.INFO): This line configures the logging system to display log messages with the INFO level or higher. It ensures that log messages of sufficient importance are shown during the execution of the code.

- logger = logging.getLogger(__name__): This line creates a logger object named logger specifically for the current module. It allows for logging messages within the module, which can help in tracking the program's progress, debugging issues, and monitoring activities.

Step 3: Data Loading Function (load_data)

In this step, we define the load_data function for loading Tesla stock price data from a CSV file stored on S3 on AWS. This data contains date, open price, high price, low price, volume, and closing price of Tesla stocks for the period from February 26, 2019, to February 23, 2024:

```
[In]: def load_data(file_path: str) -> DataFrame:
          """
          Load stock price data from a CSV file using
          SparkSession.
          """
          try:
              spark = (SparkSession.builder
                      .appName("StockPricePrediction")
                      .getOrCreate())
              df = spark.read.csv(
                      file_path,
                      header=True,
                      inferSchema=True
              )
              return df
          except Exception as e:
              raise RuntimeError(
                  f"Error loading data from {file_path}: {e}"
              )
```

CHAPTER 4   DEEP LEARNING WITH PYTORCH FOR REGRESSION

Let's examine the use of SparkSession within the function along with the process of data loading and exception handling:

Using SparkSession:

- SparkSession is a main entry point for accessing Spark functionality and working with data in Spark. In the load_data function, SparkSession is used to create a Spark application context, which allows for efficient parallel processing of large datasets distributed across multiple nodes in a cluster.

- By using SparkSession, the code leverages the distributed computing capabilities of Apache Spark, enabling scalable and high-performance data processing.

Data Loading Process:

- Inside the load_data function, the SparkSession is created using the SparkSession.builder method. This sets up the configuration for the Spark application, such as the application name.

- The function then reads the CSV file using the spark.read.csv method into a DataFrame.

- The header=True argument specifies that the first row of the CSV file contains column names, while inferSchema=True infers the schema of the DataFrame from the data.

- The loaded DataFrame is then returned by the function.

CHAPTER 4 DEEP LEARNING WITH PYTORCH FOR REGRESSION

Exception Handling:

- The code uses a try-except block to handle potential exceptions that may occur during the data loading process. Exceptions could arise due to various reasons such as missing files, permission issues, or data format inconsistencies.

- If an exception occurs during the execution of the code within the try block, it is caught by the except block. The except clause then raises a RuntimeError, providing an informative error message indicating the source of the error.

Step 4: Data Preprocessing Function (preprocess_data)

In this step, we define the preprocess_data function, which is responsible for preprocessing the loaded Tesla stock price data:

```
[In]: def preprocess_data(df: DataFrame) -> DataFrame:
          """
          Preprocess the data by assembling feature
          vectors using
          VectorAssembler and scaling them using
          StandardScaler.
          """
          assembler = VectorAssembler(
              inputCols=['Open', 'High', 'Low', 'Volume'],
              outputCol='features'
          )
          df = assembler.transform(df)

          scaler = StandardScaler(
              inputCol="features",
              outputCol="scaled_features",
```

CHAPTER 4   DEEP LEARNING WITH PYTORCH FOR REGRESSION

```
            withStd=True,
            withMean=True
    )
    scaler_model = scaler.fit(df)
    df = scaler_model.transform(df)
    df = df.select('scaled_features', 'Close')
    return df
```

The preprocess_data function is tasked with preparing the loaded Tesla stock price data for model training. Preprocessing is a crucial step in machine learning pipelines, as it helps improve model performance by transforming the raw data into a suitable format for training.

The preprocessing steps involve feature vector assembly and feature scaling:

Feature Vector Assembly:

- The first step in preprocessing with PySpark is assembling feature vectors, which are collections of numerical values representing different features of the data. In this case, they represent the Tesla stock open price, high price, low price, and volume. (The target variable, closing price, is not included.)

- The VectorAssembler from PySpark is utilized to concatenate these feature columns into a single feature vector column, enabling smoother processing and facilitating input into the PyTorch machine learning model.

Feature Scaling:

- Once the feature vectors are assembled, the next step is feature scaling. Feature scaling is the process of standardizing or normalizing the feature values to a

CHAPTER 4 DEEP LEARNING WITH PYTORCH FOR REGRESSION

similar scale. This is important because features with larger magnitudes can dominate those with smaller magnitudes during model training.

- The StandardScaler from PySpark is employed to scale the feature vectors. It standardizes the feature values by subtracting the mean and dividing by the standard deviation, ensuring that each feature has a mean of 0 and a standard deviation of 1.

- The scaled feature vectors are stored in a new column named scaled_features, which replaces the original feature vectors in the DataFrame.

In terms of the implementation details of the preprocess_data function

- Inside the function, the VectorAssembler is initialized with the inputCols parameter specifying the columns to assemble into the feature vector and the outputCol parameter specifying the name of the output column.

- The transform method of the VectorAssembler is then called on the DataFrame to assemble the feature vectors.

- Similarly, the StandardScaler is initialized with the inputCol parameter specifying the column containing the feature vectors and the outputCol parameter specifying the name of the output column for scaled features.

- The fit method of the StandardScaler is applied to the DataFrame to compute summary statistics (mean and standard deviation) for feature scaling, and then the transform method is called to scale the feature vectors.

231

CHAPTER 4   DEEP LEARNING WITH PYTORCH FOR REGRESSION

By performing feature vector assembly and feature scaling, the preprocess_data function prepares the Tesla stock price data for subsequent model training using PyTorch. These preprocessing steps help ensure that the data is appropriately formatted and scaled for optimal performance of the PyTorch machine learning model.

Step 5: DataLoader Creation Function (create_data_loader)

In this step, we define the create_data_loader function, which plays a critical role in creating DataLoader objects for both training and testing datasets:

```
[In]: def create_data_loader(
          features,
          labels,
          batch_size=32,
          num_workers=4) -> DataLoader:
      """
      Convert the preprocessed data into PyTorch
      tensors and
      create DataLoader objects for both the training
      and test
      sets.

      Args:
          features (Tensor): The input features.
          labels (Tensor): The corresponding labels.
          batch_size (int):
              The number of samples per batch to load.
              Defaults to
              32.
          num_workers (int):
              The number of subprocesses to use for data
              loading. Defaults to 4.
```

CHAPTER 4 DEEP LEARNING WITH PYTORCH FOR REGRESSION

```
    Returns:
        DataLoader: A PyTorch DataLoader for the dataset.
    """
    dataset = TensorDataset(features, labels)
    return DataLoader(
        dataset,
        batch_size=batch_size,
        shuffle=True
    )
```

The create_data_loader function is responsible for converting preprocessed data into PyTorch tensors and creating DataLoader objects. These DataLoader objects are essential for efficiently feeding data to the machine learning model during both the training and testing phases.

The function takes preprocessed features and labels as inputs and converts them into a PyTorch TensorDataset, which is a dataset wrapping tensors. It then creates DataLoader objects for both the training and testing datasets using the TensorDataset. The num_workers parameter in the DataLoader allows for parallel data loading, which can significantly speed up the data preparation phase, particularly when working with large datasets or complex preprocessing steps.

DataLoader is a PyTorch utility that provides efficient data loading by automatically batching the data and shuffling it if necessary. The DataLoader plays a crucial role in managing data batches during model training. Instead of feeding the entire dataset into the model at once, DataLoader splits the dataset into batches of specified sizes. This batching process allows for more efficient computation, particularly when dealing with large datasets that may not fit into memory all at once. It also enables the use of mini-batch gradient descent, which updates model parameters based on subsets of the data, leading to faster convergence and better generalization.

CHAPTER 4   DEEP LEARNING WITH PYTORCH FOR REGRESSION

The DataLoader is also important for shuffling data. During model training, it's essential to shuffle the data to prevent the model from learning the order or patterns specific to the dataset.

DataLoader provides an option to shuffle the data, ensuring that each batch seen by the model during training contains a random selection of samples from the dataset. This randomness helps prevent the model from overfitting to the training data and improves its ability to generalize to unseen data.

In terms of the implementation details of the create_data_loader function

- Inside the function, the preprocessed features and labels are combined into a PyTorch TensorDataset using the TensorDataset constructor.

- DataLoader objects for both the training and testing datasets are created by passing the TensorDataset along with the specified batch size and shuffle parameters to the DataLoader constructor.

- The function then returns the DataLoader objects, which can be used to iterate over batches of data during model training and evaluation.

By creating DataLoader objects, the create_data_loader function enables efficient data loading and management during the training and testing phases of the machine learning pipeline. DataLoader's ability to handle data batching and shuffling contributes to improved training efficiency, model performance, and generalization capability.

Step 6: Model Training Function (train_model)
In this step, we create the train_model function, which is responsible for training the neural network model using the training data:

```
[In]: def train_model(
          model,
          train_loader,
          criterion,
          optimizer,
          num_epochs
      ):
          """
          Train the model on the training data using the
          DataLoader
          and the defined loss function and optimizer.
          Iterate over
          the data for a specified number of epochs,
          calculate the
          loss, and update the model parameters.
          """
          for epoch in range(num_epochs):
              for inputs, labels in train_loader:
                  optimizer.zero_grad()
                  outputs = model(inputs)
                  loss = criterion(outputs, labels.
                  unsqueeze(1))
                  loss.backward()
                  optimizer.step()
          logger.info(
              f"Epoch [{epoch + 1}/{num_epochs}],"
              f"Loss: {loss.item():.4f}"
          )
```

The train_model function is tasked with training the neural network model on the training data. Training a neural network involves iteratively updating the model's parameters (weights and biases) based on the difference between the predicted outputs and the actual targets.

Let's explain this process in more detail:

Iterative Training:

- The function iterates over the training dataset for a specified number of epochs. Each epoch represents a complete pass through the entire training dataset.

- Within each epoch, the function iterates over mini-batches of data. This process is facilitated by the DataLoader object, which automatically batches the data.

- For each mini-batch, the model makes predictions using the current parameter values (forward pass), computes the loss between the predicted outputs and the actual targets, and then updates the model parameters to minimize the loss (backward pass).

Parameter Updates Using Backpropagation:

- Backpropagation is a key algorithm for training neural networks. It involves computing the gradients of the loss function with respect to the model's parameters, which represent how much each parameter contributes to the error.

- After computing the gradients, the optimizer (in this case, the Adam optimizer) adjusts the model's parameters in the direction that minimizes the loss.

- This process is repeated for each mini-batch in the training dataset and over multiple epochs, gradually improving the model's performance by minimizing the loss function.

CHAPTER 4 DEEP LEARNING WITH PYTORCH FOR REGRESSION

Regarding the specific details of the train_model function implementation

- Inside the function, there is a nested loop structure where the outer loop iterates over epochs and the inner loop iterates over mini-batches of data.

- For each mini-batch, the optimizer's zero_grad method is called to reset the gradients of all model parameters to zero.

- The model's forward method is then invoked to make predictions on the input data, followed by computing the loss using the specified loss function (criterion) and the actual labels.

- After computing the loss, the backward method is called to compute the gradients of the loss with respect to the model parameters.

- Finally, the optimizer's step method is called to update the model parameters based on the computed gradients, effectively performing one optimization step.

By implementing the iterative training process and parameter updates using backpropagation, the train_model function enables the PyTorch neural network model to learn from the training data and improve its performance over successive epochs. This iterative optimization process is fundamental to the success of neural network training in machine learning applications.

CHAPTER 4   DEEP LEARNING WITH PYTORCH FOR REGRESSION

Step 7: Model Evaluation Function (evaluate_model)
In this step, we create the evaluate_model function, which evaluates the trained neural network model on the test data:

```
[In]: def evaluate_model(
          model,
          test_loader,
          criterion
      ) -> Tuple[float, float]:
          """
          Evaluate the trained model on the test data to assess its
          performance.
          Calculate the test loss and additional evaluation metrics
          such as the R-squared score.
          """
          with torch.no_grad():
              model.eval()
              predictions = []
              targets = []
              test_loss = 0.0
              for inputs, labels in test_loader:
                  outputs = model(inputs)
                  loss = criterion(outputs, labels.unsqueeze(1))
                  test_loss += loss.item() * inputs.size(0)
                  predictions.extend(outputs.squeeze().tolist())
                  targets.extend(labels.tolist())
              test_loss /= len(test_loader.dataset)
              predictions = torch.tensor(predictions)
```

CHAPTER 4   DEEP LEARNING WITH PYTORCH FOR REGRESSION

```
targets = torch.tensor(targets)
ss_res = torch.sum((targets - predictions) ** 2)
ss_tot = torch.sum((targets - torch.
mean(targets)) ** 2)
r_squared = 1 - ss_res / ss_tot
logger.info(f"Test Loss: {test_loss:.4f}")
logger.info(f"R-squared Score: {r_squared:.4f}")
return test_loss, r_squared.item()
```

The evaluate_model function assesses the performance of the trained neural network model on the unseen test data. Evaluation is essential to understand how well the model generalizes to new, unseen samples and to identify any potential issues such as overfitting.

The function computes the test loss, which measures the discrepancy between the model's predictions and the actual targets in the test dataset. This is typically done using a loss function such as Mean Squared Error (MSE) or Mean Absolute Error (MAE). In this case, the MSE loss function is used. The total test loss is computed by averaging the individual losses across all samples in the test dataset.

In addition to the test loss, the function calculates an additional evaluation metric, the R-squared score (also known as the coefficient of determination). The R-squared score measures the proportion of the variance in the target variable (close prices in this case) that is predictable from the independent variables (features) in the model. It ranges from 0 to 1, with higher values indicating better model fit. A score of 1 indicates a perfect fit, while a score of 0 indicates that the model's predictions are no better than simply using the mean of the target variable.

Inside the evaluate_model function, the model is set to evaluation mode using model.eval(). The function iterates over mini-batches of data in the test DataLoader, making predictions using the trained model and computing the test loss for each mini-batch. It also accumulates the predictions and actual targets to calculate the R-squared score after iterating over all mini-batches.

CHAPTER 4   DEEP LEARNING WITH PYTORCH FOR REGRESSION

The R-squared score is computed using the formula

```
[In]: r_squared = 1 - ss_res / ss_tot
```

where ss_res is the sum of squares of residuals (differences between predictions and targets) and ss_tot is the total sum of squares.

The logger.info logs the test loss and R-squared score to provide insights into the model's performance on the test dataset. These metrics help in assessing the model's predictive capabilities and identifying areas for improvement.

Step 8: Main Function (main)
In this step, we'll discuss the main function, which serves as the entry point for executing the code:

```
[In]: def main(
        data_file_path: str,
        num_epochs: int = 100,
        batch_size: int = 32,
        learning_rate: float = 0.001
    ):
        """
        Main function for training and evaluating a deep learning
        model for Tesla stock price prediction.

        Args:
            data_file_path (str): The path to the CSV file
                containing stock price data.
            num_epochs (int): The number of epochs for
            training the
                model. Defaults to 100.
            batch_size (int): The batch size for data
            loading during
                training. Defaults to 32.
```

240

CHAPTER 4   DEEP LEARNING WITH PYTORCH FOR REGRESSION

```
    learning_rate (float): The learning rate for the
        optimizer. Defaults to 0.001.
    num_workers (int): The number of subprocesses
    to use for
        data loading. Defaults to 4.

Raises: FileNotFoundError: If the specified data
file is not
    found.
"""
try:
    subprocess.run([
        "aws", "s3", "cp",
        "s3://instance1bucket/TSLA_stock.csv",
        "/home/ubuntu/airflow/dags/"
    ])

    df = load_data(data_file_path)
    df = preprocess_data(df)

    train_df, test_df = df.randomSplit([0.8, 0.2],
    seed=42)

    train_features, train_labels = \
        spark_df_to_tensor(train_df)
    test_features, test_labels = spark_df_to_
    tensor(test_df)

    train_loader = create_data_loader(
        train_features,
        train_labels,
        batch_size=batch_size,
        num_workers=num_workers
    )
```

```python
    test_loader = create_data_loader(
        test_features,
        test_labels,
        batch_size=batch_size,
        num_workers=num_workers
    )

    input_size = train_features.shape[1]
    output_size = 1
    model = nn.Sequential(
        nn.Linear(input_size, 64),
        nn.ReLU(),
        nn.Linear(64, 32),
        nn.ReLU(),
        nn.Linear(32, 16),
        nn.ReLU(),
        nn.Linear(16, output_size)
    )
    criterion = nn.MSELoss()
    optimizer = optim.Adam(
        model.parameters(),
        lr=learning_rate
    )

    train_model(
        model,
        train_loader,
        criterion,
        optimizer,
        num_epochs
    )
```

CHAPTER 4   DEEP LEARNING WITH PYTORCH FOR REGRESSION

```
        evaluate_model(model, test_loader, criterion)
        torch.save(model.state_dict(), 'trained_
        model.pth')
    except FileNotFoundError:
        logger.error(f"Data file not found at {data_
        file_path}")
```

Let's explore the function's role and break it down into individual actions, including data loading, preprocessing, model initialization, training, saving, and evaluation:

- The main function orchestrates the entire process of Tesla stock price prediction, from data loading to model evaluation. It serves as the central coordinator for the machine learning pipeline.

- The main function begins by copying data from an AWS S3 bucket to a local directory on EC2 using the subprocess.run method. This step is a prerequisite for data loading. Next, it loads the Tesla stock price data from a CSV file using the load_data function, passing the file path as an argument.

- Following data loading, the main function initializes the neural network model using PyTorch. It defines the model architecture, specifying the number of input features, which is dynamically inferred from the dataset, and the output size, which is fixed at 1 for predicting a single value (stock price). Additionally, the model includes three hidden layers with 64, 32, and 16 units, respectively, each followed by a rectified linear unit (ReLU) activation function.

- Once initialized, the main function proceeds to train the model using the train_model function. It specifies parameters such as the number of epochs, batch size, learning rate, loss function, and optimizer. During training, the model's parameters are iteratively updated based on the training data to minimize the loss function.

- After training, the main function saves the trained model's parameters to a file using the torch.save method. This allows the model to be reused or deployed for making predictions on new data without needing to retrain it.

- Subsequently, the main function evaluates the trained model's performance on the test data using the evaluate_model function. It computes metrics such as test loss and the R-squared score to assess how well the model generalizes to unseen data.

- The main function includes error handling mechanisms, such as catching file not found errors during data loading using a try–except block. This ensures that the code handles unexpected issues and provides informative error messages for debugging purposes.

By encapsulating these stages within the main function, the code achieves a modular and organized structure, making it easier to understand and maintain. The main function acts as a central coordinator, integrating the different components of the machine learning pipeline to accomplish the task of Tesla stock price prediction. In Chapter 8, we explore further enhancements to this process by deploying the code through Apache Airflow.

CHAPTER 4   DEEP LEARNING WITH PYTORCH FOR REGRESSION

Step 9: Helper Function (spark_df_to_tensor)
In this step, we create the spark_df_to_tensor helper function, which converts Spark DataFrame columns to PyTorch tensors:

```
[In]: def spark_df_to_tensor(
            df: DataFrame
    ) -> Tuple[torch.Tensor, torch.Tensor]:
        """
        Converts Spark DataFrame columns to PyTorch tensors.
        """
        features = torch.tensor(
            np.array(
                df.rdd.map(lambda x: x.scaled_features.toArray())
                .collect()
            ),
            dtype=torch.float32
        )
        labels = torch.tensor(
            np.array(
                df.rdd.map(lambda x: x.Close).collect()
            ),
            dtype=torch.float32)
        return features, labels
```

The spark_df_to_tensor function serves the purpose of converting Spark DataFrame columns to PyTorch tensors. This conversion is necessary because PyTorch tensors are the primary data structure used for representing and manipulating data within neural network models.

The function takes a Spark DataFrame as input, containing the preprocessed features and labels. It extracts the feature vectors and labels from the DataFrame and converts them into NumPy arrays using the collect method on the DataFrame's RDD (Resilient Distributed Dataset).

CHAPTER 4   DEEP LEARNING WITH PYTORCH FOR REGRESSION

The NumPy arrays are then converted into PyTorch tensors using the torch.tensor constructor, which creates tensors from array-like objects. Finally, the function returns the feature tensors and label tensors as a tuple, ready to be used as inputs for training or evaluating the neural network model.

In terms of the spark_df_to_tensor function's implementation details

- Inside the function, the feature vectors are extracted from the DataFrame by mapping each row to its corresponding feature vector array using a lambda function.

- The toArray() method is used to convert the Spark vector column to a NumPy array.

- Similarly, the labels are extracted from the DataFrame using the collect method and converted to a NumPy array.

- Both the feature and label arrays are then converted to PyTorch tensors using the torch.tensor constructor.

- Finally, the feature and label tensors are returned as a tuple.

By providing a convenient interface for converting Spark DataFrame columns to PyTorch tensors, the spark_df_to_tensor function facilitates the integration of PySpark-based data processing pipelines with PyTorch-based machine learning models. This enables data flow between the data preprocessing stage (performed using PySpark) and the model training/evaluation stage (performed using PyTorch), without the need for manual conversion or compatibility issues. Additionally, converting the data to tensors ensures compatibility with PyTorch's tensor-based operations and automatic differentiation capabilities, which are essential for training neural network models. Finally, this mechanism of converting Spark

CHAPTER 4 DEEP LEARNING WITH PYTORCH FOR REGRESSION

DataFrame columns to PyTorch tensors simplifies the data processing workflow and enables efficient training and evaluation of neural network models on large-scale datasets.

Step 10: Execution
The code in this step is typically placed at the end of the script, and it serves as the entry point for executing the code when the script is run directly. It calls the main function with the specified data file path as an argument, initiating the entire process of Tesla stock price prediction using the PyTorch machine learning technique:

```
[In]: if __name__ == "__main__":
          data_file_path = "/home/ubuntu/airflow/dags/TSLA_stock.csv"
          main(data_file_path)
```

In Chapter 8, we demonstrate how to optimize this code deployment process using Apache Airflow, a powerful workflow management platform. We highlight Airflow's advanced task scheduling and orchestration capabilities, which eliminate the necessity for the traditional main function. Instead, we leverage Airflow's Directed Acyclic Graph (DAG) approach, where each node represents a specific task in our data pipeline. Airflow automates task dependencies, enables parallel execution, facilitates retries, and offers comprehensive monitoring and logging features. This enhances scalability and improves operational efficiency, paving the way for a streamlined and automated pipeline.

To run the code, we save it as "tesla_stock_price_prediction.py" and execute it in a JupyterLab terminal (alternatively, we could have run it directly on the EC2 instance). Since we have created an environment named "myenv," we activate it as follows:

```
[In]: source /home/ubuntu/myenv/bin/activate
```

CHAPTER 4 DEEP LEARNING WITH PYTORCH FOR REGRESSION

After activating the environment, we execute the code using Python 3 with the following command:

[In]: python3 tesla_stock_price_prediction.py

The output from the code shows three main metrics:

- Loss value for each epoch during the training process
- Test loss
- R-squared score after the training is completed

Training Loss:

Epoch	Loss
1/100	48533.29
2/100	40968.77
3/100	26662.96
4/100	13865.01
5/100	3026.32
6/100	1682.71
7/100	1077.39
8/100	475.19
9/100	1194.83
10/100	1361.66
...	...
100/100	11.12

Test Loss: 20.54
R-squared Score: 0.998

Let's explain these metrics one by one:

Train Loss:

The value of the loss function computed during training represents how well the model is performing on the training data. Lower values indicate better performance.

As shown in the sample table above for the first ten epochs, the loss declines with each epoch. For instance, the loss value for the first epoch is 48,533, while for the tenth epoch, it's about 1,362. By the time the model reaches epoch 100, the training loss is approximately 11 or 0.02% of the initial loss. This indicates that the model has achieved a relatively low error on the training data after the 100th epoch, suggesting that it has effectively learned the patterns present in the training data.

Test Loss:

The test loss value is computed on the test set (20% of the total dataset in this case), providing an estimate of how well the model generalizes to unseen data. The test loss, as shown in the output above, is about 21. This is reasonably close to the training loss, indicating that the model is not overfitting the training data excessively and is generalizing well to unseen data. The fact that the test loss is slightly higher than the training loss is expected and reasonable, as it's common for the test loss to be slightly higher due to the model needing to generalize to unseen data.

R-squared Score:

The R-squared score (calculated on the test set) is a metric that represents the goodness of fit of the model, indicating the proportion of the variance in the dependent variable that is predictable from the independent variables. A value close to 1 indicates a good fit.

The model achieved an impressive R-squared score of 0.998, indicating that it explains 99.8% of the variance in the dependent variable. This high level of explanatory power demonstrates a strong correlation between the predicted and actual values, suggesting that the model effectively captures the underlying patterns in the data.

CHAPTER 4   DEEP LEARNING WITH PYTORCH FOR REGRESSION

To further validate the model's performance, we can compare the actual and predicted values. The code below achieves this goal by first comparing the top ten actual and predicted values in tabular format and then graphing the actual and predicted values for all Tesla stock prices.

Note This code depends on the tesla_stock_price_prediction.py module we ran earlier for training and evaluating the model. The code preprocesses the data by scaling the features using the preprocess_data function imported from the tesla_stock_price_prediction.py module and loads the pre-trained model weights generated by that module.

Let's begin by importing the necessary modules: PyTorch library, NumPy, Matplotlib, and load_data and preprocess_data functions from the tesla_stock_price_prediction module:

```
[In]: import torch
[In]: import numpy as np
[In]: import matplotlib.pyplot as plt
[In]: from tesla_stock_price_prediction import (
          load_data,
          preprocess_data
      )
```

We then define a main function, which performs the following tasks:

- Loads test data from a CSV file containing stock price data

- Preprocesses the data by scaling the features using the preprocess_data function imported from the tesla_stock_price_prediction module

CHAPTER 4　DEEP LEARNING WITH PYTORCH FOR REGRESSION

- Defines a neural network model architecture using PyTorch

- Loads pre-trained model weights from a file

- Makes predictions on the preprocessed test data using the loaded model

- Prints the top ten actual and predicted stock prices side by side

- Plots all actual and predicted values on a line chart for visualization

```
[In]: def main():
    """
    Load test data, preprocess it, and make predictions
    using a pre-trained model.

    This function loads test data from a CSV file,
    preprocesses it, loads a pre-trained PyTorch model,
    makes predictions on the test data, and visualizes
    the actual vs. predicted stock prices.
    """
    test_data_file_path = \
    "/home/ubuntu/airflow/dags/TSLA_stock.csv"

    test_data = load_data(test_data_file_path)
    test_data = preprocess_data(test_data)

    test_features = \
        test_data.select('scaled_features').collect()
```

CHAPTER 4   DEEP LEARNING WITH PYTORCH FOR REGRESSION

```
test_features = np.array(
    [
        row.scaled_features.toArray()
        for row in test_features
    ],
    dtype=np.float64
)

input_size = 4
output_size = 1
model = torch.nn.Sequential(
    torch.nn.Linear(input_size, 64),
    torch.nn.ReLU(),
    torch.nn.Linear(64, 32),
    torch.nn.ReLU(),
    torch.nn.Linear(32, 16),
    torch.nn.ReLU(),
    torch.nn.Linear(16, output_size)
)

model.load_state_dict(
    torch.load(
        '/home/ubuntu/airflow/dags/trained_model.pth',
        map_location=torch.device('cpu')
    )
)
model.eval()

test_features_tensor = torch.tensor(
    test_features,
    dtype=torch.float64
)
```

CHAPTER 4 DEEP LEARNING WITH PYTORCH FOR REGRESSION

```
model = model.double()

with torch.no_grad():
    predictions = model(
        test_features_tensor
    ).squeeze().numpy()

actual_labels = test_data.select('Close').collect()
actual_labels = np.array(
    [row.Close for row in actual_labels],
    dtype=np.float64
)

actual_labels = actual_labels[::-1]
predictions = predictions[::-1]

print("Top 10 Actual vs Predicted Values:")
print("Actual\tPredicted")
for i in range(10):
    print(f"{actual_labels[i]:.4f}\
    t{predictions[i]:.4f}")

plt.figure(figsize=(10, 6))
plt.plot(actual_labels, label='Actual')
plt.plot(predictions, label='Predicted')
plt.xlabel('Index')
plt.ylabel('Price')
plt.title(
    'Actual vs Predicted Tesla Stock Prices'
    '(Feb 26, 2019 to Feb 23, 2024)'
)
```

253

CHAPTER 4   DEEP LEARNING WITH PYTORCH FOR REGRESSION

```
        plt.legend()
        plt.grid(True)
        plt.show()
[In]: if __name__ == "__main__":
        main()
```

The code defines a function named main() that encompasses the process of predicting Tesla stock prices using a pre-trained PyTorch model:

- The function begins by defining its purpose through a docstring, stating that it loads test data from a CSV file, preprocesses it, loads a pre-trained PyTorch model, makes predictions on the test data, and visualizes the actual vs. predicted stock prices.

- First, it sets the file path for the test data CSV file. Then, it loads and preprocesses the test data using custom functions (load_data and preprocess_data), converting it into a format suitable for the model.

- Next, the neural network model architecture is defined using PyTorch's Sequential API, comprising several linear layers with ReLU activation functions.

- The pre-trained model weights are loaded from a file using torch.load() and model.load_state_dict(). The model is then put into evaluation mode using model.eval().

- The test data features are converted into a PyTorch tensor and cast to the appropriate data type (torch.float64). The model is also converted to double precision using model = model.double().

CHAPTER 4　DEEP LEARNING WITH PYTORCH FOR REGRESSION

- Predictions are made on the test data using the pre-trained model. The actual labels for the test data are retrieved, and both the actual and predicted values are reversed to align with the correct chronological order.

- The top ten actual vs. predicted values are printed.

- Finally, the actual and predicted stock prices are visualized using Matplotlib, with the x-axis representing the index and the y-axis representing the price. The plot title indicates the time range of the data being visualized.

- Lastly, the main() function is executed only if the script is run directly as the main program (if __name__ == "__main__":). This ensures that the code within the function is executed when the script is executed directly, but not when it's imported as a module into another script.

There are two outputs from the code:

- A table displaying the top ten actual and predicted target values

- A plot graphing all the actual and predicted target values

255

CHAPTER 4   DEEP LEARNING WITH PYTORCH FOR REGRESSION

Starting with the table, we have added a % change column, even though this isn't produced by the code, to provide an idea of the differences between the actual and predicted values:

Date	Actual	Predicted	% change
2/23/24	192	196	2%
2/22/24	197	196	-1%
2/21/24	195	197	1%
2/20/24	194	196	1%
2/16/24	200	202	1%
2/15/24	200	196	-2%
2/14/24	189	187	-1%
2/13/24	184	186	1%
2/12/24	188	192	2%
2/09/24	194	193	-1%

The data shows that the maximum difference between the predicted and actual values is abs(2%), indicating a close match.

Moving to the plot, Figure 4-1 illustrates the actual and predicted values for all Tesla stock prices (1,258 observations). The proximity of the two lines confirms the model's strong performance.

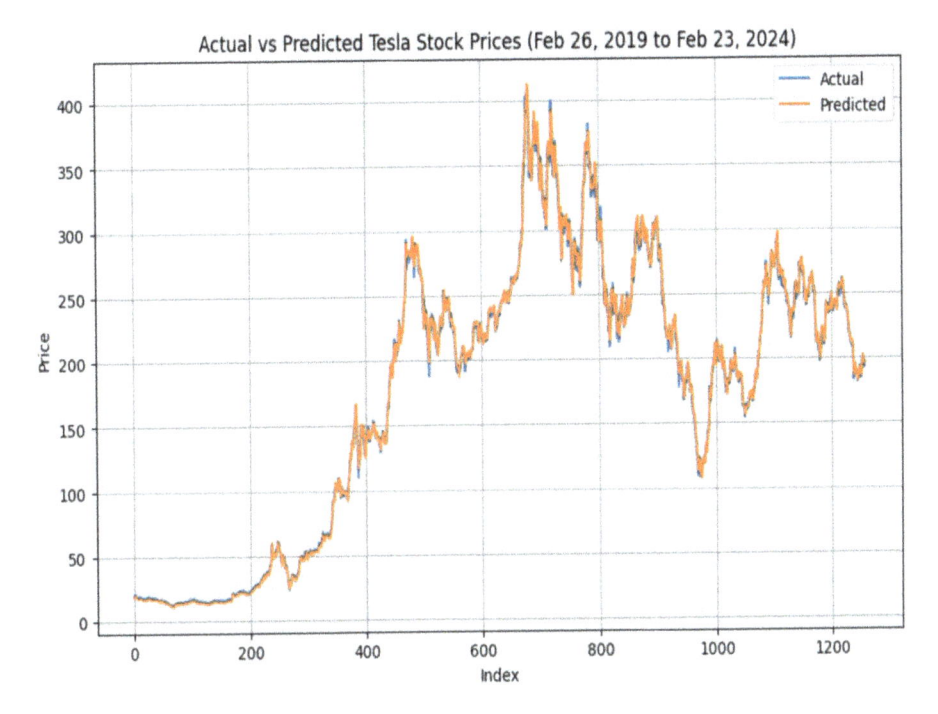

Figure 4-1. Actual and Predicted Tesla Stock Prices

Bringing It All Together

In this section, we consolidate the various components of the code examined in this chapter to construct a comprehensive codebase. There are three subsections: exploring the Tesla stock dataset, predicting Tesla stock price with PySpark and PyTorch, and plotting actual and predicted values side by side.

CHAPTER 4　DEEP LEARNING WITH PYTORCH FOR REGRESSION

Exploring the Tesla Stock Dataset

The script below copies the TSLA (Tesla) stock price CSV file from an S3 bucket to the local directory "/home/ubuntu/airflow/dags/", loads the data into a Spark DataFrame, and then proceeds to print the first five observations from this DataFrame:

```
[In]: import subprocess
[In]: from pyspark.sql import SparkSession
[In]: import logging

[In]: def load_data(file_path: str):
        """
        Load stock price data from a CSV file using
        SparkSession.
        """
        spark = (SparkSession.builder
                .appName("StockPricePrediction")
                .getOrCreate())
        df = spark.read.csv(
                file_path,
                header=True,
                inferSchema=True
        )
        return df

[In]: def copy_and_print_data():
        """
        Copy a CSV file from an S3 bucket to a local
        directory and
        print the first 5 observations.
        """
```

```
        s3_bucket_path = "s3://instance1bucket/TSLA_
        stock.csv"
        local_file_path = "/home/ubuntu/airflow/dags/TSLA_
        stock.csv"
        logging.basicConfig(level=logging.INFO)
        logger = logging.getLogger(__name__)

        try:
            subprocess.run(
                [
                    "aws", "s3", "cp",
                    s3_bucket_path,
                    local_file_path
                ],
                check=True
            )

            df = load_data(local_file_path)
            print("First 5 observations of the DataFrame:")
            df.show(5)

        except FileNotFoundError:
            logger.error(f"Data file not found at {s3_
            bucket_path}")
        except subprocess.CalledProcessError as e:
            logger.error(f"Error copying data from S3: {e}")
[In]: if __name__ == "__main__":
        copy_and_print_data()
```

CHAPTER 4   DEEP LEARNING WITH PYTORCH FOR REGRESSION

Predicting Tesla Stock Price with PySpark and PyTorch

The script below performs the following tasks:

- Loads stock price data from a CSV file using SparkSession

- Preprocesses the data by assembling feature vectors and scaling them

- Splits the preprocessed data into training and testing sets

- Converts the data into PyTorch tensors and loads them into DataLoader objects for batch processing

- Defines, trains, and evaluates a deep learning model for predicting Tesla stock prices

- Saves the trained model for future use

```
[In]: import logging
[In]: import subprocess
[In]: from typing import Tuple
[In]: import numpy as np
[In]: import torch
[In]: import torch.nn as nn
[In]: import torch.optim as optim
[In]: from torch.utils.data import DataLoader, TensorDataset
[In]: from pyspark.sql import SparkSession
[In]: from pyspark.ml.feature import VectorAssembler, StandardScaler
[In]: from pyspark.sql import DataFrame
[In]: logging.basicConfig(level=logging.INFO)
```

CHAPTER 4 DEEP LEARNING WITH PYTORCH FOR REGRESSION

```
[In]: logger = logging.getLogger(__name__)

[In]: def load_data(file_path: str) -> DataFrame:
          """
          Load stock price data from a CSV file using
          SparkSession.
          """
          try:
              spark = (SparkSession.builder
                      .appName("StockPricePrediction")
                      .getOrCreate())
              df = spark.read.csv(
                      file_path,
                      header=True,
                      inferSchema=True
              )
              return df
          except Exception as e:
              raise RuntimeError(
                  f"Error loading data from {file_path}: {e}"
              )

[In]: def preprocess_data(df: DataFrame) -> DataFrame:
          """
          Preprocess the data by assembling feature
          vectors using
          VectorAssembler and scaling them using
          StandardScaler.
          """
          assembler = VectorAssembler(
              inputCols=['Open', 'High', 'Low', 'Volume'],
              outputCol='features'
          )
```

261

CHAPTER 4   DEEP LEARNING WITH PYTORCH FOR REGRESSION

```
    df = assembler.transform(df)

    scaler = StandardScaler(
        inputCol="features",
        outputCol="scaled_features",
        withStd=True,
        withMean=True
    )
    scaler_model = scaler.fit(df)
    df = scaler_model.transform(df)
    df = df.select('scaled_features', 'Close')
    return df
```

```
[In]: def create_data_loader(
        features,
        labels,
        batch_size=32,
        num_workers=4) -> DataLoader:
        """
        Convert the preprocessed data into PyTorch
        tensors and
        create DataLoader objects for both the training
        and test
        sets.

        Args:
            features (Tensor): The input features.
            labels (Tensor): The corresponding labels.
            batch_size (int):
                The number of samples per batch to load.
                Defaults to
                32.
            num_workers (int):
```

CHAPTER 4 DEEP LEARNING WITH PYTORCH FOR REGRESSION

            The number of subprocesses to use for data
            loading.
            Defaults to 4.

        Returns:
            DataLoader: A PyTorch DataLoader for the dataset.
        """

        dataset = TensorDataset(features, labels)
        return DataLoader(
            dataset,
            batch_size=batch_size,
            shuffle=True
)

```
[In]: def train_model(
        model,
        train_loader,
        criterion,
        optimizer,
        num_epochs
    ):
        """
        Train the model on the training data using the
        DataLoader and the defined loss function and
        optimizer. Iterate over the data for a specified
        number of epochs, calculate the loss, and update the
        model parameters.
        """
        for epoch in range(num_epochs):
            for inputs, labels in train_loader:
                optimizer.zero_grad()
```

CHAPTER 4 DEEP LEARNING WITH PYTORCH FOR REGRESSION

```
                outputs = model(inputs)
                loss = criterion(outputs, labels.
                unsqueeze(1))
                loss.backward()
                optimizer.step()
        logger.info(
            f"Epoch [{epoch + 1}/{num_epochs}],"
            f"Loss: {loss.item():.4f}"
        )
[In]: def evaluate_model(
        model,
        test_loader,
        criterion
    ) -> Tuple[float, float]:
        """
        Evaluate the trained model on the test data to assess
        its performance.
        Calculate the test loss and additional evaluation
        metrics such as the R-squared score.
        """
        with torch.no_grad():
            model.eval()
            predictions = []
            targets = []
            test_loss = 0.0
            for inputs, labels in test_loader:
                outputs = model(inputs)
                loss = criterion(outputs, labels.unsqueeze(1))
                test_loss += loss.item() * inputs.size(0)
                predictions.extend(outputs.squeeze().tolist())
```

```
            targets.extend(labels.tolist())
        test_loss /= len(test_loader.dataset)
        predictions = torch.tensor(predictions)
        targets = torch.tensor(targets)
        ss_res = torch.sum((targets - predictions) ** 2)
        ss_tot = torch.sum((targets - torch.mean(targets)) ** 2)
        r_squared = 1 - ss_res / ss_tot
        logger.info(f"Test Loss: {test_loss:.4f}")
        logger.info(f"R-squared Score: {r_squared:.4f}")
        return test_loss, r_squared.item()
```

```
[In]: def main(
        data_file_path: str,
        num_epochs: int = 100,
        batch_size: int = 32,
        learning_rate: float = 0.001
    ):
        """
        Main function for training and evaluating a deep
        learning model for Tesla stock price prediction.

        Args:
            data_file_path (str): The path to the CSV file
                containing stock price data.
            num_epochs (int): The number of epochs for
            training the model. Defaults to 100.
            batch_size (int): The batch size for data loading
            during training. Defaults to 32.
            learning_rate (float): The learning rate for the
                optimizer. Defaults to 0.001.
```

CHAPTER 4 DEEP LEARNING WITH PYTORCH FOR REGRESSION

```
        num_workers (int): The number of subprocesses to
        use for data loading. Defaults to 4.

    Raises: FileNotFoundError: If the specified data file
    is not found.
    """

    try:
        subprocess.run(
            [
                "aws", "s3", "cp",
                "s3://instance1bucket/TSLA_stock.csv",
                "/home/ubuntu/airflow/dags/"
            ]
        )

        df = load_data(data_file_path)
        df = preprocess_data(df)

        train_df, test_df = df.randomSplit([0.8, 0.2],
        seed=42)

        train_features, train_labels = \
          spark_df_to_tensor(train_df)
        test_features, test_labels = spark_df_to_
        tensor(test_df)

        train_loader = create_data_loader(
            train_features,
            train_labels,
            batch_size=batch_size,
            num_workers=num_workers
        )
```

```python
test_loader = create_data_loader(
    test_features,
    test_labels,
    batch_size=batch_size,
    num_workers=num_workers
)

input_size = train_features.shape[1]
output_size = 1
model = nn.Sequential(
    nn.Linear(input_size, 64),
    nn.ReLU(),
    nn.Linear(64, 32),
    nn.ReLU(),
    nn.Linear(32, 16),
    nn.ReLU(),
    nn.Linear(16, output_size)
)
criterion = nn.MSELoss()
optimizer = optim.Adam(
    model.parameters(),
    lr=learning_rate
)

train_model(
    model,
    train_loader,
    criterion,
    optimizer,
    num_epochs
)
```

CHAPTER 4 DEEP LEARNING WITH PYTORCH FOR REGRESSION

```
            evaluate_model(model, test_loader, criterion)
            torch.save(model.state_dict(), 'trained_model.pth')
        except FileNotFoundError:
            logger.error(f"Data file not found at {data_
            file_path}")
```
```
[In]: def spark_df_to_tensor(df: DataFrame) -> Tuple[torch.
      Tensor, torch.Tensor]:
          """
          Converts Spark DataFrame columns to PyTorch tensors.
          """
          features = torch.tensor(
              np.array(
                  df.rdd.map(lambda x: x.scaled_features.
                  toArray())
                  .collect()
              ),
              dtype=torch.float32)
          labels = torch.tensor(
              np.array(
                  df.rdd.map(lambda x: x.Close).collect()),
                  dtype=torch.float32
              )
          return features, labels
```
```
[In]: if __name__ == "__main__":
          data_file_path = "/home/ubuntu/airflow/dags/TSLA_
          stock.csv"
          main(data_file_path)
```

CHAPTER 4   DEEP LEARNING WITH PYTORCH FOR REGRESSION

Comparing and Plotting Actual and Predicted Values

The script below performs the following:

- Loads test data from a CSV file containing stock price data

- Preprocesses the data by scaling the features using the preprocess_data function imported from the tesla_stock_price_prediction module

- Defines a neural network model architecture using PyTorch

- Loads pre-trained model weights from a file

- Makes predictions on the preprocessed test data using the loaded model

- Prints the top ten actual and predicted stock prices side by side

- Plots all actual and predicted values on a line chart for visualization

```
[In]: import torch
[In]: import numpy as np
[In]: import matplotlib.pyplot as plt
[In]: from tesla_stock_price_prediction import (
        load_data,
        preprocess_data
    )
```

CHAPTER 4   DEEP LEARNING WITH PYTORCH FOR REGRESSION

```
[In]: def main():
          """
          Load test data, preprocess it, and make predictions
          using a pre-trained model.

          This function loads test data from a CSV file,
          preprocesses it, loads a pre-trained PyTorch model,
          makes predictions on the test data, and visualizes
          the actual vs. predicted stock prices.
          """
          test_data_file_path = \
          "/home/ubuntu/airflow/dags/TSLA_stock.csv"

          test_data = load_data(test_data_file_path)
          test_data = preprocess_data(test_data)

          test_features = \
             test_data.select('scaled_features').collect()

          test_features = np.array(
              [
                  row.scaled_features.toArray()
                  for row in test_features
              ],
              dtype=np.float64)

          input_size = 4
          output_size = 1
          model = torch.nn.Sequential(
              torch.nn.Linear(input_size, 64),
              torch.nn.ReLU(),
              torch.nn.Linear(64, 32),
              torch.nn.ReLU(),
```

```python
    torch.nn.Linear(32, 16),
    torch.nn.ReLU(),
    torch.nn.Linear(16, output_size)
)
model.load_state_dict(
    torch.load(
        '/home/ubuntu/airflow/dags/'
        'trained_model.pth',
        map_location=torch.device('cpu')
    )
)
model.eval()

test_features_tensor = torch.tensor(
    test_features,
    dtype=torch.float64
)

model = model.double()

with torch.no_grad():
    predictions = model(
        test_features_tensor
    ).squeeze().numpy()

actual_labels = test_data.select('Close').collect()
actual_labels = np.array(
    [
        row.Close for row in actual_labels
    ],
    dtype=np.float64
)
```

CHAPTER 4 DEEP LEARNING WITH PYTORCH FOR REGRESSION

```
    actual_labels = actual_labels[::-1]
    predictions = predictions[::-1]

    print("Top 10 Actual vs Predicted Values:")
    print("Actual\tPredicted")
    for i in range(10):
        print(f"{actual_labels[i]:.4f}
        \t{predictions[i]:.4f}")

    plt.figure(figsize=(10, 6))
    plt.plot(actual_labels, label='Actual')
    plt.plot(predictions, label='Predicted')
    plt.xlabel('Index')
    plt.ylabel('Price')
    plt.title(
        'Actual vs Predicted Tesla Stock Prices'
        '(Feb 26, 2019 to Feb 23, 2024)'
    )
    plt.legend()
    plt.grid(True)
    plt.show()
```

```
[In]: if __name__ == "__main__":
        main()
```

Summary

In this chapter, we began our exploration of deep learning using PyTorch. Our focus was on regression, a fundamental task in machine learning aimed at predicting continuous numerical values. We discussed the advantages of deep learning for regression, such as its ability to capture complex patterns and relationships in data, particularly useful for datasets with nonlinearities and interactions. However, we also acknowledged the black-box nature of these models, which can make interpretation difficult. More importantly, we highlighted the computational intensity of deep learning algorithms, a challenge that can be mitigated by using Amazon Web Services (AWS) as a solution for scalable cloud computing infrastructure, essential for deploying and running deep learning models efficiently.

We demonstrated PyTorch's application for regression by building and evaluating a neural network to predict Tesla stock prices, showing strong performance with low MSE and high R-squared score. Additionally, we briefly discussed early stopping and overfitting, topics we will explore further in Chapter 9.

In the next chapter, we will demonstrate how to follow similar steps using TensorFlow, mirroring the process employed with PyTorch. We will utilize the same Tesla stock dataset for model comparison, allowing for a direct evaluation of the performance and characteristics of models implemented in both frameworks.

CHAPTER 5

Deep Learning with TensorFlow for Regression

In the previous chapter, we began our journey of exploring deep learning models for regression tasks. We trained and evaluated a PyTorch model and used it to predict Tesla's daily stock prices. In this chapter, we demonstrate how to replicate that same process with TensorFlow. We'll use the same dataset to directly compare the performance of models in both frameworks. Additionally, we will leverage the distributed computing capabilities of PySpark to preprocess this dataset.

It's worth noting that the advantages and disadvantages of deep learning algorithms, compared with traditional predictive models like linear regression, as discussed in the previous chapter, remain relevant in the current chapter.

The chapter begins with a reminder of what the Tesla stock dataset looks like, followed by the construction, training, and evaluation of the TensorFlow regression model. It then brings everything together by combining the code snippets discussed throughout the chapter before concluding with a summary of the key takeaways.

CHAPTER 5   DEEP LEARNING WITH TENSORFLOW FOR REGRESSION

The Dataset

We will use the same daily historical dataset of Tesla stock obtained from Yahoo! Finance. This dataset was downloaded in CSV format and then stored on Amazon Web Services (AWS) S3 within a bucket named "instance1bucket" from the following link:

https://finance.yahoo.com/quote/TSLA/history/

We explored this dataset in great detail in Chapter 3 and used it to train and evaluate a PyTorch deep learning regression model in Chapter 4. As a reminder, below are the top five observations from the dataset:

Date	Open	High	Low	Close	Volume
2/23/24	195.31	197.57	191.50	191.97	78670300
2/22/24	194.00	198.32	191.36	197.41	92739500
2/21/24	193.36	199.44	191.95	194.77	103844000
2/20/24	196.13	198.60	189.13	193.76	104545800
2/16/24	202.06	203.17	197.40	199.95	111173600

This dataset spans from February 26, 2019, to February 23, 2024, encompassing 1,258 observations. Each trading day records real-time fluctuations in Tesla stock price. These observations provide insight into key indicators—Open, High, Low, Close, and Volume—depicting price movements over a specific timeframe.

Here are their definitions:

- Open: The opening price of Tesla stock on the given date

- High: The highest price of Tesla stock reached during the trading session on the given date

- Low: The lowest price of Tesla stock reached during the trading session on the given date

- Close: The closing price of Tesla stock on the given date

- Volume: The trading volume, representing the total number of shares traded on the given date

Predicting Tesla Stock Price with TensorFlow

In this section, we utilize the dataset outlined in the preceding section to implement a deep learning algorithm aimed at forecasting Tesla's stock prices. The dataset comprises key features, including Open, High, Low, and Volume, with the Close price serving as the target variable. By integrating these features into our predictive model, we can effectively capture both fundamental market dynamics, such as opening sentiment and trading activity, and technical aspects like price volatility and trading range, all of which are critical for precise stock price predictions.

It's essential to recognize that the purpose of our model is illustrative rather than exhaustive. While the current feature set provides valuable insights, the inclusion of additional features like technical indicators, fundamental data, or market sentiment could further augment our model's predictive capabilities. However, such enhancements fall beyond the scope of this chapter.

To facilitate data processing, we leverage the distributed computing capabilities of PySpark, while TensorFlow is utilized for constructing and training the neural network model. The Python code facilitates an end-to-end workflow, starting with data loading and culminating in model evaluation. In Chapter 8, we will demonstrate how this code can be deployed using Apache Airflow.

CHAPTER 5 DEEP LEARNING WITH TENSORFLOW FOR REGRESSION

In this chapter, we have embraced an object-oriented approach to our methodology. This marks an advancement from the previous chapter, as we have introduced a class called TSLARegressor. This acts as a centralized hub for organizing and orchestrating the various functionalities essential for conducting regression analysis on TSLA stock prices. Within this class, each method is dedicated to specific tasks within the regression workflow, including data loading, feature preprocessing, model creation and training, performance evaluation, and prediction generation.

While the functional approach used in the previous chapter is still effective, the transition to an object-oriented methodology offers several benefits. By encapsulating these functionalities within a class, we enhance the modularity, readability, and reusability of our code. This structured approach simplifies the development and maintenance of our regression analysis workflow. Additionally, leveraging classes enables us to achieve better code organization and abstraction, facilitating easier management and scalability as our project evolves.

However, the modeling steps remain the same as in the previous chapter, comprising the following:

- Imports
- Logging setup
- Data loading
- Data preprocessing
- Data loader creation
- Model training
- Model evaluation
- Main
- Execution

CHAPTER 5 DEEP LEARNING WITH TENSORFLOW FOR REGRESSION

Let's examine the code and delve into these steps in detail, beginning with the imports.

Step 1: Imports

```
[In]: import logging
[In]: import numpy as np
[In]: import tensorflow as tf
[In]: from pyspark.sql import SparkSession
[In]: from pyspark.ml.feature import VectorAssembler, StandardScaler
[In]: from sklearn.metrics import r2_score
[In]: from tensorflow.keras.models import Sequential
[In]: from tensorflow.keras.layers import Dense
[In]: import matplotlib.pyplot as plt
```

- import logging: This import is used to enable logging functionality in the code, allowing us to log information, warnings, and errors during execution. In our upcoming code, the logging module will be used to log information about the test loss and R2 score in the main() function. Additionally, it will be used to log errors in case an exception occurs.

- import numpy as np: This import aliases the NumPy module to np, making it easier to reference. In our code, NumPy will be used for array manipulation and mathematical operations. Specifically, it will be used in the convert_to_numpy() method of the TSLARegressor class to convert Spark DataFrames to NumPy arrays for further processing.

CHAPTER 5   DEEP LEARNING WITH TENSORFLOW FOR REGRESSION

- import tensorflow as tf: This is our Tensorflow deep learning framework. It is extensively used throughout the code for building, training, and evaluating our neural network model. It is used to create a Sequential model and its layers, compile the model with an optimizer and loss function, train the model, and evaluate its performance.

- from pyspark.sql import SparkSession: This imports the SparkSession class from the pyspark.sql module, which is necessary for creating a SparkSession object. In the code, SparkSession is used to create a SparkSession object for Spark-related operations, such as reading data from CSV files and creating Spark DataFrames.

- from pyspark.ml.feature import VectorAssembler, StandardScaler: This line imports the VectorAssembler and StandardScaler classes from the pyspark.ml.feature module, which are used for feature engineering and preprocessing. In the code, these classes are used in the preprocess_data() method to assemble feature vectors and standardize features before training the model.

- from sklearn.metrics import r2_score: This imports the r2_score function from the sklearn.metrics module, which is used to calculate the R-squared score, a metric for evaluating the performance of regression models. In the code, r2_score is used in the predict_and_evaluate() method to calculate the R2 score of the trained model.

- from tensorflow.keras.models import Sequential: This imports the Sequential class from the tensorflow.keras.models module, which is used to create a sequential neural network model in TensorFlow Keras. In the code, Sequential is used to define the architecture of the neural network model in the create_and_train_model() method.

- from tensorflow.keras.layers import Dense: This imports the Dense layer class from the tensorflow.keras.layers module, which is used to add fully connected layers to the neural network model. In the code, Dense layers are used to define the hidden layers of the neural network model in the create_and_train_model() method.

- import matplotlib.pyplot as plt: This imports the pyplot module from the Matplotlib library, which is used for creating visualizations such as plots and charts. In the code, pyplot is used in the plot_actual_vs_predicted() function to create a scatter plot of actual vs. predicted values.

Step 2: Class Definition

In this step, we define a class that encapsulates various methods for data loading, preprocessing, model creation, training, evaluation, and prediction:

```
[In]: class TSLARegressor:
        """
        Class for TSLA stock price regression.
        """
        def __init__(self):
            """
            Initialize TSLARegressor object.
            """
            self.logger = logging.getLogger(__name__)
            self.spark = (SparkSession.builder
```

CHAPTER 5   DEEP LEARNING WITH TENSORFLOW FOR REGRESSION

```
                .appName("TSLA_Regression")
                .getOrCreate())
        self.input_shape = None
        self.model = None
```

The code defines a Python class named TSLARegressor, which serves as a blueprint for creating objects that can perform regression tasks on TSLA (Tesla) stock prices.

Let's break down the code:

- class TSLARegressor:: This line initiates the definition of the class named TSLARegressor.

- """ Class for TSLA stock price regression. """: This is a docstring (documentation string) that provides a brief description of the class's purpose, which is to perform regression analysis on TSLA stock prices.

- def __init__(self):: This is the initializer method (constructor) for the TSLARegressor class. It is called automatically when an object of the class is created.

- """ Initialize TSLARegressor object. """: This is the docstring for the __init__ method, providing a brief description of its purpose.

- self.logger = logging.getLogger(__name__): This line creates an instance attribute named logger for each object of the TSLARegressor class. It assigns a logger object from the logging module to the logger attribute. This logger is used for logging information, warnings, and errors during the execution of the code.

- self.spark = SparkSession.builder.appName("TSLA_Regression").getOrCreate(): This line creates an instance attribute named spark for each object of the

CHAPTER 5   DEEP LEARNING WITH TENSORFLOW FOR REGRESSION

TSLARegressor class. It creates a SparkSession object using SparkSession.builder, sets the application name to "TSLA_Regression" using appName(), and retrieves an existing SparkSession or creates a new one using getOrCreate(). This SparkSession is used for Spark-related operations, such as reading data from CSV files and creating Spark DataFrames.

- self.input_shape = None: This line creates an instance attribute named input_shape and initializes it to None. This attribute is intended to store the shape of the input data for the neural network model.

- self.model = None: This line creates an instance attribute named model and initializes it to None. This attribute is intended to store the trained neural network model.

In the next steps, we define a series of methods, within the TSLARegressor class, for data loading, preprocessing, model creation, training, evaluation, and prediction.

Step 3: preprocess_data Method Definition
In this step, we define the preprocess_data() method within the TSLARegressor class. The model's features undergo transformation as they are assembled from the loaded data using PySpark's VectorAssembler, which consolidates multiple columns into a single feature vector. Subsequently, the features are scaled using StandardScaler to standardize each feature's values, ensuring a mean of 0 and a standard deviation of 1:

```
[In]: def preprocess_data(
        self,
        file_path,
        feature_cols=['Open', 'High', 'Low', 'Volume'],
        train_ratio=0.8,
```

CHAPTER 5 DEEP LEARNING WITH TENSORFLOW FOR REGRESSION

```
        seed=42
):
    """
    Preprocesses the data.

    Parameters:
        - file_path (str): Path to the CSV file.
        - feature_cols (list): List of feature columns.
        - train_ratio (float): Ratio of training data.
        - seed (int): Random seed for splitting data.

    Returns:
        - tuple: Tuple containing train and test
        DataFrames.
    """
    df = self.spark.read.csv(
            file_path,
            header=True,
            inferSchema=True
        )
    assembler = VectorAssembler(
        inputCols=feature_cols,
        outputCol='features'
    )
    df = assembler.transform(df)
    scaler = StandardScaler(
        inputCol='features',
        outputCol='scaled_features',
        withMean=True,
        withStd=True
    )
```

CHAPTER 5 DEEP LEARNING WITH TENSORFLOW FOR REGRESSION

```
scaler_model = scaler.fit(df)
df = scaler_model.transform(df)
train_df, test_df = df.randomSplit(
    [train_ratio, 1-train_ratio],
    seed=seed
)
return train_df, test_df
```

This method prepares the data for training the TensorFlow neural network model.

Below is a breakdown of what it does:

- Reading Data: The method reads the CSV file located at file_path into a Spark DataFrame using PySpark's read.csv() method. The header=True argument indicates that the first row contains column names, while inferSchema=True enables automatic schema inference.

- Feature Assembly: Using PySpark's VectorAssembler, the method combines the specified feature columns (['Open', 'High', 'Low', 'Volume']) into a single feature vector column named 'features'. This consolidation facilitates the processing of features as a single entity.

- Feature Scaling: The feature vector column 'features' is scaled using PySpark's StandardScaler to ensure that each feature has a mean of 0 and a standard deviation of 1. This normalization step helps in preventing features with large values from dominating the training process and ensures that the features contribute equally to the model.

CHAPTER 5   DEEP LEARNING WITH TENSORFLOW FOR REGRESSION

- Data Splitting: Finally, the method splits the DataFrame into training and test sets using Spark's randomSplit() method. The train_ratio parameter determines the ratio of training data, while the seed parameter ensures reproducibility by providing a random seed for data splitting.

- Return: The method returns a tuple containing the training and test DataFrames.

Step 4: convert_to_numpy Method Definition

In this step, we define the convert_to_numpy method, which is responsible for converting Spark DataFrames, train_df and test_df, containing training and test data, respectively, into NumPy arrays:

```
[In]: def convert_to_numpy(self, train_df, test_df):
        """
        Convert Spark DataFrames to numpy arrays.

        Parameters:
            - train_df (DataFrame): Spark DataFrame
              containing
              training data.
            - test_df (DataFrame): Spark DataFrame containing
              test data.

        Returns:
            - tuple: Tuple containing train and test
              numpy arrays
              for features and labels.
        """
        train_features = np.array(
            train_df.select('scaled_features')
            .rdd.map(lambda x: x.scaled_features.toArray())
```

CHAPTER 5 DEEP LEARNING WITH TENSORFLOW FOR REGRESSION

```
        .collect()
    )
    train_labels = np.array(
        train_df.select('Close')
        .rdd.map(lambda x: x.Close).collect())
    test_features = np.array(
        test_df.select('scaled_features')
        .rdd.map(lambda x: x.scaled_features.toArray())
        .collect()
    )
    test_labels = np.array(
        test_df.select('Close')
        .rdd.map(lambda x: x.Close).collect()
    )
    return train_features, train_labels, \
            test_features, test_labels
```

The method convert_to_numpy facilitates compatibility with TensorFlow by converting Spark DataFrames into NumPy arrays, as TensorFlow typically operates with NumPy data structures. It takes two parameters: train_df and test_df, which are Spark DataFrames containing training and test data, respectively.

The method returns a tuple containing four NumPy arrays: train_features, train_labels, test_features, and test_labels. The conversion process inside the method is as follows:

- train_features: Converts the 'scaled_features' column of the training DataFrame (train_df) into a NumPy array
- train_labels: Converts the 'Close' column of the training DataFrame (train_df) into a NumPy array

287

CHAPTER 5   DEEP LEARNING WITH TENSORFLOW FOR REGRESSION

- test_features: Converts the 'scaled_features' column of the test DataFrame (test_df) into a NumPy array
- test_labels: Converts the 'Close' column of the test DataFrame (test_df) into a NumPy array

In terms of implementation, the method uses Spark's select method to extract specific columns from the DataFrames. It then converts these columns into RDDs (Resilient Distributed Datasets) to enable parallel processing. The map function is applied to each row of the RDD to extract the values of the selected columns, which are then converted into NumPy arrays using the np.array function. Finally, the NumPy arrays are collected from the RDDs and returned as part of a tuple.

Step 5: create_and_train_model Method Definition
Moving on, we define the create_and_train_model method, which is responsible for building and training a neural network model using TensorFlow's Keras API:

```
[In]: def create_and_train_model(
        self,
        train_features,
        train_labels,
        epochs=100,
        batch_size=32
    ):
        """
        Create and train the neural network model.

        Parameters:
            - train_features (ndarray): Training features.
            - train_labels (ndarray): Training labels.
            - epochs (int): Number of epochs for training.
            - batch_size (int): Batch size for training.
```

CHAPTER 5 DEEP LEARNING WITH TENSORFLOW FOR REGRESSION

```
Returns:
    - None
"""
self.input_shape = (train_features.shape[1],)
self.model = Sequential([
    Dense(64, activation='relu',
    input_shape=self.input_shape),
    Dense(32, activation='relu'),
    Dense(16, activation='relu'),
    Dense(1)
])
self.model.compile(optimizer='adam', loss='mse')
self.model.fit(
    train_features,
    train_labels,
    epochs=epochs,
    batch_size=batch_size
)
```

The create_and_train_model method is responsible for creating the TensorFlow neural network model with a specified architecture and training it using the provided training features and labels.

Below is a breakdown of the method's parameters and what it accomplishes:

Parameters:

- train_features (ndarray): NumPy array containing the training features.
- train_labels (ndarray): NumPy array containing the corresponding training labels.

CHAPTER 5 DEEP LEARNING WITH TENSORFLOW FOR REGRESSION

- epochs (int): Number of epochs (iterations over the entire dataset) for training. The default is set to 100.
- batch_size (int): Number of samples processed per gradient update. The default is set to 32.

Returns:

- None

Model Architecture:

- The method initializes a sequential model (Sequential), which is a linear stack of layers. Four dense layers are added to the model with different configurations:
    - The first layer (Dense(64, activation='relu', input_shape=self.input_shape)) has 64 units with the ReLU activation function and takes the input shape defined by the number of features.
    - The subsequent layers (Dense(32, activation='relu'), Dense(16, activation='relu')) have 32 and 16 units, respectively, both with the ReLU activation function.
    - The last layer (Dense(1)) has a single unit, representing the output layer.

Compilation:

- The model is compiled using the Adam optimizer (optimizer='adam') and the Mean Squared Error loss function (loss='mse').

Training:

- The compiled model is trained (self.model.fit) using the training features and labels. The training process runs for the specified number of epochs (epochs) and uses the specified batch size (batch_size).

CHAPTER 5   DEEP LEARNING WITH TENSORFLOW FOR REGRESSION

Hyperparameter Selection:

- Epochs: The choice of 100 epochs was made based on common practices in neural network training, providing a balance between learning the data patterns and avoiding overfitting. While no extensive experimentation was conducted, 100 epochs is a reasonable starting point that typically allows the model to converge.

- Batch Size: A batch size of 32 was selected as it is a widely accepted standard that balances computational efficiency and model accuracy. It is a commonly used default that works well in most cases, offering a good trade-off between training speed and generalization.

Step 6: evaluate_model Method Definition

In this step, we create the evaluate_model method, within the TSLARegressor class, which is responsible for evaluating the trained neural network model using test data:

```
[In]: def evaluate_model(self, test_features, test_labels):
          """
          Evaluate the trained model.

          Parameters:
              - test_features (ndarray): Test features.
              - test_labels (ndarray): Test labels.

          Returns:
              - float: Test loss.
          """
          test_loss = self.model.evaluate(
```

291

```
            test_features,
            test_labels
    )
    return test_loss
```

The evaluate_model method evaluates the trained neural network model by computing the loss on the test data.

Here's an explanation of the method's parameters and what it does:

Parameters:

- test_features (ndarray): NumPy array containing the test features

- test_labels (ndarray): NumPy array containing the corresponding test labels

Returns:

- float: Test loss, indicating how well the model performs on the test data

Evaluation:

- The method uses the evaluate function of the trained model (self.model) to compute the loss on the test data.

- It passes the test features and labels to the evaluate function, which calculates the loss based on the model's predictions and the true labels.

Test Loss:

- The computed test loss is returned as the output of the method. This value provides insight into the model's performance on unseen data. A lower test loss indicates better performance, as it signifies that the model's predictions are closer to the true labels.

CHAPTER 5 DEEP LEARNING WITH TENSORFLOW FOR REGRESSION

Step 7: predict_and_evaluate Method Definition

In this step, we create the predict_and_evaluate method, which serves the purpose of predicting values using the trained neural network model and then evaluating its performance based on R-squared score:

```
[In]: def predict_and_evaluate(self, test_features, test_labels):
        """
        Predict and evaluate the neural network model.

        Parameters:
            - test_features (ndarray): Test features.
            - test_labels (ndarray): Test labels.

        Returns:
            - float: R2 score.
        """
        test_predictions = self.model.predict(test_features)
        r2_score_value = r2_score(test_labels, test_predictions)
        return r2_score_value
```

This method predicts values using the trained neural network model (self.model) on the test features and evaluates its performance using the R-squared score.

Let's break down the method's parameters and functionality:

Parameters:

- test_features (ndarray): NumPy array containing the test features

- test_labels (ndarray): NumPy array containing the corresponding test labels

293

CHAPTER 5　DEEP LEARNING WITH TENSORFLOW FOR REGRESSION

Returns:

- float: R-squared score, which measures the proportion of the variance in the dependent variable (test labels) that is predictable from the independent variable (test features) in the model

Prediction:

- The method first predicts values for the test features using the predict function of the trained model (self.model).
- The predictions are stored in the variable test_predictions.

Evaluation (R-squared Score):

- The method then calculates the R-squared score by comparing the predicted values (test_predictions) with the true test labels (test_labels).
- The r2_score function from the Scikit-Learn library is used for this purpose.

R-squared Score Calculation:

- R-squared score indicates the proportion of the variance in the dependent variable that is predictable from the independent variable.
- A higher R-squared score (closer to 1) indicates that a larger proportion of the variance in the dependent variable is predictable from the independent variable, suggesting a better fit of the model to the data.

CHAPTER 5　DEEP LEARNING WITH TENSORFLOW FOR REGRESSION

Return Value:

- The computed R-squared score is returned as the output of the method, providing a measure of the model's predictive performance on the test data.

Step 8: plot_actual_vs_predicted Function Definition

We are now outside the TSLARegressor class. We create the plot_actual_vs_predicted function, which plots the actual vs. predicted values of the test data using Matplotlib:

```
[In]: def plot_actual_vs_predicted(tsla_regressor, test_
          features, test_labels):
          """
          Plot actual vs. predicted values.

          Parameters:
              - tsla_regressor (TSLARegressor):
                  Instance of TSLARegressor class.
              - test_features (ndarray): Test features.
              - test_labels (ndarray): Test labels.

          Returns:
              - None
          """
          test_predictions = \
              tsla_regressor.model.predict(test_features)

          print("Actual    Predicted")
          for actual, predicted in zip(test_labels, test_
          predictions):
              print(f"{actual:.2f}    {predicted[0]:.2f}")

          plt.figure(figsize=(10, 6))
          plt.scatter(test_labels, test_predictions,
          color='blue')
```

295

CHAPTER 5   DEEP LEARNING WITH TENSORFLOW FOR REGRESSION

```
        plt.plot(
            [test_labels.min(), test_labels.max()],
            [test_labels.min(), test_labels.max()],
            'k--', lw=3
        )
        plt.xlabel('Actual')
        plt.ylabel('Predicted')
        plt.title('Actual vs. Predicted Values')
        plt.show()
```

Let's break down what the function accomplishes:

Parameters:

- tsla_regressor (TSLARegressor): An instance of the TSLARegressor class, containing the trained neural network model (model)

- test_features (ndarray): NumPy array containing the test features

- test_labels (ndarray): NumPy array containing the corresponding test labels

Returns:

- None

Predictions:

- The function first predicts the values of the test features using the trained neural network model (tsla_regressor.model.predict(test_features)).

- The predicted values are stored in the variable test_predictions.

CHAPTER 5 DEEP LEARNING WITH TENSORFLOW FOR REGRESSION

Printing Actual vs. Predicted Values:

- The function prints the actual and predicted values side by side for comparison.
- It iterates over each pair of actual and predicted values using a loop and prints them formatted to two decimal places.

Plotting Actual vs. Predicted Values:

- The function creates a scatter plot to visualize the relationship between the actual and predicted values.
- The actual values are plotted on the x-axis (test_labels), and the predicted values are plotted on the y-axis (test_predictions).
- Additionally, a diagonal dashed line is plotted to represent the ideal scenario where actual and predicted values are identical.

Plot Customization:

- The function customizes the plot by setting labels for the x-axis and y-axis and providing a title.
- The plot is displayed using plt.show().

Visualization:

- This visualization allows for a visual assessment of how well the model's predictions align with the actual values.
- A tight clustering of points along the diagonal dashed line indicates a strong correlation between actual and predicted values, suggesting good model performance.

CHAPTER 5   DEEP LEARNING WITH TENSORFLOW FOR REGRESSION

Step 9: Main Function Definition

In this step, we create the main function, which serves as the main entry point to orchestrate the entire workflow of the script. It takes a single parameter, file_path, which represents the path to the CSV file containing the data to be processed:

```
[In]: def main(file_path):
          """
          Main function to orchestrate the entire workflow.
          Parameters:
              - file_path (str): Path to the CSV file.
          Returns:
              - None
          """
          try:
              tsla_regressor = TSLARegressor()
              train_df, test_df = \
                tsla_regressor.preprocess_data(file_path)
              train_features, train_labels, test_features, \
              test_labels = tsla_regressor.convert_to_numpy(
                  train_df,
                  test_df
              )
              tsla_regressor.create_and_train_model(
                  train_features,
                  train_labels
              )
              test_loss = tsla_regressor.evaluate_model(
                  test_features,
                  test_labels
              )
```

CHAPTER 5 DEEP LEARNING WITH TENSORFLOW FOR REGRESSION

```
    r2_score_value = tsla_regressor.predict_and_
    evaluate(
        test_features,
        test_labels
    )
    logging.info("Test Loss: {}".format(test_loss))
    logging.info("R2 Score: {}".format(r2_
    score_value))

    plot_actual_vs_predicted(
        tsla_regressor,
        test_features,
        test_labels
    )
except Exception as e:
    logging.error(f"An error occurred: {e}")
```

The function begins by instantiating an object of the TSLARegressor class named tsla_regressor. It then calls the preprocess_data method of the TSLARegressor instance to preprocess the data from the specified CSV file. The resulting data is split into training and testing sets. The convert_to_numpy method is invoked to convert the Spark DataFrames containing the training and testing data into NumPy arrays for further processing. Next, the create_and_train_model method is called to create and train the neural network model using the training features and labels. After training the model, the evaluate_model method is used to evaluate the model's performance on the test data, computing the test loss. Additionally, the predict_and_evaluate method is called to make predictions on the test data and compute the R2 score. Finally, the function logs the test loss and R2 score using the logging module. If an exception occurs during the execution of the workflow, it is caught, and an error message is logged using the logging module.

CHAPTER 5　DEEP LEARNING WITH TENSORFLOW FOR REGRESSION

Step 10: Conditional Block

In this final step, the conditional block of code checks if the script is being run directly as the main program:

```
[In]: if __name__ == "__main__":
          logging.basicConfig(level=logging.INFO)
          file_path = "/home/ubuntu/airflow/dags/TSLA_stock.csv"
          main(file_path)
```

- if __name__ == "__main__":: This condition checks if the script is being executed directly, rather than being imported as a module into another script.

- logging.basicConfig(level=logging.INFO): This line configures the logging system to display log messages at the INFO level or higher. This ensures that log messages with severity level INFO or higher are displayed.

- file_path = "/home/ubuntu/airflow/dags/TSLA_stock.csv": This line defines the file path to the CSV file containing the TSLA stock data.

- main(file_path): This line calls the main function with the specified file_path as an argument. The main function orchestrates the entire workflow of the script, including preprocessing the data, training the model, evaluating its performance, and plotting actual vs. predicted values.

To run the code, we saved it as "tesla_stock_price_prediction.py" and executed it in a Jupyter Lab terminal (we could have run it directly on the EC2 instance). Since we have created an environment named "myenv," we activated it as follows:

CHAPTER 5 DEEP LEARNING WITH TENSORFLOW FOR REGRESSION

[In]: source /home/ubuntu/myenv/bin/activate

Following environment activation, we executed the code using Python 3 with the following command:

[In]: python3 tesla_stock_price_prediction.py

Now, let's examine the output from the code.

The first output displays the training loss, test loss, and the R2 score obtained during the training and evaluation of the TSLA stock price regression model:

Training Loss:

Epoch	Loss
1	41,731
2	40,362
3	39,658
4	33,450
5	19,520
6	5,492
7	2,532
8	1,984
9	1,657
10	1,231
...	...
100	14

Test Loss: 12.11
R2 Score: 0.998

CHAPTER 5   DEEP LEARNING WITH TENSORFLOW FOR REGRESSION

We'll now explain these metrics one by one:

Training Loss:

- It displays the training loss for each epoch during the training process.

- Each row represents an epoch and the corresponding loss value for that epoch.

- As the training progresses, the loss decreases. This indicates that the model is learning to better fit the training data. Initially, the loss is relatively high, but it decreases rapidly with each epoch until it converges to a lower value. In our output, the training loss decreases significantly over the epochs from 41,731 to 14, indicating that the model is learning and improving its fit to the training data.

Test Loss:

- It represents the loss (Mean Squared Error) calculated on the test dataset after training the model.

- The test loss indicates how well the trained model generalizes to unseen data.

- In our output, the test loss is about 12, which means, on average, the model's predictions on the test data are off by this amount (measured in the same units as the target variable, in USD for stock prices).

R2 Score:

- R2 score (also known as the coefficient of determination) is a statistical measure that represents the proportion of the variance in the dependent variable (stock prices in this case) that is predictable from the independent variables (features).

CHAPTER 5   DEEP LEARNING WITH TENSORFLOW FOR REGRESSION

- It ranges from 0 to 1, with 1 indicating a perfect fit.
- In our output, the R2 score is 0.998, which means that the model can explain approximately 99.8% of the variance in the test data, indicating a very high level of predictive accuracy.

In conclusion, the output suggests that the model has been trained effectively, with the training loss decreasing over epochs and the evaluation metrics (test loss and R2 score) indicating good performance on unseen data.

The second output from the code is a comparison between the top ten actual TSLA stock prices and the prices predicted by the regression model:

Actual	Predicted	% Difference
118.85	117.87	-1%
123.22	122.63	0%
283.15	282.61	0%
284.8	281.55	-1%
34.57	35.09	2%
34.03	34.17	0%
281.52	280.22	0%
283.48	282.09	0%
143.75	138.58	-4%
209.14	211.78	1%

We have added the percentage difference between the actual and predicted values for easy interpretation. If the predicted value is lower than the actual value, the percentage difference will be negative, indicating an underestimation. If the predicted value is higher than the actual value,

CHAPTER 5 DEEP LEARNING WITH TENSORFLOW FOR REGRESSION

the percentage difference will be positive, indicating an overestimation. If the predicted value is exactly the same as the actual value, the percentage difference will be 0%.

In our output, for some entries, such as the second row (123.22 actual vs. 122.63 predicted), the predicted value is slightly lower than the actual value, resulting in a very small negative percentage difference (less than –1%), which is rounded to 0%. Similarly, in the fifth row (34.57 actual vs. 35.09 predicted), the predicted value is slightly higher than the actual value, resulting in a positive percentage difference of approximately 2%.

Overall, there is a maximum difference of –4% with 50% of the predicted values aligning with the actual values (approx. 0% difference) indicating a close match.

The final output is a scatter plot to visualize the relationship between the actual and predicted values for all observations. In Figure 5-1, the actual values are plotted on the x-axis (test_labels), and the predicted values are plotted on the y-axis (test_predictions). Additionally, a diagonal dashed line is plotted to represent the ideal scenario where actual and predicted values are identical.

CHAPTER 5   DEEP LEARNING WITH TENSORFLOW FOR REGRESSION

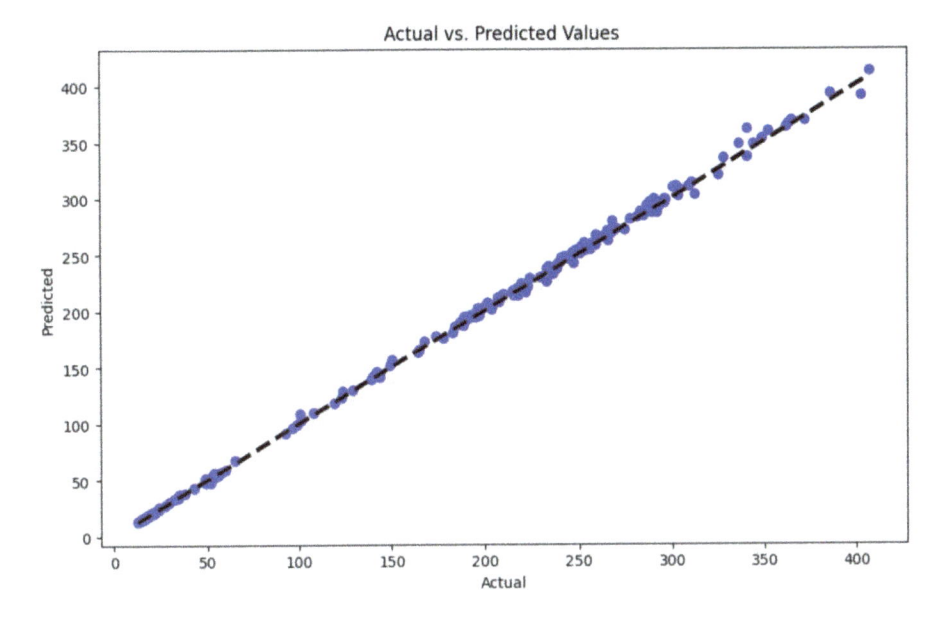

Figure 5-1. Actual vs. Predicted Values

This graph allows for a visual assessment of how well the model's predictions align with the actual values. We see a tight clustering of points along the diagonal dashed line indicating a strong correlation between actual and predicted values, suggesting good model performance.

TensorFlow vs. PyTorch

Leveraging results from the previous chapter where we employed PyTorch and the current chapter where TensorFlow was utilized, we can now conduct a comparison between the performances of the two models. By analyzing training loss, test loss, and R2 score, we can gain insights into how the two algorithms perform in predicting TSLA stock prices.

CHAPTER 5   DEEP LEARNING WITH TENSORFLOW FOR REGRESSION

The table below compares the training loss for PyTorch and TensorFlow for the first ten epochs and the 100th epoch:

Epoch	PyTorch	TensorFlow
1/100	48,533	41,731
2/100	40,969	40,362
3/100	26,663	39,658
4/100	13,865	33,450
5/100	3,026	19,520
6/100	1,683	5,492
7/100	1,077	2,532
8/100	475	1,984
9/100	1,195	1,657
10/100	1,362	1,231
...	...	
100/100	11	14

We can make the following remarks based on this data:

- PyTorch generally starts with higher initial training loss values compared with TensorFlow.

- Both frameworks show a consistent decrease in training loss over epochs, indicating effective optimization.

- PyTorch and TensorFlow exhibit similar trends in loss reduction, with both eventually converging to lower values.

CHAPTER 5   DEEP LEARNING WITH TENSORFLOW FOR REGRESSION

- In some epochs, there are minor differences in the loss values between PyTorch and TensorFlow, but overall, the trends are comparable.

- Toward the end of training, both frameworks achieve low training loss values, indicating that the models have converged effectively.

Turning to test loss and R-squared score, we have the data below:

Metric	PyTorch	TensorFlow
Test loss	20.54	12.11
R-squared	0.998	0.998

The table shows that the test loss obtained from the PyTorch model is 20.54, while the TensorFlow model achieves a lower test loss of 12.11. Interestingly, both models exhibit an identical R-squared value of approximately 0.998, indicating an exceptionally high level of explanatory power and predictive accuracy in capturing the variance of the test data. Despite the difference in test loss, the comparable R-squared values suggest that both frameworks produce models with similar predictive performance in explaining the TSLA stock prices.

While this comparison is largely apples-to-apples, as we used the same Tesla stock dataset, there are practical differences in implementing the models in PyTorch and TensorFlow that extend beyond the data and model architecture. In TensorFlow, the use of a static computational graph requires that the model's structure be defined upfront. This approach allows TensorFlow to optimize the graph for performance, making it particularly well-suited for production environments where efficiency is critical. However, this static nature can make the development process more rigid, as any modifications to the model require the graph to be redefined. This can slow down the iterative process of model refinement, especially in research settings where flexibility is key.

On the other hand, PyTorch's dynamic computational graph allows for real-time modifications, offering a more flexible and intuitive development experience. This dynamism makes PyTorch highly advantageous during the experimentation phase, where models are frequently adjusted and tested. The ability to alter the graph on-the-fly simplifies debugging, allowing developers to inspect and modify the model's behavior at runtime, leading to quicker iterations and problem-solving.

Additionally, the usability and ecosystem of each framework play a crucial role. TensorFlow, particularly with the advancements made since TensorFlow 2.0 and the continued integration of Keras as its high-level API, has become increasingly accessible and user-friendly. The framework's comprehensive ecosystem, including tools like TensorFlow Serving for deployment, makes it a robust choice for end-to-end machine learning pipelines, from development to production.

In contrast, PyTorch's straightforward interface, coupled with its seamless integration with Python libraries like NumPy, provides a more intuitive experience for developers, especially those new to deep learning. Its ecosystem is rapidly evolving, with tools like PyTorch Lightning and TorchServe further expanding its capabilities. It still excels in research settings, where its flexibility and ease of use are significant advantages.

Ultimately, while both frameworks are powerful and capable of delivering high-performance models, the choice between TensorFlow and PyTorch often comes down to the specific needs of the project. TensorFlow's static graph and robust deployment tools make it a strong candidate for production environments, whereas PyTorch's dynamic graph and intuitive interface are particularly well-suited for research and experimentation.

Bringing It All Together

In this section, we integrate the diverse elements of the code explored throughout this chapter to develop a cohesive codebase. We also provide more context on why specific layers and activation functions were chosen for the model to help readers understand the design choices. Additionally, we include an environment.yml file to make it easy for readers to replicate the environment.

Layers and Activation Functions

Layer Selection:
The model uses a sequence of dense layers with 64, 32, and 16 neurons. This gradual reduction in neurons helps the model to start by capturing complex patterns with more neurons in the initial layers, while the subsequent layers focus on refining and reducing the data complexity. This design helps in preventing overfitting by gradually narrowing down the focus of the model as it learns.

Activation Function—ReLU:
ReLU (rectified linear unit) is used as the activation function in the hidden layers due to its efficiency in handling nonlinear data and avoiding the vanishing gradient problem. ReLU is known for speeding up the convergence during training, making it a preferred choice in many deep learning models, particularly for tasks involving complex data like stock prices.

Output Layer:
The output layer consists of a single neuron with no activation function (linear). This choice is typical in regression models where the goal is to predict a continuous value. The linear activation ensures that the output can range across all possible values, which is necessary for accurately predicting stock prices.

CHAPTER 5 DEEP LEARNING WITH TENSORFLOW FOR REGRESSION

environment.yml

This file specifies the versions of the Python packages used in the project, allowing readers to easily recreate the same setup.

Below is the configuration:

```
name: tsla_regression
channels:
  - defaults
dependencies:
  - python=3.12.3
  - numpy=1.26.4
  - tensorflow=2.17.0
  - pyspark=3.5.2
  - scikit-learn=1.5.1
  - matplotlib=3.9.2
```

TensorFlow Regression Code

Below is the code for building, training, and evaluating the TensorFlow regression model to predict Tesla stock prices:

```
[In]: import logging
[In]: import numpy as np
[In]: import tensorflow as tf
[In]: from pyspark.sql import SparkSession
[In]: from pyspark.ml.feature import VectorAssembler, StandardScaler
[In]: from sklearn.metrics import r2_score
[In]: from tensorflow.keras.models import Sequential
[In]: from tensorflow.keras.layers import Dense
[In]: import matplotlib.pyplot as plt
[In]: class TSLARegressor:
```

CHAPTER 5   DEEP LEARNING WITH TENSORFLOW FOR REGRESSION

```
"""
Class for TSLA stock price regression.
"""
def __init__(self):
    """
    Initialize TSLARegressor object.
    """
    self.logger = logging.getLogger(__name__)
    self.spark = (SparkSession.builder
                  .appName("TSLA_Regression")
                  .getOrCreate())
    self.input_shape = None
    self.model = None
```

```
[In]: def preprocess_data(
        self,
        file_path,
        feature_cols=['Open', 'High', 'Low', 'Volume'],
        train_ratio=0.8,
        seed=42
    ):
        """
        Preprocesses the data.

        Parameters:
            - file_path (str): Path to the CSV file.
            - feature_cols (list): List of feature columns.
            - train_ratio (float): Ratio of training data.
            - seed (int): Random seed for splitting data.
```

311

```
    Returns:
        - tuple: Tuple containing train and test
          DataFrames.
    """
    df = self.spark.read.csv(
            file_path,
            header=True,
            inferSchema=True
    )
    assembler = VectorAssembler(
        inputCols=feature_cols,
        outputCol='features'
    )
    df = assembler.transform(df)
    scaler = StandardScaler(
        inputCol='features',
        outputCol='scaled_features',
        withMean=True,
        withStd=True
    )
    scaler_model = scaler.fit(df)
    df = scaler_model.transform(df)
    train_df, test_df = df.randomSplit(
        [train_ratio, 1-train_ratio],
        seed=seed
    )
    return train_df, test_df
```

```
[In]: def convert_to_numpy(self, train_df, test_df):
    """
    Convert Spark DataFrames to numpy arrays.
```

CHAPTER 5 DEEP LEARNING WITH TENSORFLOW FOR REGRESSION

```
Parameters:
    - train_df (DataFrame):
        Spark DataFrame containing training data.
    - test_df (DataFrame):
        Spark DataFrame containing test data.

Returns:
    - tuple:
        Tuple containing train and test numpy
        arrays for
        features and labels.
"""
train_features = np.array(
    train_df.select('scaled_features')
    .rdd.map(lambda x: x.scaled_features.toArray())
    .collect()
)
train_labels = np.array(
    train_df.select('Close')
    .rdd.map(lambda x: x.Close).collect()
)
test_features = np.array(
    test_df.select('scaled_features')
    .rdd.map(lambda x: x.scaled_features.toArray())
    .collect()
)
test_labels = np.array(
    test_df.select('Close')
    .rdd.map(lambda x: x.Close).collect()
)
return train_features, train_labels, \
       test_features, test_labels
```

CHAPTER 5   DEEP LEARNING WITH TENSORFLOW FOR REGRESSION

```
[In]: def create_and_train_model(
          self,
          train_features,
          train_labels,
          epochs=100,
          batch_size=32
      ):
          """
          Create and train the neural network model.

          Parameters:
              - train_features (ndarray): Training features.
              - train_labels (ndarray): Training labels.
              - epochs (int): Number of epochs for training.
              - batch_size (int): Batch size for training.

          Returns:
              - None
          """
          self.input_shape = (train_features.shape[1],)
          self.model = Sequential([
              Dense(64, activation='relu',
              input_shape=self.input_shape),
              Dense(32, activation='relu'),
              Dense(16, activation='relu'),
              Dense(1)
          ])
          self.model.compile(optimizer='adam', loss='mse')
          self.model.fit(
              train_features,
              train_labels,
```

CHAPTER 5 DEEP LEARNING WITH TENSORFLOW FOR REGRESSION

```
            epochs=epochs,
            batch_size=batch_size
        )
[In]: def evaluate_model(self, test_features, test_labels):
        """
        Evaluate the trained model.

        Parameters:
            - test_features (ndarray): Test features.
            - test_labels (ndarray): Test labels.

        Returns:
            - float: Test loss.
        """
        test_loss = self.model.evaluate(
            test_features,
            test_labels
        )
        return test_loss
[In]: def predict_and_evaluate(self, test_features,
      test_labels):
        """
        Predict and evaluate the neural network model.

        Parameters:
            - test_features (ndarray): Test features.
            - test_labels (ndarray): Test labels.

        Returns:
            - float: R2 score.
        """
```

CHAPTER 5   DEEP LEARNING WITH TENSORFLOW FOR REGRESSION

```
        test_predictions = self.model.predict(
            test_features
        )
        r2_score_value = r2_score(
            test_labels,
            test_predictions
        )
        return r2_score_value
[In]: def plot_actual_vs_predicted(
        tsla_regressor,
        test_features,
        test_labels
    ):
        """
        Plot actual vs. predicted values.

        Parameters:
            - tsla_regressor (TSLARegressor):
                    Instance of TSLARegressor class.
            - test_features (ndarray): Test features.
            - test_labels (ndarray): Test labels.

        Returns:
            - None
        """
        test_predictions = tsla_regressor.model.predict(
            test_features
        )

        print("Actual   Predicted")
        for actual, predicted in zip(
```

CHAPTER 5   DEEP LEARNING WITH TENSORFLOW FOR REGRESSION

```
        test_labels,
        test_predictions
    ):
        print(f"{actual:.2f}    {predicted[0]:.2f}")

    plt.figure(figsize=(10, 6))
    plt.scatter(test_labels, test_predictions,
    color='blue')
    plt.plot(
        [test_labels.min(), test_labels.max()],
        [test_labels.min(), test_labels.max()],
        'k--', lw=3
    )
    plt.xlabel('Actual')
    plt.ylabel('Predicted')
    plt.title('Actual vs. Predicted Values')
    plt.show()
```

```
[In]: def main(file_path):
    """
    Main function to orchestrate the entire workflow.

    Parameters:
        - file_path (str): Path to the CSV file.

    Returns:
        - None
    """
    try:
        tsla_regressor = TSLARegressor()
        train_df, test_df = tsla_regressor.
        preprocess_data(
            file_path
```

```
        )
        train_features, train_labels, test_features, \
        test_labels = tsla_regressor.convert_to_numpy(
            train_df,
            test_df
        )
        tsla_regressor.create_and_train_model(
            train_features,
            train_labels
        )
        test_loss = tsla_regressor.evaluate_model(
            test_features,
            test_labels
        )
        r2_score_value = tsla_regressor.predict_and_
        evaluate(
            test_features,
            test_labels
        )

        logging.info("Test Loss: {}".format(test_loss))
        logging.info("R2 Score: {}".format(r2_
        score_value))

        plot_actual_vs_predicted(
            tsla_regressor,
            test_features,
            test_labels
        )
    except Exception as e:
        logging.error(f"An error occurred: {e}")
```

```
[In]: if __name__ == "__main__":
          logging.basicConfig(level=logging.INFO)
          file_path = "/home/ubuntu/airflow/dags/TSLA_
          stock.csv"
          main(file_path)
```

Summary

In this chapter, we continued our exploration of deep learning models for regression tasks. We constructed, trained, and assessed a TensorFlow deep learning model, using it to forecast Tesla's daily stock prices. Although the deep learning regression model was not intended for commercial purposes, it demonstrated significant performance, achieving an R-squared value of over 99%. This performance is comparable to the results obtained in the previous chapter using PyTorch for predictions.

In the upcoming chapter, we will shift our focus to classification tasks. We will delve into how PyTorch can be utilized to classify categorical variables, marking a new phase in our exploration of deep learning.

CHAPTER 6

Deep Learning with PyTorch for Classification

In Chapters 4 and 5, we explored deep learning for regression tasks. In this chapter and the next, we shift our focus to classification, another fundamental task in deep learning. Regression and classification are distinct types of machine learning tasks: regression aims to predict continuous numerical values from input features, such as predicting Tesla stock prices as demonstrated in the previous chapters. In contrast, classification involves assigning input data into predefined categories or classes. In this chapter, our objective is to predict the probability of a woman being diagnosed with diabetes based on attributes such as the number of pregnancies, glucose levels, blood pressure, Body Mass Index (BMI), age, and family history of diabetes, also known as the diabetes pedigree function.

We leverage Spark for data preprocessing to demonstrate its distributed computing capabilities in handling large datasets. Although our dataset is relatively small, the concepts and code are applicable to larger datasets, making this exercise valuable for production environments. The small dataset allows us to delve into important

CHAPTER 6   DEEP LEARNING WITH PYTORCH FOR CLASSIFICATION

topics such as data augmentation, transfer learning, resampling, model complexity, data quality, and data representation—all critical aspects of deep learning.

In developing the code for this project, we followed key modeling steps: data preprocessing, model definition, model training, and model evaluation. We also incorporated additional functionalities, such as copying data from an S3 bucket to a local directory on EC2 for processing and handling errors and log messages for debugging and monitoring throughout the modeling process. Furthermore, we evaluated the model using K-fold cross-validation to ensure that our conclusions are not dependent on a single test dataset.

The chapter begins with an examination of the diabetes dataset, followed by the construction, training, and evaluation of the PyTorch classification model. We then introduce K-fold cross-validation, consolidate the individual codes discussed throughout the chapter, and conclude with a summary of the key takeaways.

The Dataset

The dataset used for this project is the well-known Pima Indians Diabetes Dataset. Pima refers to the Pima people, who are a group of Native American tribes indigenous to the southwestern United States, primarily living in what is now Arizona.

The dataset contains 768 records. Each record represents a Pima Indian woman and includes various health-related attributes, along with a target variable (outcome) indicating whether the woman developed diabetes. This dataset is suitable for classification tasks aimed at predicting diabetes outcomes based on well-defined attributes such as the number of pregnancies, glucose levels, blood pressure, Body Mass Index (BMI), age, and family history of diabetes (diabetes pedigree function).

CHAPTER 6   DEEP LEARNING WITH PYTORCH FOR CLASSIFICATION

The dataset was originally created by the National Institute of Diabetes and Digestive and Kidney Diseases. For this project, we sourced it from Kaggle, a platform designed for data scientists. The following are the contributor's name, approximate upload date, dataset name, site name, and the URL from which we downloaded a copy of the CSV file and stored it in an S3 bucket on AWS:

> Title: Pima Indians Diabetes Database
>
> Source: Kaggle
>
> URL: `www.kaggle.com/uciml/pima-indians-diabetes-database`
>
> Contributor: UCI Machine Learning
>
> Date: 2016

The Pima Indians Diabetes Dataset was previously explored in our book, *Distributed Machine Learning with PySpark: Migrating Effortlessly from Pandas and Scikit-Learn* (Apress, November 2023), where logistic regression was employed to address this classification problem. While logistic regression serves as a robust method for binary classification, deep learning introduces new opportunities to capture complex patterns in data that simpler models may overlook.

The PySpark code in this section performs several types of data exploration on this dataset, including the following:

- Handling Missing Values: The code counts the number of missing values for each column in the dataset and displays the counts. This helps in understanding the extent of missing data and potentially deciding on strategies for dealing with it.

CHAPTER 6   DEEP LEARNING WITH PYTORCH FOR CLASSIFICATION

- Counting Zero Values: It counts the number of zero values for each column in the dataset. This exploration can reveal potential issues, such as missing or incorrect data, especially in features where zero values might not be valid (e.g., blood pressure, glucose levels).

- Data Summary: The code provides a summary of the dataset. It calculates statistics like mean, standard deviation, min, max, and quartiles for numeric columns. This summary helps in understanding the distribution and range of values for different features.

- Counting Outcome Values: It counts the occurrences of each outcome value in the dataset. This exploration helps understand the distribution of diabetes cases in the dataset.

- Feature Distributions: The code plots histograms for the distribution of relevant features. Visualizing feature distributions can provide insights into the data's underlying patterns, potential outliers, and whether the data is skewed.

- Calculating Feature-Target Correlation: It calculates the correlation between each feature and the target variable (Outcome). This exploration helps in understanding which features might be more strongly correlated with the target variable, which is crucial for predictive modeling tasks like classification.

Let's go through the code step by step, starting with the import statements. Once we have explained the code, we will turn to the output.

CHAPTER 6　DEEP LEARNING WITH PYTORCH FOR CLASSIFICATION

Step 1: Imports

```
[In]: from pyspark.sql import SparkSession
[In]: from pyspark.sql.functions import col, sum as pyspark_sum
[In]: import boto3
```

Let's break down the import statements and where they are used in the upcoming code:

- from pyspark.sql import SparkSession: This line imports the SparkSession class from the pyspark.sql module. SparkSession serves as the entry point to Spark, providing access to various functionalities for distributed data processing. In the upcoming code, it's used to create a Spark session object named spark, which enables interaction with Spark's DataFrame API for data loading and manipulation.

- from pyspark.sql.functions import col, sum as pyspark_sum: This line imports specific functions from the pyspark.sql.functions module. col is a function used to reference a column in a DataFrame, and sum is an aggregation function to calculate the sum of values in a column. These functions are used in various methods of the PimaDatasetExplorer class for data preprocessing, handling missing values, and calculating correlations between features and the target variable.

- import boto3: This line imports the boto3 module, which is the official AWS SDK for Python. Boto3 provides an easy-to-use interface to interact with various AWS services, such as S3 and EC2, programmatically. In our code, boto3 is used in the

CHAPTER 6 DEEP LEARNING WITH PYTORCH FOR CLASSIFICATION

copy_file_from_s3() function to interact with AWS S3 services for downloading the diabetes.csv file from an S3 bucket to a local path on the EC2 instance.

Step 2: Class Definition

In this step, we define a class named PimaDatasetExplorer, which is designed for exploring and analyzing the Pima Indians Diabetes Dataset using PySpark:

```
[In]: class PimaDatasetExplorer:
          """
          Class for exploring and analyzing the Pima Indian
          Diabetes
          dataset using PySpark.
          """

          def __init__(self, file_path):
              """
              Initializes the PimaDatasetExplorer object.

              Args:
                  file_path (str): The path to the
                  dataset file.
              """
              self.file_path = file_path
              self.data = None
```

After this, we define a series of methods within the PimaDatasetExplorer class, starting with the load_data method.

CHAPTER 6　DEEP LEARNING WITH PYTORCH FOR CLASSIFICATION

Step 3: load_data Method Definition

The load_data method is defined within the class and is responsible for loading the dataset from the specified file path into a PySpark DataFrame:

```
[In]: def load_data(self):
          """
          Loads the dataset from the specified file path into a
          PySpark DataFrame.
          """
          try:
              self.data = spark.read.csv(
                  self.file_path,
                  header=True,
                  inferSchema=True
              )
              self.data.cache()
          except Exception as e:
              print(
                  "An error occurred while loading the
                  data:", str(e)
              )
              self.data = None
```

The method begins with a docstring that describes its purpose, which is to load the dataset. Within a try block, the method reads a CSV file located at the path specified by self.file_path into a PySpark DataFrame (self.data). The header=True argument indicates that the first row of the CSV file contains column names, and inferSchema=True instructs Spark to automatically infer the data types of each column during the loading process.

If loading is successful, the DataFrame (self.data) is cached using the cache() method. Caching stores the DataFrame contents in memory or disk, reducing the need to recompute transformations or actions on the DataFrame in subsequent operations, which can improve performance

327

CHAPTER 6   DEEP LEARNING WITH PYTORCH FOR CLASSIFICATION

(it may not be noticeable with our relatively small dataset, but it's a good practice for large ones).

In case an exception occurs during the loading process, the method catches the exception using an except block. It prints an error message indicating that an error occurred while loading the data, along with the specific error message obtained from the exception (str(e)).

Finally, if an error occurs, the data attribute of the object is set to None to indicate that the dataset was not successfully loaded. This ensures that subsequent operations can check whether the dataset was loaded successfully before proceeding with further analysis or processing.

Step 4: preprocess_data Method Definition

In this step, we define a method named preprocess_data within the class, which, as indicated in the docstring, performs preprocessing steps on the data:

```
[In]: def preprocess_data(self):
          """
          Performs preprocessing steps on the dataset.
          """
          self.load_data()
```

- The method takes self as its first parameter, referring to the instance of the class.

- """"Performs preprocessing steps on the dataset."""": This docstring provides a brief description of what the preprocess_data method does.

- self.load_data(): This line invokes the load_data method, which loads the dataset into memory as a PySpark DataFrame.

CHAPTER 6   DEEP LEARNING WITH PYTORCH FOR CLASSIFICATION

Step 5: handle_missing_values Method Definition

The handle_missing_values method is defined within the PimaDatasetExplorer class. It's designed to count the number of missing values for each column in the dataset and display these counts:

```
[In]: def handle_missing_values(self):
        """
        Counts the number of missing values for each
        column in
        the dataset and displays the counts.
        """
        if self.data is not None:
            missing_counts = self.data.select(
                [
                    pyspark_sum(
                        col(c).isNull().cast("int")
                    ).alias(c)
                    for c in self.data.columns
                ]
            )
            print("Missing Value Counts:")
            missing_counts.show()
        else:
            print("Data not loaded.")
```

The method starts with a docstring, which clarifies the purpose of the method. In this case, the purpose is to count the number of missing values for each column in the dataset and display the counts. It then checks whether the dataset has been loaded by verifying if self.data is not None. If the dataset is loaded, it proceeds to count missing values.

To count missing values, the method constructs a DataFrame named missing_counts. It employs list comprehension to iterate over each column (c) in the dataset, utilizing PySpark functions to determine if values

329

CHAPTER 6 DEEP LEARNING WITH PYTORCH FOR CLASSIFICATION

are null. Specifically, it uses isNull() to identify null values, cast("int") to convert Boolean results into integers (0 or 1), and pyspark_sum to calculate the sum of missing values for each column. Each column's missing value count is aliased with its original column name.

Following the counting process, the method prints a header indicating that missing value counts will be displayed and then proceeds to display the missing value counts DataFrame using the show() method. Each row corresponds to a column in the dataset, and the value in each row represents the count of missing values in that column.

In case the dataset is not loaded (i.e., self.data is None), the method prints a message stating that the data is not loaded, informing the user of the necessity to load the dataset before performing missing value analysis. This ensures clarity regarding the state of the dataset and prevents analysis on unloaded data.

Step 6: count_zeros Method Definition

At this stage, we define the count_zeros method, also within the PimaDatasetExplorer class, which is aimed at tallying the occurrences of zero values for each column in the dataset and then presenting these counts (we will explain the purpose for this when we discuss the output):

```
[In]: def count_zeros(self):
          """
          Counts the number of zero values for each
          column in the
          dataset and displays the counts.
          """
          if self.data is not None:
              zero_counts = self.data.select([
                  pyspark_sum((col(c) == 0).cast("int")).
                  alias(c)
                  for c in self.data.columns
              ])
```

CHAPTER 6   DEEP LEARNING WITH PYTORCH FOR CLASSIFICATION

```
            print("Zero Counts:")
            zero_counts.show()
    else:
        print("Data not loaded.")
```

After the docstring, the method begins by checking whether the dataset has been loaded. This check is performed by verifying if self.data is not None. If the dataset is loaded, the method proceeds with the zero value count. For counting zero values, the method constructs a DataFrame called zero_counts. It utilizes list comprehension to iterate over each column (c) in the dataset. Within this comprehension, PySpark functions are employed to assess if values are equal to zero. Specifically, (col(c) == 0) evaluates whether values in a column equal zero, cast("int") converts Boolean results into integers (0 or 1), and pyspark_sum computes the sum of zero values for each column. Each column's zero value count is aliased with its original column name.

After the count operation, the method prints a header announcing that zero value counts will be displayed. It then proceeds to display the zero value counts DataFrame using the show() method. Each row in the displayed DataFrame corresponds to a column in the dataset, with the value in each row indicating the count of zero values in that column.

If the dataset is not loaded (i.e., self.data is None), the method prints a message indicating that the data is not loaded. This message serves as a reminder to load the dataset before attempting zero value analysis.

Step 7: data_summary Method Definition

The data_summary method within the PimaDatasetExplorer class provides a summary of the relevant features in the diabetes dataset:

```
[In]: def data_summary(self):
         """
         Displays a summary of the dataset, excluding certain
         columns, and filtering out rows with zero values in
         specific columns.
```

331

CHAPTER 6   DEEP LEARNING WITH PYTORCH FOR CLASSIFICATION

```
    Note: The 'SkinThickness' and 'Insulin' columns are
    excluded from the summary because they contain a
    large number of zero values, which are considered
    invalid because these measurements should not be zero
    in a living person and could
    skew the results.
    """
    if self.data is not None:
        print(
            "Data Summary (
            excluding 'Outcome',
            'SkinThickness',
            'Insulin'
            ):"
        )
        columns_to_exclude = [
            'Outcome',
            'SkinThickness',
            'Insulin'
        ]
        summary_cols = [
            c for c in self.data.columns
            if c not in columns_to_exclude
        ]
        filtered_data = self.data.filter(
            (col("Glucose") != 0)
            & (col("BloodPressure") != 0)
            & (col("BMI") != 0)
        )
```

CHAPTER 6 DEEP LEARNING WITH PYTORCH FOR CLASSIFICATION

```
        filtered_data.select(summary_cols).
        describe().show()
    else:
        print("Data not loaded.")
```

Let's dissect the method step by step:

- It starts with a docstring describing its purpose.

- It then verifies whether the dataset has been loaded, ensuring that self.data is not None. If the dataset is loaded, the method proceeds with generating the data summary.

- To generate the summary, the method first prints a header indicating that a summary of the dataset will be displayed. It then defines a list named columns_to_exclude, containing the names of columns to be excluded from the summary ('Outcome', 'SkinThickness', 'Insulin').

- Next, the method constructs a list named summary_cols using list comprehension. This list contains all columns from the dataset (self.data.columns) except those specified in columns_to_exclude.

- Subsequently, the method filters the dataset (self.data) to exclude rows where specific columns ('Glucose', 'BloodPressure', 'BMI') have zero values. This is achieved using the filter() method in PySpark, with conditions ensuring that none of the specified columns have zero values.

- After filtering the dataset, the method selects only the columns specified in summary_cols and generates descriptive statistics using the describe()

333

CHAPTER 6　DEEP LEARNING WITH PYTORCH FOR CLASSIFICATION

method. These statistics provide insights into the central tendency, dispersion, and shape of the data distribution.

- Finally, the method displays the summary by invoking the show() method on the generated summary DataFrame.

- In case the dataset is not loaded (i.e., self.data is None), the method prints a message indicating that the data is not loaded.

Step 8: count_outcome Method Definition

Here, we define the count_outcome method, which is part of the PimaDatasetExplorer class. This serves to tally the occurrences of each outcome value in the diabetes dataset:

```
[In]: def count_outcome(self):
        """
        Counts the occurrences of each outcome value in the
        dataset.
        """
        if self.data is not None:
            print("Outcome Counts:")
            outcome_counts = self.data.filter(
                (col("Glucose") != 0)
                & (col("BloodPressure") != 0)
                & (col("BMI") != 0)
            ).groupBy("Outcome").count()
            outcome_counts.show()
        else:
            print("Data not loaded.")
```

Let's break it down:

- The method begins with a docstring, which provides a brief explanation of its purpose.
- It then checks whether the dataset has been loaded, ensuring that self.data is not None. If the dataset is loaded, the method proceeds with counting the outcome occurrences.
- To count the occurrences of each outcome value, the method constructs a DataFrame named outcome_counts. It filters the dataset to exclude rows where specific columns ('Glucose', 'BloodPressure', 'BMI') have zero values (invalid data). This filtering is achieved using the filter() method in PySpark, ensuring that none of the specified columns have zero values.
- After filtering the dataset, the method groups the data by the Outcome column (target variable) and computes the count of occurrences for each outcome value using the groupBy() and count() functions.
- Once the counts are computed, the method prints a header indicating that outcome counts will be displayed. It then proceeds to display the outcome counts DataFrame using the show() method. Each row in the displayed DataFrame represents an outcome value, with the corresponding count indicating the number of occurrences of that outcome value in the dataset.
- In case the dataset is not loaded (i.e., self.data is None), the method prints a message indicating that the data is not loaded.

CHAPTER 6   DEEP LEARNING WITH PYTORCH FOR CLASSIFICATION

Step 9: explore_feature_distributions Method Definition
In this step, we define the explore_feature_distributions method, which is designed to analyze and visualize the distribution of each feature in the dataset through histograms:

```
[In]: def explore_feature_distributions(self):
          """
          Plots histograms for the distribution of each
          feature in the
          dataset.
          """
          if self.data is not None:
              print("Feature Distributions:")
              for column in self.data.columns:
                  if column not in ['SkinThickness',
                  'Insulin']:
                      print(f"Feature: {column}")
                      if column in [
                          "Glucose",
                          "BloodPressure",
                          "BMI"
                      ]:
                          filtered_data = self.data.filter(
                          (col(column) != 0)
                          & (col("Glucose") != 0)
                          & (col("BloodPressure") != 0)
                          & (col("BMI") != 0)
                          )
```

CHAPTER 6　DEEP LEARNING WITH PYTORCH FOR CLASSIFICATION

```
        else:
            filtered_data = self.data
        plot_data=filtered_data.select(column).
        toPandas()
        plot_data.plot(kind='hist', title=column)
    else:
        print("Data not loaded.")
```

Let's break it down:

- Within the method, the first step is to check whether the dataset has been loaded by verifying if self.data is not None. This ensures that the method proceeds only when the dataset is available for analysis.

- Upon confirming that the dataset is loaded, the method starts exploring the feature distributions. It begins by printing a header indicating that feature distributions will be examined.

- Next, the method iterates over each column in the dataset. For each column, it checks whether it is not in the exclusion list, which includes columns like 'SkinThickness' and 'Insulin'. This exclusion ensures that these specific columns are not included in the analysis of feature distributions.

- For each feature being analyzed, the method prints out a label indicating the name of the feature. It then proceeds to check if the feature is among the specified columns: 'Glucose', 'BloodPressure', and 'BMI'. If so, it filters the dataset to exclude rows where any of these columns have zero values.

CHAPTER 6 DEEP LEARNING WITH PYTORCH FOR CLASSIFICATION

- If the feature is not among the specified columns, the method utilizes the unfiltered dataset for analysis.

- Finally, the method first converts the selected data to a Pandas DataFrame using the toPandas() method and then plots a histogram for the selected feature using the plot(kind='hist') method. This generates a histogram displaying the distribution of values for the feature under analysis.

- In case the dataset is not loaded (i.e., self.data is None), the method prints a message indicating that the data is not loaded.

Step 10: calculate_feature_target_correlation Method Definition

In this step, we create the calculate_feature_target_correlation method, which is tasked with computing the correlation between each feature and the target variable (Outcome):

```
[In]: def calculate_feature_target_correlation(self):
        """
        Calculates the correlation between each feature and
        the target variable (Outcome).

        Note: The 'SkinThickness' and 'Insulin' columns are
        excluded from the correlation calculations because
        they contain a large number of zero values, which are
        considered invalid since these measurements should
        not be zero in a living person and could skew the
        results.
        """
```

```python
        if self.data is not None:
            print(
                "Feature-Target Correlation "
                "(excluding 'SkinThickness', 'Insulin'):"
            )
            for column in self.data.columns:
                if column not in [
                    'Outcome',
                    'SkinThickness',
                    'Insulin'
                ]:
                    correlation = self.data.filter(
                        (col(column) != 0)
                        & (col("Glucose") != 0)
                        & (col("BloodPressure") != 0)
                        & (col("BMI") != 0)
                    ).stat.corr(column, 'Outcome')

                    print(f"{column}: {correlation}")
        else:
            print("Data not loaded.")
```

The explanations of this code are given below:

- After the docstring, a check is performed to ensure that the dataset has been loaded (self.data is not None). This ensures that the correlation computation proceeds only when the dataset is available.

- Once the dataset is loaded, the method proceeds to compute the feature–target correlation. It begins by printing a header indicating that the feature–target correlation will be calculated.

CHAPTER 6   DEEP LEARNING WITH PYTORCH FOR CLASSIFICATION

- Next, the method iterates over each column in the dataset. For each column, it checks whether the column is not in the exclusion list ('Outcome', 'SkinThickness', 'Insulin'). This exclusion ensures that these specific columns are not included in the correlation analysis.

- For each feature being analyzed, the method computes the correlation between that feature and the target variable ('Outcome'). This computation is performed by filtering the dataset to exclude rows where the feature or any of the specified columns ('Glucose', 'BloodPressure', 'BMI') have zero values. The stat.corr() method is then invoked to compute the correlation between the feature and the target variable.

- The computed correlation value is then printed out, along with the name of the feature, providing insight into the strength and direction of the relationship between each feature and the target variable.

- In case the dataset is not loaded (i.e., self.data is None), the method prints a message indicating that the data is not loaded.

Step 11: copy_file_from_s3 Function Definition

Working outside the PimaDatasetExplorer class, we now define the copy_file_from_s3 function. This is responsible for copying the diabetes.csv file from an AWS S3 bucket to a local path on the EC2 instance:

```
[In]: def copy_file_from_s3(bucket_name, file_key, local_path):
      """
      Copies CSV file from an AWS S3 bucket to a
      local path.
```

CHAPTER 6   DEEP LEARNING WITH PYTORCH FOR CLASSIFICATION

```
    Args:
        bucket_name (str): The name of the S3 bucket.
        file_key (str): The key of the file in the
        S3 bucket.
        local_path (str):
            The local path where the file will be copied.
    """
    try:
        s3 = boto3.client('s3')
        s3.download_file(bucket_name, file_key,
        local_path)
        print(f"File {file_key} downloaded to
        {local_path}")
    except Exception as e:
        print(f"An error occurred while
        downloading file",
            f"{file_key}: {str(e)}")
```

Within the copy_file_from_s3 function

- A docstring provides documentation of the method's purpose and its arguments.

- A try-except block is utilized to handle potential exceptions that may occur during the file copying process.

- Inside the try block, the method initiates a connection to the AWS S3 service using the boto3.client() function from the boto3 library. This establishes a client to interact with the S3 service.

CHAPTER 6 DEEP LEARNING WITH PYTORCH FOR CLASSIFICATION

- The function then uses the download_file method of the S3 client to copy the file from the specified S3 bucket (bucket_name) with the given key (file_key) to the specified local path (local_path).

- If the file copying process is successful, a message is printed confirming that the file has been downloaded to the specified local path.

- In case an exception occurs during the file copying process, the except block captures the exception and prints an error message indicating that an error occurred while downloading the file, along with the specific error message provided by the exception (str(e)). This ensures that the user is notified of any issues that arise during the file copying process, facilitating troubleshooting and error resolution.

Step 12: Main

In this final step, we define the if __name__ == "__main__": block. This is a common conditional statement in Python that allows code to be executed only when the script is run directly, not when it's imported as a module. It ensures that the following code is executed only when the script is run as the main program:

```
[In]: if __name__ == "__main__":
          spark = (SparkSession.builder
                   .appName("PimaDatasetExplorer")
                   .getOrCreate())
          local_file_path = "/home/ubuntu/airflow/dags/diabetes.csv"
          s3_bucket_name = "instance1bucket"
          s3_file_key = "diabetes.csv"
          copy_file_from_s3(
```

CHAPTER 6   DEEP LEARNING WITH PYTORCH FOR CLASSIFICATION

```
        s3_bucket_name,
        s3_file_key,
        local_file_path
    )
    explorer = PimaDatasetExplorer(local_file_path)
    explorer.preprocess_data()
    explorer.handle_missing_values()
    explorer.count_zeros()
    explorer.data_summary()
    explorer.count_outcome()
    explorer.explore_feature_distributions()
    explorer.calculate_feature_target_correlation()
```

Inside this block

- The SparkSession class creates a Spark session with the specified application name (or retrieves an existing one if available).

- local_file_path is defined as the local path where the dataset file will be stored.

- s3_bucket_name is defined as the name of the S3 bucket from which the dataset file will be copied.

- s3_file_key is defined as the key of the dataset file in the S3 bucket.

- The copy_file_from_s3 function is called to copy the dataset file from the specified S3 bucket to the local file path on the EC2 instance.

- An instance of the PimaDatasetExplorer class is created with the local file path as the argument.

343

- The preprocess_data method of the PimaDatasetExplorer instance is called to load and preprocess the dataset.

- Various methods of the PimaDatasetExplorer instance are called to explore and analyze the dataset:

  - handle_missing_values: Counts missing values for each column

  - count_zeros: Counts zero values for each column

  - data_summary: Displays a summary of the dataset

  - count_outcome: Counts occurrences of each outcome value

  - explore_feature_distributions: Plots histograms for feature distributions

  - calculate_feature_target_correlation: Computes feature-target correlations

This block of code orchestrates the entire process of loading, preprocessing, and analyzing the dataset, demonstrating the functionality of the PimaDatasetExplorer class.

To run the code, we saved it as "diabetes_data_exploration.py" and executed it in a JupyterLab terminal (we could have run it directly on the EC2 instance). Since we have created an environment named "myenv," we activated it as follows:

```
[In]: source /home/ubuntu/myenv/bin/activate
```

CHAPTER 6 DEEP LEARNING WITH PYTORCH FOR CLASSIFICATION

Following environment activation, we executed the code using Python 3 with the following command:

[In]: python3 diabetes_data_exploration.py

It's time now to examine the output from the code.

The first output is a confirmation message indicating that the file "diabetes.csv" was successfully downloaded to the specified local path "/home/ubuntu/airflow/dags/diabetes.csv". This confirms that the "copy_file_from_s3" function executed without any errors:

[Out]: File diabetes.csv downloaded to /home/ubuntu/airflow/dags/diabetes.csv

The second output is the missing value counts generated by the handle_missing_values() method:

CHAPTER 6 DEEP LEARNING WITH PYTORCH FOR CLASSIFICATION

Pregnancies	Glucose	BloodPressure	SkinThickness	Insulin	BMI	DiabetesPedigreeFunction	Age	Outcome
0	0	0	0	0	0	0	0	0

The output shows that all counts are zero, indicating that there are no missing values present in any of the columns.

The third output, generated by the count_zeros() method, represents the counts of zero values for each column in the dataset:

CHAPTER 6 DEEP LEARNING WITH PYTORCH FOR CLASSIFICATION

Pregnancies	Glucose	BloodPressure	SkinThickness	Insulin	BMI	DiabetesPedigreeFunction	Age	Outcome
111	5	35	227	374	11	0	0	500

The output indicates that there are no cases with 0 values for DiabetesPedigreeFunction and Age, but there are many cases with 0 values for the other variables. For example, 111 cases have 0 values for the Pregnancies variable, which makes sense since some females in the sample have no kids. Additionally, 500 cases have 0 values for the Outcome variable, indicating that these women do not have diabetes. However, it is illogical for Glucose, BloodPressure, SkinThickness, Insulin, or BMI to have 0 values.

There are three common ways to deal with invalid readings: exclude columns or features with 0 values, exclude rows with 0 values, or impute 0 values with mean or average values. For this chapter, we have chosen options 1 and 2:

- Option 1: Exclude SkinThickness and Insulin as the number of cases with 0 values is too large (227 and 374, respectively). Excluding rows instead of columns would make the sample too small.

- Option 2: Exclude rows with 0 values for Glucose, BloodPressure, and BMI. The number of invalid cases is not too large, so excluding rows won't significantly impact the sample size.

The next output, generated by the data_summary() method, provides a summary of statistical measures for the relevant features in the dataset after dealing with invalid readings, i.e. (excluding 'Outcome', 'SkinThickness', 'Insulin'):

CHAPTER 6   DEEP LEARNING WITH PYTORCH FOR CLASSIFICATION

Summary	Pregnancies	Glucose	Blood Pressure	BMI	Diabetes PedigreeFunction	Age
count	724	724	724	724	724	724
mean	3.87	121.88	72.40	32.47	0.47	33.35
stddev	3.36	30.75	12.38	6.89	0.33	11.77
min	0	44	24	18.2	0.078	21
max	17	199	122	67.1	2.42	81

Here's a breakdown of the summary statistics for each feature:

- Pregnancies: Count, 724; mean, 3.87; standard deviation, 3.36; minimum, 0; maximum, 17

- Glucose: Count, 724; mean, 121.88; standard deviation, 30.75; minimum, 44; maximum, 199

- BloodPressure: Count, 724; mean, 72.40; standard deviation, 12.38; minimum, 24; maximum, 122

- BMI: Count, 724; mean, 32.47; standard deviation, 6.89; minimum, 18.2; maximum, 67.1

- DiabetesPedigreeFunction: Count, 724; mean, 0.47; standard deviation, 0.33; minimum, 0.078; maximum, 2.42

- Age: Count, 724; mean, 33.35; standard deviation, 11.77; minimum, 21; maximum, 81

These summary statistics provide insights into the central tendency, dispersion, and range of values for each feature, aiding in the understanding of the dataset's characteristics. In fact, upon closer inspection, there are some values that appear suspicious in the data. For instance

- Glucose: The minimum value is 44, which seems unusually low for a glucose measurement. While it's possible for glucose levels to fall within this range, values below 70 mg/dL are generally considered hypoglycemic and could indicate an issue with the data.

- BloodPressure: The minimum value is 24, which is extremely low and unlikely to be a valid blood pressure reading. Typically, blood pressure values below 90/60 mmHg are considered hypotensive and could indicate an error in the data collection process.

- BMI: The minimum value is 18.2, which falls within the healthy BMI range but is at the lower end. It's worth investigating whether these low BMI values are valid or if they might be outliers or errors in the data.

- DiabetesPedigreeFunction: The minimum value is 0.078, which seems unusually low for a genetic risk factor associated with diabetes. While it's possible for this value to be low, further investigation may be warranted to ensure its accuracy.

- Age: The minimum value is 21, which is within a reasonable range. However, depending on the context of the dataset, it's worth verifying if this minimum age is appropriate for the population being studied.

- Pregnancies: Finally, the maximum number of pregnancies in the data is 17. While it's biologically possible for a woman to have 17 pregnancies, it's extremely rare and would be considered an outlier in most populations.

CHAPTER 6   DEEP LEARNING WITH PYTORCH FOR CLASSIFICATION

These suspicious values could potentially be outliers or errors in the data and would normally warrant further investigation and data validation to ensure the integrity of the dataset. For the purpose of our illustration, we will proceed with utilizing this dataset without expecting excellent performance. However, this demonstrates how data exploration, especially descriptive statistics, can identify potential issues. It also highlights the concept of data quality in machine learning.

The next output, generated by the count_outcome() method, shows the number of cases where the outcome is 1 (indicating diabetes) and where the outcome is 0 (indicating no diabetes):

Outcome	count
1	249
0	475

In this dataset, there are 249 cases with an outcome of 1 (diabetes) and 475 cases with an outcome of 0 (no diabetes). An unbalanced dataset, such as this one, with significantly more instances of one outcome (0 in this case) than the other, can have implications in machine learning tasks.

The next output is the correlation coefficient between each feature with valid readings and the target variable (Outcome) generated by the calculate_feature_target_correlation() method:

Feature	Correlation Value
Pregnancies	0.261
Glucose	0.488
BloodPressure	0.167
BMI	0.299
DiabetesPedigreeFunction	0.185
Age	0.246

CHAPTER 6   DEEP LEARNING WITH PYTORCH FOR CLASSIFICATION

The correlation coefficient measures the strength and direction of the linear relationship between two variables. A positive correlation indicates that as one variable increases, the other variable also tends to increase, while a negative correlation indicates that as one variable increases, the other variable tends to decrease.

In deep learning classification tasks, understanding the correlation between features and the target variable can be helpful for several reasons:

- Feature Selection: Features with higher correlation values are more likely to have a stronger influence on the prediction of the target variable. In feature selection processes, such as in neural network architecture design, selecting features that are highly correlated with the target variable can lead to more effective models with better predictive performance. Given the correlation coefficients above, Glucose has the strongest impact on the model, followed by BMI, Pregnancies, Age, DiabetesPedigreeFunction, and BloodPressure.

- Model Interpretability: Knowing which features have higher correlations with the target variable can provide insights into the underlying relationships between input features and the target variable. This information can help in interpreting the trained model's predictions and understanding which features are most important for classification. This is especially important given that deep learning models are often considered "black box."

- Model Optimization: In the training process, understanding feature-target correlations can guide optimization strategies. For example, during the optimization of deep learning models using techniques

CHAPTER 6 DEEP LEARNING WITH PYTORCH FOR CLASSIFICATION

like gradient descent, features with higher correlations may receive more weight updates, potentially leading to faster convergence and improved model performance.

- Feature Engineering: Feature engineering plays a crucial role in improving model performance. By leveraging the information about feature–target correlations, data scientists can make informed decisions about feature transformations, interactions, or the creation of new features to enhance the model's ability to capture relevant patterns in the data.

The final output, generated by the explore_feature_distributions() method, represents histograms generated for various columns of the diabetes dataset (Figures 6-1 to 6-6).

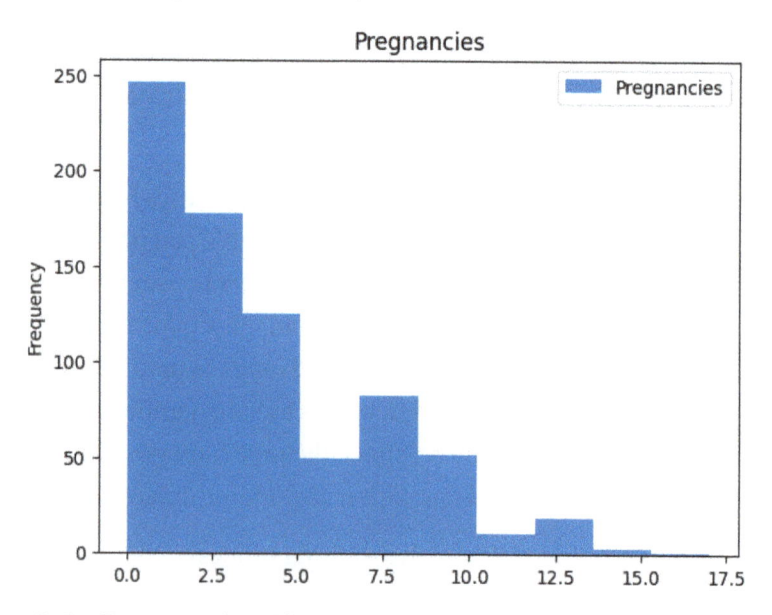

Figure 6-1. Pregnancies Histogram

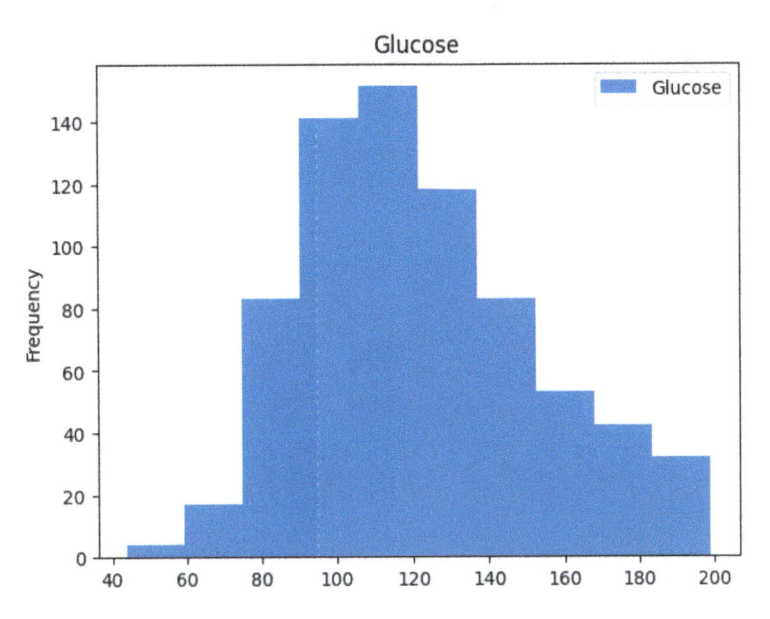

CHAPTER 6 DEEP LEARNING WITH PYTORCH FOR CLASSIFICATION

Figure 6-2. Glucose Histogram

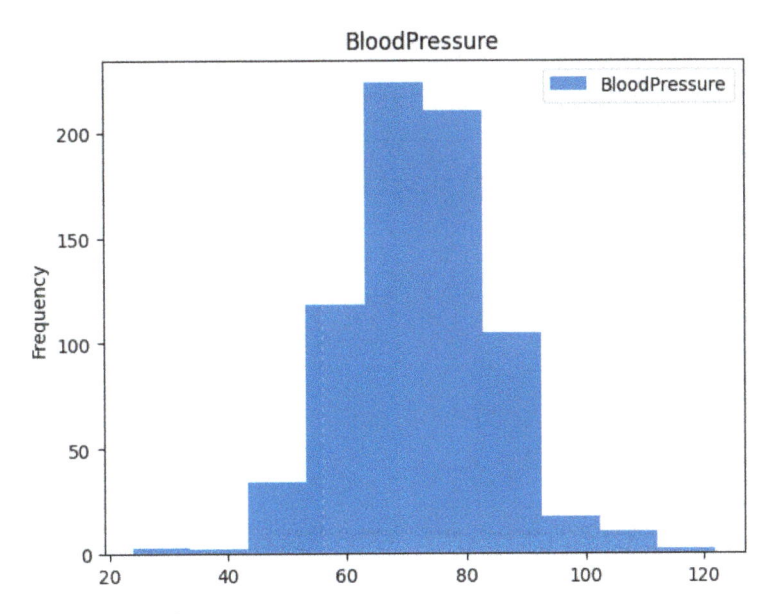

Figure 6-3. BloodPressure Histogram

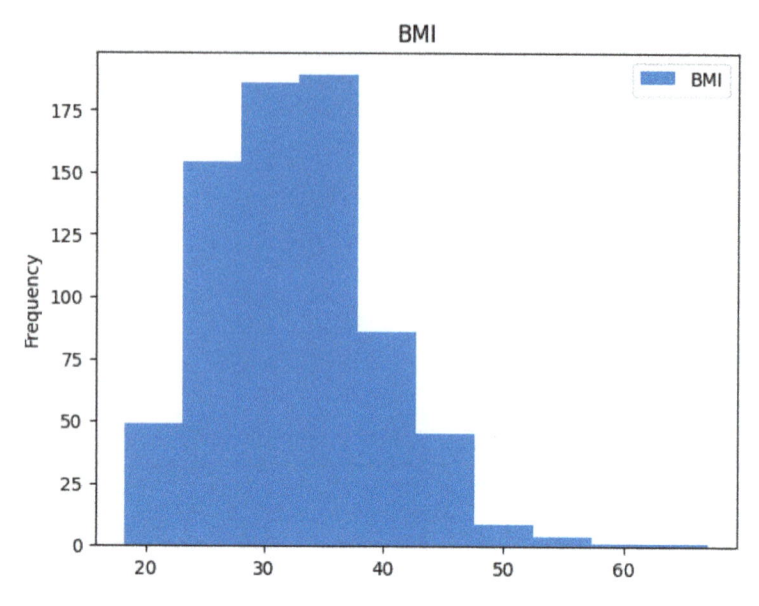

CHAPTER 6   DEEP LEARNING WITH PYTORCH FOR CLASSIFICATION

Figure 6-4. *BMI Histogram*

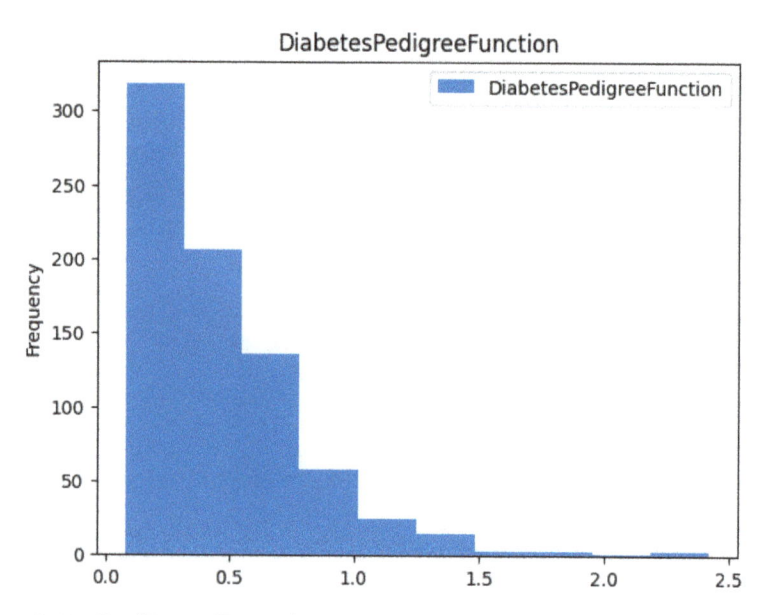

Figure 6-5. *Pedigree Function Histogram*

CHAPTER 6 DEEP LEARNING WITH PYTORCH FOR CLASSIFICATION

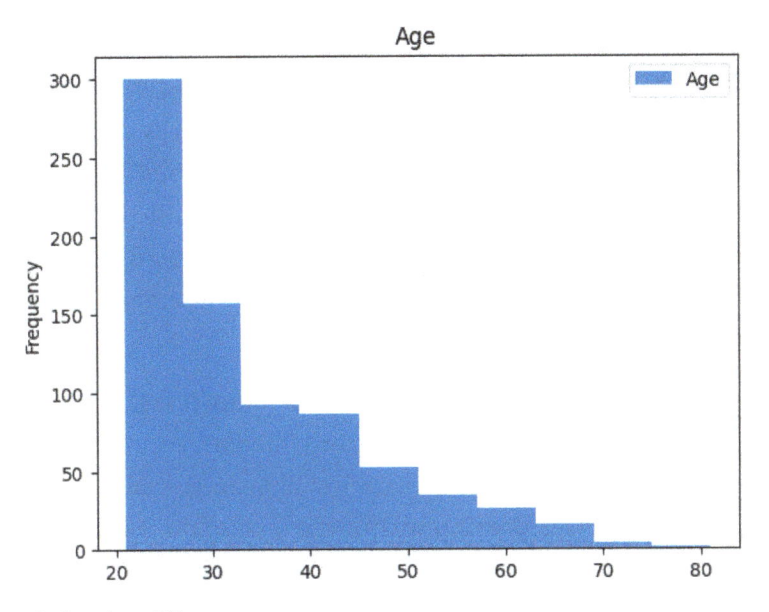

Figure 6-6. Age Histogram

The histograms indicate that Glucose and BloodPressure appear to be somewhat normally distributed, while the other features are skewed to the right. A right-skewed distribution (also known as positively skewed) means that the majority of the data points are clustered toward the left side of the histogram, with a tail extending toward the right. In other words, the distribution is asymmetric, with a longer tail on the right side.

Having skewed data can impact the performance and training dynamics of deep learning models in a number of ways:

- Model Bias: Skewed data can introduce bias in the trained model, as the model may become more biased toward the majority class (in classification tasks) or toward the predominant patterns in the data. This can result in poorer generalization performance, especially for minority classes or less frequent patterns.

- Model Learning: Deep learning models trained on skewed data may prioritize learning features and patterns that are more prevalent in the majority class, potentially ignoring or under-representing minority classes or less common patterns. As a result, the model may struggle to generalize well to new, unseen data points, particularly those from under-represented classes.

- Loss Function Imbalance: In classification tasks, imbalanced classes can lead to an imbalance in the loss function during training. This imbalance can cause the model to prioritize minimizing the loss for the majority class at the expense of the minority class, further exacerbating the bias in the trained model.

In reality, addressing skewed data in deep learning often involves employing specialized sampling strategies, such as oversampling minority classes or undersampling majority classes. These strategies aim to balance the class distribution in the training data and mitigate the impact of skewness on model training.

While addressing these issues falls beyond the scope of this book, when evaluating the performance of our deep learning model trained on this skewed data, we will use appropriate evaluation metrics that account for class imbalance, namely, precision, recall, and F1 score. These metrics provide a more comprehensive assessment of model performance across all classes, rather than relying solely on accuracy.

CHAPTER 6   DEEP LEARNING WITH PYTORCH FOR CLASSIFICATION

Predicting Diabetes with PyTorch

In this section, we build, train, and evaluate a neural network model and use it to predict the likelihood of a Pima Indian woman having diabetes using the dataset we have just explored. We use six features: number of pregnancies, blood pressure, BMI, glucose level, diabetes pedigree function, and age. These features are important factors in predicting the presence of diabetes.

We leverage the distributed computing capabilities of Spark for data preprocessing and PyTorch for developing the deep learning model.

The code has been organized into the following steps:

1. Data Preprocessing:

    - Load the diabetes dataset from a CSV file on AWS S3 using Spark.

    - Filter out invalid records in specific columns (Glucose, BloodPressure, BMI).

    - Assemble feature columns into a single vector using VectorAssembler.

    - Scale the assembled features using StandardScaler.

    - Split the preprocessed data into training and testing datasets.

2. Model Definition:

    - Define the architecture of the neural network model for diabetes classification using PyTorch. The model consists of multiple fully connected layers with ReLU activation functions.

CHAPTER 6   DEEP LEARNING WITH PYTORCH FOR CLASSIFICATION

3. Model Training:

    - Initialize the model trainer with the defined neural network model, loss function, optimizer, and training data loader.

    - Train the neural network model on the training dataset for a specified number of epochs.

    - Update the model parameters based on the optimization process to minimize the loss.

4. Model Evaluation:

    - Evaluate the trained model using the testing dataset.

    - Calculate various performance metrics such as accuracy, precision, recall, F1 score, ROC-AUC score, and confusion matrix to assess the model's effectiveness in classifying diabetes.

5. Additional Functionality:

    - Copy a data file from an S3 bucket to an EC2 local directory for processing, demonstrating integration with AWS services.

    - Handle errors and log messages for debugging and monitoring purposes throughout the modeling process.

These steps encapsulate an end-to-end process of building, training, and evaluating a neural network model for diabetes classification, leveraging Spark for data preprocessing and PyTorch for deep learning model development.

Before proceeding with any task, the first step is to import the necessary libraries.

CHAPTER 6 DEEP LEARNING WITH PYTORCH FOR CLASSIFICATION

Step 1: Imports

```
[In]: import boto3
[In]: import logging
[In]: import numpy as np
[In]: import torch
[In]: import torch.nn as nn
[In]: import torch.optim as optim
[In]: from torch.utils.data import DataLoader, TensorDataset
[In]: from sklearn.metrics import (
        accuracy_score,
        precision_score,
        recall_score,
        f1_score,
        confusion_matrix,
        roc_auc_score
    )
[In]: from pyspark.sql import SparkSession
[In]: from pyspark.ml.feature import (
        VectorAssembler,
        StandardScaler
    )
[In]: from pyspark.sql.functions import col
```

Below is a breakdown of each import statement and its purpose along with where it will be used in the upcoming code:

- import boto3:
    - This statement imports the boto3 library, which is the Amazon Web Services (AWS) SDK for Python. It's used for interacting with S3 to download a file. In our code, this import is used in the copy_file_from_s3 function to interact with Amazon S3 and download a file from the specified S3 bucket.

CHAPTER 6   DEEP LEARNING WITH PYTORCH FOR CLASSIFICATION

- import logging:
  - This imports the built-in logging module in Python, which provides a framework for displaying log messages from Python programs. Throughout the code, logging is used to handle logging messages, including errors, warnings, and informational messages. It's used for debugging and monitoring purposes.
- import numpy as np:
  - This imports the NumPy library and aliases it as np. NumPy is a powerful library for numerical computing in Python. In our code, various numerical computations are performed using NumPy arrays, such as data preprocessing and transforming data into NumPy arrays.
- import torch:
  - This imports the PyTorch library, which is an open source machine learning framework developed by Facebook's AI Research lab. In our code, PyTorch is used for working with tensors and neural network operations, including model definition, training, and evaluation.
- import torch.nn as nn:
  - This imports the neural network module nn from PyTorch. In the code, the nn module is used to define the architecture of the neural network model for diabetes classification. It provides building blocks for creating neural network layers and activation functions.

CHAPTER 6 DEEP LEARNING WITH PYTORCH FOR CLASSIFICATION

- import torch.optim as optim:

    - This imports the optimization module optim from PyTorch. The optim module will be used to specify the optimization algorithm and parameters for training the neural network model. It includes implementations of various optimization algorithms like Adam and SGD.

- from torch.utils.data import DataLoader, TensorDataset:

    - These imports are used to handle data loading and batching during model training. DataLoader and TensorDataset will be used to load and batch data efficiently during model training.

- from sklearn.metrics import accuracy_score, precision_score, recall_score, f1_score, confusion_matrix, roc_auc_score:

    - These import specific performance metrics from the Scikit-Learn (sklearn) library. In the code, these metrics are used in the ModelEvaluator class to calculate performance metrics such as accuracy, precision, recall, F1 score, confusion matrix, and ROC-AUC score for evaluating the trained model.

- from pyspark.sql import SparkSession:

    - This imports the SparkSession class from the pyspark.sql module. It will be used in the main execution part of the code to create a Spark session for working with structured data using the Spark DataFrame API.

CHAPTER 6   DEEP LEARNING WITH PYTORCH FOR CLASSIFICATION

- from pyspark.ml.feature import VectorAssembler, StandardScaler:
  - These import specific feature engineering tools from the pyspark.ml.feature module. VectorAssembler and StandardScaler will be used in the DataPreprocessor class to perform feature engineering tasks such as assembling feature columns into a single vector and scaling the features.
- from pyspark.sql.functions import col:
  - This imports the col function from the pyspark.sql.functions module. It will be used in the DataPreprocessor class to reference DataFrame columns by name, enabling convenient column operations within Spark SQL queries, particularly for filtering records based on specific column values.

After these imports, we move on to Step 2 in our modeling process, which is data preprocessing.

Step 2: Data Preprocessing

For this modeling step, we define a DataPreprocessor class and its preprocess method. This code loads the data, performs filtering for invalid rows, assembles features, scales the features, and splits the data into training and testing datasets:

```
[In]: class DataPreprocessor:
        """
        Preprocesses the diabetes dataset.
        """
```

CHAPTER 6 DEEP LEARNING WITH PYTORCH FOR CLASSIFICATION

```python
def __init__(self, spark, data_file_path):
    """
    Initializes the DataPreprocessor.

    Args:
        spark (pyspark.sql.SparkSession):
            The Spark session.
        data_file_path (str):
            The path to the data file.
    """
    self.spark = spark
    self.data_file_path = data_file_path

def preprocess(self):
    """
    Performs data preprocessing.

    Returns:
        tuple:
            A tuple containing train and test
            DataFrames.
    """
    try:
        diabetes_df = self.spark.read.csv(
            self.data_file_path,
            header=True,
            inferSchema=True
        )
        diabetes_df = diabetes_df.filter(
            (col("Glucose") != 0)
            & (col("BloodPressure") != 0)
            & (col("BMI") != 0)
        )
```

CHAPTER 6 DEEP LEARNING WITH PYTORCH FOR CLASSIFICATION

```python
        feature_cols = [
            "Pregnancies",
            "Glucose",
            "BloodPressure",
            "BMI",
            "DiabetesPedigreeFunction",
            "Age",
        ]
        assembler = VectorAssembler(
            inputCols=feature_cols,
            outputCol="features"
        )
        diabetes_df = assembler.
        transform(diabetes_df)
        scaler = StandardScaler(
            inputCol="features",
            outputCol="scaled_features"
        )
        scaler_model = scaler.fit(diabetes_df)
        diabetes_df = \
          scaler_model.transform(diabetes_df)
        train_df, test_df = diabetes_df.randomSplit(
            [0.8, 0.2],
            seed=42
        )
        return train_df, test_df
    except Exception as e:
        logging.error(
            f"Error occurred during data"
            f"preprocessing: {str(e)}"
        )
        raise e
```

CHAPTER 6   DEEP LEARNING WITH PYTORCH FOR CLASSIFICATION

The code defines a Python class named DataPreprocessor, which encapsulates functionality for preprocessing the diabetes dataset. Within this class, there's a constructor method __init__ that initializes the class instance with two parameters: spark, representing a Spark session, and data_file_path, indicating the path to the data file. The core preprocessing functionality is implemented in the preprocess method of the class. This method first reads the CSV data file into a Spark DataFrame (diabetes_df) using Spark's read.csv method, specifying that the file has a header row and inferring the schema. It then filters the DataFrame to remove records where certain columns (Glucose, BloodPressure, and BMI) have zero values, as these are biologically implausible values and are likely to indicate missing or incorrect data. Keeping these values could negatively impact the model's accuracy and predictive power. Next, a list of feature columns is defined, including features such as pregnancies, glucose levels, blood pressure, Body Mass Index (BMI), diabetes pedigree function, and age.

Subsequently, a VectorAssembler object is created to assemble these feature columns into a single vector column named "features". This assembled feature column is then added to the DataFrame. After feature assembly, a StandardScaler object is instantiated to scale the features, ensuring that they have similar scales for improved model performance. This is particularly important for neural networks, which are sensitive to the scale of input data. The scaler is fitted to the DataFrame to compute summary statistics, and the DataFrame is transformed using the fitted scaler to obtain scaled features. Finally, the preprocessed data is split into training and testing sets using an 80–20 split ratio, with 80% of the data allocated for training and 20% for testing. The use of a seed=42 in the randomSplit method ensures that the data is split in the same way every time the code is run, enhancing the reproducibility of the results. The training and testing DataFrames are returned as a tuple from the preprocess method.

CHAPTER 6   DEEP LEARNING WITH PYTORCH FOR CLASSIFICATION

In case any exceptions occur during the preprocessing steps, the code handles them within a try–except block. If an exception is caught, an error message is logged using Python's logging.error method, containing details of the exception.

Step 3: Model Definition

In this step, we define the architecture of the neural network model for diabetes classification using PyTorch. The architecture specifies the layers and activation functions of the model:

```
[In]: class DiabetesClassifierModel(nn.Module):
          """
          Neural network model for diabetes classification.
          """

          def __init__(self, input_size, output_size):
              """
              Initializes the DiabetesClassifierModel.

              Args:
                  input_size (int): The input size.
                  output_size (int): The output size.
              """
              super(DiabetesClassifierModel, self).__init__()
              self.model = nn.Sequential(
                  nn.Linear(input_size, 64),
                  nn.ReLU(),
                  nn.Linear(64, 32),
                  nn.ReLU(),
                  nn.Linear(32, 16),
                  nn.ReLU(),
                  nn.Linear(16, output_size),
              )
```

```
def forward(self, x):
    """
    Defines the forward pass of the model.

    Args:
        x (torch.Tensor): The input tensor.

    Returns:
        torch.Tensor: The output tensor.
    """
    return self.model(x)
```

The code defines a Python class named DiabetesClassifierModel, which serves as a neural network model for diabetes classification. Within this class, there are two main methods: __init__ and forward.

The __init__ method acts as the constructor and is responsible for initializing the model. It takes two parameters: input_size, representing the size of the input features, and output_size, indicating the size of the output (i.e., the number of classes for classification).

Inside the __init__ method, the class inherits from nn.Module, which is a base class for all neural network modules in PyTorch. The neural network architecture is defined using PyTorch's nn.Sequential container, which simplifies the model definition by automatically connecting layers in the order they are defined. This container allows for a clear, linear arrangement of layers, making the model easier to read and maintain.

The architecture consists of several fully connected (linear) layers with decreasing output sizes from input_size to output_size. After each linear layer, a rectified linear unit (ReLU) activation function is applied. The ReLU function introduces nonlinearity into the model, which is crucial for the network to learn complex patterns in the data. Without this nonlinearity, the model would simply be a linear combination of inputs, which could severely limit its predictive power.

CHAPTER 6   DEEP LEARNING WITH PYTORCH FOR CLASSIFICATION

This model uses three hidden layers with 64, 32, and 16 neurons, respectively. While this architecture provides a solid foundation, it might not be optimal and could lead to overfitting or underfitting, depending on the dataset size and complexity. The absence of regularization techniques, such as dropout or L2 regularization, means that the model might overfit the training data, capturing noise rather than general patterns. Conversely, it might underfit the data if the architecture is too simple to capture the underlying structure of the dataset. Additionally, the current training approach splits the dataset once into training and testing sets, which may introduce variability and impact model performance. These potential issues—including the need for techniques like dropout, early stopping, L1/L2 regularization, and cross-validation—will be addressed in Chapter 9. Chapter 9 will also demonstrate how to leverage PyTorch's learning rate schedulers to further enhance model performance and generalization.

The forward method defines the forward pass of the model, specifying how input data is processed through the neural network to produce output predictions. It takes a single parameter, x, which represents the input tensor containing the features. Within the forward method, the input tensor x is passed through the neural network model (self.model). The output tensor produced by the model is then returned as the result of the forward pass.

Step 4: Model Training
In this step, we define the ModelTrainer class and its train method. The code initializes the model trainer with the neural network model, loss function, optimizer, and training data loader. It trains the model by updating its parameters based on the optimization process:

```
[In]: class ModelTrainer:
          """
          Trainer for the diabetes classifier model.
          """
```

CHAPTER 6   DEEP LEARNING WITH PYTORCH FOR CLASSIFICATION

```
def __init__(self, model, criterion, optimizer,
    train_loader):
    """
    Initializes the ModelTrainer.

    Args:
        model (torch.nn.Module):
            The neural network model.
        criterion (torch.nn.Module):
            The loss function.
        optimizer (torch.optim.Optimizer):
            The optimizer.
        train_loader (torch.utils.data.DataLoader):
            The training data loader.
    """
    self.model = model
    self.criterion = criterion
    self.optimizer = optimizer
    self.train_loader = train_loader

def train(self, epochs=100, lr=0.01):
    """
    Trains the model.

    Args:
        epochs (int, optional):
            The number of epochs. Defaults to 100.
        lr (float, optional):
            The learning rate. Defaults to 0.01.
    """
    try:
        for epoch in range(epochs):
            self.model.train()
```

371

CHAPTER 6   DEEP LEARNING WITH PYTORCH FOR CLASSIFICATION

```
                for inputs, targets in self.train_loader:
                    self.optimizer.zero_grad()
                    outputs = self.model(inputs)
                    loss = self.criterion(
                        outputs.squeeze(),
                        targets
                    )
                    loss.backward()
                    self.optimizer.step()
                if (epoch + 1) % 10 == 0:
                    logging.info(
                        f"Epoch [{epoch + 1}/{epochs}],"
                        f"Loss: {loss.item():.4f}"
                    )
        except Exception as e:
            logging.error(
                f"Error occurred during model training:"
                f"{str(e)}"
            )
            raise e
```

The ModelTrainer class is designed to facilitate the training process of the diabetes classifier model. Its primary role is to manage the training loop, executing the forward and backward passes of the model on the training data.

In the __init__ method, the trainer is initialized with four essential components: the neural network model (model), the loss function (criterion), the optimizer (optimizer), and the training data loader (train_loader). These components are crucial for training the model effectively. The method serves as a constructor for the ModelTrainer class, ensuring that all necessary components are properly set up before training begins. Each argument is documented with its expected type and purpose, enhancing code readability and maintainability.

CHAPTER 6 DEEP LEARNING WITH PYTORCH FOR CLASSIFICATION

The train method is the heart of the ModelTrainer class, responsible for executing the training loop. Within this method, the model undergoes training for a specified number of epochs, with each epoch consisting of multiple iterations over the training dataset. During each iteration, the model is set to training mode (self.model.train()). This step is important because certain layers, like dropout or batch normalization, behave differently during training compared with evaluation mode. For example, dropout randomly disables some neurons during training to prevent overfitting, but all neurons are active during evaluation.

Gradients are then reset (self.optimizer.zero_grad()). Zeroing the gradients is necessary because, by default, gradients accumulate in PyTorch. Without this step, previous gradient information would interfere with the current backpropagation process, leading to incorrect updates to the model's parameters.

The forward pass is executed (outputs = self.model(inputs)), where the model processes the input data and produces output predictions. The loss is then calculated (loss = self.criterion(outputs.squeeze(), targets)) by comparing the model's predictions to the actual target values. The backward pass (loss.backward()) computes the gradients of the loss with respect to the model parameters, and the optimizer step (self.optimizer.step()) updates the model parameters based on these gradients.

Additionally, the method logs the loss at regular intervals (every ten epochs) to monitor the training progress. Logging the loss is useful for diagnosing issues such as vanishing gradients or overfitting early in training, allowing for adjustments if necessary.

Exception handling is implemented to catch and log any errors that may occur during the training process. This ensures that issues are identified and reported, facilitating debugging and model refinement.

CHAPTER 6 DEEP LEARNING WITH PYTORCH FOR CLASSIFICATION

Step 5: Model Evaluation

In this step, we define the ModelEvaluator class and its evaluate method. This code evaluates the trained model using the testing dataset. It calculates performance metrics such as accuracy, precision, recall, F1 score, ROC-AUC score, and confusion matrix to assess the model's effectiveness:

```
[In]: class ModelEvaluator:
          """
          Evaluator for the diabetes classifier model.
          """

          def __init__(self, model, test_loader, y_test_np):
              """
              Initializes the ModelEvaluator.

              Args:
                  model (torch.nn.Module):
                      The neural network model.
                  test_loader (torch.utils.data.DataLoader):
                      The test data loader.
                  y_test_np (numpy.ndarray):
                      The ground truth labels for test data.
              """
              self.model = model
              self.test_loader = test_loader
              self.y_test_np = y_test_np

          def evaluate(self):
              """
              Evaluates the model.
              """
              try:
                  with torch.no_grad():
```

CHAPTER 6   DEEP LEARNING WITH PYTORCH FOR CLASSIFICATION

```
        self.model.eval()
        predictions = []
        probabilities = []
        for inputs, _ in self.test_loader:
            outputs = self.model(inputs)
            probabilities.extend(
                torch.sigmoid(outputs)
                .squeeze().tolist()
            )
            predictions.extend(
                outputs.squeeze().tolist()
            )
y_pred = np.array(
    [1 if pred > 0.5 else 0
      for pred in probabilities]
)

accuracy = accuracy_score(
                self.y_test_np,
                y_pred
            )
precision = precision_score(
    self.y_test_np,
    y_pred
)
recall = recall_score(self.y_test_np, y_pred)
f1 = f1_score(self.y_test_np, y_pred)

logging.info(f"Accuracy: {accuracy:.4f}")
logging.info(f"Precision: {precision:.4f}")
logging.info(f"Recall: {recall:.4f}")
logging.info(f"F1 Score: {f1:.4f}")
```

CHAPTER 6   DEEP LEARNING WITH PYTORCH FOR CLASSIFICATION

```
            conf_matrix = confusion_matrix(
                self.y_test_np,
                y_pred
            )
            logging.info(
                "Confusion Matrix:\n" + str(conf_matrix)
            )
            roc_auc = roc_auc_score(
                self.y_test_np, probabilities
            )
            logging.info(f"ROC-AUC Score: {roc_auc:.4f}")
        except Exception as e:
            logging.error(
                f"Error occurred during model "
                f"evaluation: {str(e)}"
            )
            raise e
```

The ModelEvaluator class serves as an evaluator for the diabetes classifier model, responsible for assessing its performance on test data.

The class is initialized with three essential components: the neural network model (model), the test data loader (test_loader), and the ground truth labels for the test data (y_test_np). These components are crucial for evaluating the model's performance accurately. The __init__ method acts as a constructor, ensuring that all necessary elements are properly set up before evaluation begins. Each argument is documented with its expected type and purpose, enhancing code clarity and usability.

The evaluate method is the core of the ModelEvaluator class, responsible for conducting the evaluation process. Within this method

CHAPTER 6 DEEP LEARNING WITH PYTORCH FOR CLASSIFICATION

- The model is set to evaluation mode (self.model.eval()), which is important because certain layers, such as dropout, behave differently during training and evaluation.

- Predictions are generated for the test data using the trained model. The torch.no_grad() context manager is used to ensure that gradient computations are disabled, saving memory and computation time during evaluation.

To obtain final predictions

- Probabilities are calculated using the sigmoid function (torch.sigmoid(outputs)), which converts the raw output scores into probabilities.

- The final predicted labels (y_pred) are determined based on these probabilities, with a threshold of 0.5 used to classify the predictions.

Various performance metrics are then computed:

- Accuracy: Measures the proportion of correctly classified samples

- Precision: Indicates the proportion of true positive predictions among all positive predictions

- Recall: Represents the proportion of true positive predictions among all actual positives

- F1 Score: Provides a balance between precision and recall

CHAPTER 6 DEEP LEARNING WITH PYTORCH FOR CLASSIFICATION

Additionally, the method computes and logs

- Confusion Matrix: Provides a detailed breakdown of the true positives, false positives, true negatives, and false negatives

- ROC-AUC Score: Measures the model's ability to distinguish between classes across different thresholds, providing a summary of its performance

Exception handling is implemented within the evaluate method to capture and log any errors that may occur during the evaluation process. This ensures that potential issues are identified and reported, allowing for effective debugging and error resolution.

Step 6: Additional Functionality

The code in this final step includes the copy_file_from_s3 function, logging configurations, and main execution part. These codes handle additional tasks such as copying data from an AWS S3 bucket, setting up logging for monitoring, and orchestrating the main execution flow of the modeling process:

```
[In]: def copy_file_from_s3(bucket_name, file_name, local_dir):
        """
        Copies a file from S3 bucket to local directory.

        Args:
            bucket_name (str): The name of the S3 bucket.
            file_name (str):
                The name of the file to be copied.
            local_dir (str):
                The local directory where the file will be
                copied.
        """
        try:
```

```
            s3 = boto3.client("s3")
            s3.download_file(
                bucket_name,
                file_name,
                local_dir + file_name
            )
        except Exception as e:
            logging.error(
                f"Error occurred during file copying: "
                f"{str(e)}"
            )
            raise e
```

```
[In]: if __name__ == "__main__":
        logging.basicConfig(level=logging.INFO)
        spark = (SparkSession.builder
                    .appName("DiabetesClassifier")
                    .getOrCreate())
        copy_file_from_s3(
            "instance1bucket",
            "diabetes.csv",
            "/home/ubuntu/airflow/dags/"
        )

        data_preprocessor = DataPreprocessor(
            spark,
            "/home/ubuntu/airflow/dags/diabetes.csv"
        )
        train_df, test_df = data_preprocessor.preprocess()

        X_train_np = np.array(
            train_df.select("scaled_features")
            .rdd.flatMap(lambda x: x).collect(),
```

```
        dtype=np.float32,
    )
    y_train_np = np.array(
        train_df.select("Outcome")
        .rdd.flatMap(lambda x: x).collect(),
        dtype=np.float32,
    )

    train_dataset = TensorDataset(
        torch.tensor(X_train_np),
        torch.tensor(y_train_np)
    )
    train_loader = DataLoader(
        train_dataset,
        batch_size=64,
        shuffle=True
    )

    input_size = X_train_np.shape[1]
    output_size = 1
    model = DiabetesClassifierModel(
        input_size,
        output_size
    )
    criterion = nn.BCEWithLogitsLoss()
    optimizer = optim.Adam(model.parameters(), lr=0.01)
    model_trainer = ModelTrainer(
        model,
        criterion,
        optimizer, train_loader
    )
    model_trainer.train()
```

CHAPTER 6   DEEP LEARNING WITH PYTORCH FOR CLASSIFICATION

```
X_test_np = np.array(
    test_df.select("scaled_features")
    .rdd.flatMap(lambda x: x).collect(),
    dtype=np.float32,
)
y_test_np = np.array(
    test_df.select("Outcome")
    .rdd.flatMap(lambda x: x).collect()
)

test_dataset = TensorDataset(
    torch.tensor(X_test_np),
    torch.tensor(y_test_np, dtype=torch.float32)
)
test_loader = DataLoader(
    test_dataset,
    batch_size=64,
    shuffle=False
)

model_evaluator = ModelEvaluator(
    model,
    test_loader,
    y_test_np
)
model_evaluator.evaluate()
```

The code consists of a Python script responsible for copying the diabetes CSV file from an Amazon S3 bucket to a local directory on EC2, performing data preprocessing using Spark, and training and evaluating a diabetes classifier model using PyTorch.

CHAPTER 6   DEEP LEARNING WITH PYTORCH FOR CLASSIFICATION

Let's break it down:

File Copying from S3:

- The copy_file_from_s3 function is defined to copy a file from an S3 bucket to a local directory. It takes three arguments: bucket_name (the name of the S3 bucket), file_name (the name of the file to be copied), and local_dir (the local directory where the file will be copied).

- Inside the function, an S3 client is created using the boto3 library. The download_file method is then called to download the specified file from the S3 bucket to the local directory.

Data Preprocessing:

- After setting up logging, a Spark session is created using SparkSession. The session is configured with the application name "DiabetesClassifier".

- The copy_file_from_s3 function is called to copy a file named "diabetes.csv" from the S3 bucket "instance1bucket" to the local directory "/home/ubuntu/airflow/dags/".

- An instance of the DataPreprocessor class is created, specifying the Spark session and the path to the diabetes dataset file.

- The preprocess method of the DataPreprocessor instance is called to perform data preprocessing, resulting in train and test DataFrames (train_df and test_df).

CHAPTER 6 DEEP LEARNING WITH PYTORCH FOR CLASSIFICATION

Model Training and Evaluation:

- The train and test DataFrames are converted into NumPy arrays (X_train_np, y_train_np, X_test_np, y_test_np) to be used for training and evaluation.

- PyTorch's DataLoader and TensorDataset are used to create train and test datasets (train_dataset, test_dataset) from the NumPy arrays.

- The diabetes classifier model (DiabetesClassifierModel) is instantiated, along with the loss function (BCEWithLogitsLoss) and optimizer (Adam).

- An instance of the ModelTrainer class is created with the model, loss function, optimizer, and train DataLoader. The train method of the ModelTrainer instance is then called to train the model.

- Another instance of the ModelEvaluator class is created with the model, test DataLoader, and ground truth labels. The evaluate method of the ModelEvaluator instance is called to evaluate the trained model's performance.

This code demonstrates a complete pipeline for building, training, and evaluating a diabetes classifier model, incorporating file handling, data preprocessing with Spark, and deep learning with PyTorch.

To run the code, we saved it as "diabetes_classifier_pipeline.py" and executed it in a JupyterLab terminal (alternatively, we could have run it directly on the EC2 instance). Since we have created an environment named "myenv," we activated it as follows:

```
[In]: source /home/ubuntu/myenv/bin/activate
```

CHAPTER 6　DEEP LEARNING WITH PYTORCH FOR CLASSIFICATION

Following environment activation, we executed the code using Python 3 with the following command:

[In]: python3 diabetes_classifier_pipeline.py

The first output from the code is a training progress table, summarizing the training of the diabetes classifier model over multiple epochs:

Epoch	Loss
10	0.5899
20	0.4330
30	0.4608
40	0.4612
50	0.5155
60	0.4025
70	0.3406
80	0.4323
90	0.4822
100	0.3203

Each row in the training progress table represents an epoch, indicating the epoch number and the corresponding loss value. For instance, the row "Epoch 10/100, Loss 0.5899" means that after ten epochs of training, the model achieved a loss of approximately 0.59. Similarly, subsequent rows provide the loss values at different epochs until the end of the training (100 epochs in this case).

The loss values were computed using the BCEWithLogitsLoss function, which stands for Binary Cross-Entropy with Logits Loss. This function is commonly used for binary classification tasks, such as the diabetes prediction task in this scenario.

CHAPTER 6   DEEP LEARNING WITH PYTORCH FOR CLASSIFICATION

Observations based on loss values:

- Trend: Initially, there were fluctuations in the loss, but there was an overall downward trend, indicating an improvement in the model's performance. The loss increased at certain epochs before decreasing again, suggesting some instability in training.

- Magnitude: The loss values ranged from approximately 0.32 to 0.59. The decrease in loss values toward the end of training suggests that the model was converging toward a better solution.

- Optimization: The fluctuating loss values reflect the optimization process. The model adjusts its parameters to minimize the loss function, and the decreasing loss values indicate that the model's predictions were progressively aligning better with the actual target values in the training data.

The second output is the evaluation metrics table:

Metric	Value
Accuracy	0.7607
Precision	0.6585
Recall	0.6585
F1 Score	0.6585
ROC-AUC Score	0.8135

Let's explain these metrics one by one:

Accuracy: Accuracy measures the proportion of correctly classified instances out of all instances in the test dataset. Mathematically, it is calculated as the ratio of the number of correctly predicted instances to the total number of instances. For example, an accuracy of 0.7607 means that approximately 76.07% of the instances in the test dataset were correctly classified by the model.

Precision: Precision measures the proportion of true positive predictions out of all positive predictions made by the model. It focuses on the correctness of positive predictions and is calculated as the ratio of true positives to the sum of true positives and false positives. A precision score of 0.6585 indicates that approximately 65.85% of the instances predicted as positive by the model were actually positive.

Recall: Recall, also known as sensitivity, measures the proportion of true positive predictions out of all actual positive instances in the test dataset. It focuses on capturing all positive instances and is calculated as the ratio of true positives to the sum of true positives and false negatives. A recall score of 0.6585 means that approximately 65.85% of all actual positive instances were correctly identified by the model.

F1 Score: The F1 score is the harmonic or weighted mean of precision and recall, providing a balance between these two metrics. It takes into account both false positives and false negatives and is calculated as 2 * (precision * recall) / (precision + recall). An F1 score of 0.6585 indicates the overall effectiveness of the model in terms of both precision and recall, with higher scores indicating better performance.

ROC-AUC Score: The ROC-AUC score measures the model's ability to distinguish between positive and negative instances across different thresholds. A score of 0.8135 indicates a high level of distinction between the classes, with the model performing well in distinguishing between them.

CHAPTER 6   DEEP LEARNING WITH PYTORCH FOR CLASSIFICATION

The model also outputs the confusion matrix:

[[62 14]

[14 27]]

The confusion matrix provides a detailed breakdown of the model's performance:

- True Positives (TP): 27
- False Positives (FP): 14
- True Negatives (TN): 62
- False Negatives (FN): 14

This suggests that the model

- Correctly predicted 62 negative instances (true negatives)
- Incorrectly predicted 14 negative instances as positive (false positives)
- Failed to identify 14 positive instances (false negatives)
- Correctly predicted 27 positive instances (true positives)

Based on these numbers, we can replicate the calculations in the evaluation metrics table as follows:

Accuracy = (TP + TN)/(TP+TN+FP+FN)

Substituting the values:

Accuracy = (27+62)/(27+62+14+14) = 89/117
= 0.7607

CHAPTER 6   DEEP LEARNING WITH PYTORCH FOR CLASSIFICATION

Precision is calculated as TP/(TP+FP) = 27/(27+14).

Precision = 27/41 = 0.6585

Recall is calculated as TP/(TP+FN).

Substituting the values:

Recall = 27/(27+14) = 27/41 = 0.6585

F1 score = 2 × [(precision × recall)/(precision + recall)]

Substituting the values:

F1 = 2 × [(0.6585*0.6585)/(0.6585+0.6585)] = 0.6585

Based on both outputs, the loss values and evaluation metrics, the model demonstrates modest performance. One reason for this is the class imbalance and skewed features as observed during the exploration stage. Additionally, the sample size is limited, and there were potential data quality issues.

In a real-world scenario with the Pima dataset, several areas for improvement would be considered, including the following:

- Data Quality: Ensuring data quality is important for the dataset, despite its limited size. For instance, during data exploration, we observed anomalies such as the Glucose attribute's minimum value of 44, potentially indicating hypoglycemia and raising concerns about data integrity. Similarly, the BloodPressure entry displayed an exceptionally low minimum value of 24, highly unlikely for a valid blood pressure reading, suggesting potential errors in data collection. Furthermore, the DiabetesPedigreeFunction feature exhibited a minimum value of 0.078, notably low for a genetic risk factor associated with diabetes. Additionally, the Pregnancies field showed a

CHAPTER 6   DEEP LEARNING WITH PYTORCH FOR CLASSIFICATION

maximum count of 17, biologically plausible but highly uncommon. These observations underscore the need for further scrutiny and validation to ensure the reliability of the dataset.

- Resampling: Resampling techniques such as oversampling the minority class or undersampling the majority class can help balance the distribution of classes in the dataset, improving model performance and reducing bias toward the majority class.

- Data Augmentation: Despite its conventional association with image data, data augmentation can still be beneficial for the Pima dataset. By generating additional synthetic data points through transformations such as perturbing feature values or introducing noise, we can effectively enhance the size of the dataset. This augmentation diversifies the dataset, potentially improving the model's generalization ability by adding a small random variation to the glucose levels, blood pressure measurements, BMI values, or other numerical features. This variation helps in generating diverse training examples and prevents the model from relying too heavily on specific values or patterns in the data.

- Transfer Learning: While not as commonly applied in non-image tasks, transfer learning can still provide benefits for the Pima dataset. By leveraging knowledge from pre-trained models on related tasks or datasets, we can fine-tune these models on our small dataset. This adaptation helps the model capture relevant patterns more efficiently, leading to improved performance with limited training data.

- Model Complexity: Balancing model complexity is crucial, especially with small datasets like Pima. Complex models may risk overfitting, so it's essential to find the right balance between model complexity and dataset size to effectively capture underlying patterns while avoiding overfitting. For example, during exploration, we saw that Glucose and BMI were the strongest predictors. A smaller model based on these features alone may provide better results.

- Data Representation: Effective data representation is critical for small datasets like Pima. Preprocessing techniques such as feature extraction or dimensionality reduction can enhance the quality of input data and facilitate learning in deep neural networks.

- Hyperparameter Tuning: Further tuning hyperparameters such as learning rate, batch size, and network architecture can potentially improve model convergence and performance. Advanced techniques like regularization, dropout, or different optimization algorithms might also help mitigate overfitting and improve generalization. In Chapter 9, we explore some of these techniques.

- Cross-Validation: Employing K-fold cross-validation is essential for obtaining a reliable estimate of model performance, particularly with small datasets like Pima. This method reduces the risk of overfitting or underfitting due to variability in a single data split by evaluating the model across multiple folds. The section below explores this option in detail.

Enhancing Model Evaluation with Cross-Validation

K-fold cross-validation is crucial for obtaining a reliable estimate of a model's performance, particularly with limited data. By dividing the dataset into K folds and iterating the training and testing process across these folds, this technique ensures that every data point is utilized for both training and validation. This approach reduces the risks of overfitting and underfitting by averaging performance across multiple subsets, leading to a more balanced evaluation. Additionally, K-fold cross-validation maximizes the use of available data, aids in hyperparameter tuning, and helps address issues like class imbalance, ultimately resulting in models that generalize better to unseen data.

To implement K-fold cross-validation in the original code, we follow these steps:

Step 1: Importing KFold

```
[In]: from sklearn.model_selection import KFold
```

This import enables the use of the KFold class from Scikit-Learn, essential for implementing K-fold cross-validation.

Step 2: Defining the Number of Folds

```
[In]: k_folds = 5
[In]: kfold = KFold(n_splits=k_folds, shuffle=True,
      random_state=42)
```

This sets up K-fold cross-validation with five folds, shuffles the data before splitting, and uses a random seed for reproducibility.

CHAPTER 6   DEEP LEARNING WITH PYTORCH FOR CLASSIFICATION

Step 3: Looping Through Each Fold

[In]: fold_results = []

[In]: for fold, (train_idx, test_idx) in enumerate(kfold.
    split(X_np)):
        logging.info(f"Training fold {fold + 1}/{k_folds}")

        X_train, X_test = X_np[train_idx], X_np[test_idx]
        y_train, y_test = y_np[train_idx], y_np[test_idx]

        # Convert to PyTorch datasets
        train_dataset = TensorDataset(
            torch.tensor(X_train),
            torch.tensor(y_train)
)
        test_dataset = TensorDataset(
            torch.tensor(X_test),
            torch.tensor(y_test)
)

        # Create DataLoaders
        train_loader = DataLoader(
            train_dataset,
            batch_size=64,
            shuffle=True
)
        test_loader = DataLoader(
            test_dataset,
            batch_size=64,
            shuffle=False
)

```python
# Initialize the model, criterion, and optimizer
input_size = X_train.shape[1]
output_size = 1
model = DiabetesClassifierModel(input_size,
output_size)
criterion = nn.BCEWithLogitsLoss()
optimizer = optim.Adam(model.parameters(), lr=0.01)

# Train the model
model_trainer = ModelTrainer(
    model,
    criterion,
    optimizer,
    train_loader
)
model_trainer.train()

# Evaluate the model
model_evaluator = ModelEvaluator(
    model,
    test_loader,
    y_test
)
results = model_evaluator.evaluate()

# Append results to the fold_results list
fold_results.append(results)
```

In this code, K-fold cross-validation is implemented in the main function. The dataset is first preprocessed and converted into NumPy arrays (X_np for features and y_np for labels). The KFold class from Scikit-Learn is used to create the K-fold splits with k_folds = 5, meaning the dataset is divided into five folds.

CHAPTER 6   DEEP LEARNING WITH PYTORCH FOR CLASSIFICATION

Within the loop, the training and test indices are determined using the K-fold split. The corresponding data is then split into training and testing sets. PyTorch datasets are created for both the training and testing data, which are then loaded into DataLoaders to facilitate batch processing during model training.

For each fold, the model is initialized, trained using the ModelTrainer class, and evaluated using the ModelEvaluator class. The evaluation metrics for each fold—such as accuracy, precision, recall, F1 score, and ROC-AUC score—are calculated and stored in a list.

Step 4: Aggregating Results Across Folds

```
[In]: avg_accuracy = np.mean(
        [result['accuracy'] for result in fold_results]
    )
[In]: avg_precision = np.mean(
        [result['precision'] for result in fold_results])
[In]: avg_recall = np.mean(
        [result['recall'] for result in fold_results])
[In]: avg_f1 = np.mean(
        [result['f1'] for result in fold_results])
[In]: avg_roc_auc = np.mean(
        [result['roc_auc'] for result in fold_results])

[In]: logging.info(f"Average Accuracy: {avg_accuracy:.4f}")
[In]: logging.info(f"Average Precision: {avg_precision:.4f}")
[In]: logging.info(f"Average Recall: {avg_recall:.4f}")
[In]: logging.info(f"Average F1 Score: {avg_f1:.4f}")
[In]: logging.info(f"Average ROC-AUC Score: {avg_roc_auc:.4f}")
```

After all folds have been processed, the results are averaged to provide a more robust estimate of the model's performance. This averaged performance is then logged, offering a comprehensive view of how well the model generalizes across different subsets of the data. K-fold cross-

CHAPTER 6   DEEP LEARNING WITH PYTORCH FOR CLASSIFICATION

validation ensures that the model's evaluation is not dependent on a single train-test split, providing a more balanced and accurate assessment of its capabilities.

The table below shows the results of the K-fold cross validation:

Fold	Accuracy	Precision	Recall	F1 Score	ROC-AUC Score
Fold 1	0.7586	0.5833	0.6512	0.6154	0.8347
Fold 2	0.7103	0.6774	0.3962	0.500	0.7966
Fold 3	0.7586	0.6078	0.6739	0.6392	0.8048
Fold 4	0.800	0.7826	0.6545	0.7129	0.8711
Fold 5	0.7569	0.7297	0.5192	0.6067	0.8242
Average	**0.7569**	**0.6762**	**0.579**	**0.6148**	**0.8263**

The results show that the model performs consistently across the folds, with an average accuracy of 75.69%, indicating a reasonable performance given the small diabetes dataset. The other metrics—precision, recall, F1 score, and ROC-AUC—also provide insights into the model's predictive capabilities, with the ROC-AUC score averaging 0.8263, indicating good discriminative ability.

Bringing It All Together

In this section, we consolidate the individual codes discussed throughout the chapter for both exploring the Pima diabetes dataset and building, training, and evaluating the PyTorch deep learning model with and without K-fold cross validation.

CHAPTER 6   DEEP LEARNING WITH PYTORCH FOR CLASSIFICATION

Exploring the Pima Diabetes Dataset

```
[In]: from pyspark.sql import SparkSession
[In]: from pyspark.sql.functions import col, sum as pyspark_sum
[In]: import boto3

[In]: class PimaDatasetExplorer:
        """
        Class for exploring and analyzing a Pima Indian
        Diabetes
        dataset using PySpark.
        """

        def __init__(self, file_path):
            """
            Initializes the PimaDatasetExplorer object.

            Args:
                file_path (str): The path to the
                dataset file.
            """
            self.file_path = file_path
            self.data = None

[In]: def load_data(self):
        """
        Loads the dataset from the specified file path into a
        PySpark DataFrame.
        """
        try:
            self.data = spark.read.csv(
                self.file_path,
```

```python
            header=True,
            inferSchema=True
        )
        self.data.cache()
    except Exception as e:
        print(
            "An error occurred while loading the data:",
            str(e)
        )
        self.data = None

[In]: def preprocess_data(self):
    """
    Performs preprocessing steps on the dataset.
    """
    self.load_data()

[In]: def handle_missing_values(self):
    """
    Counts the number of missing values for each column
    in the dataset and displays the counts.
    """
    if self.data is not None:
        missing_counts = self.data.select([
            pyspark_sum(col(c).isNull().cast("int")).
            alias(c)
            for c in self.data.columns
        ])
        print("Missing Value Counts:")
        missing_counts.show()
```

CHAPTER 6 DEEP LEARNING WITH PYTORCH FOR CLASSIFICATION

```
        else:
            print("Data not loaded.")
[In]: def count_zeros(self):
        """
        Counts the number of zero values for each column in
        the dataset and displays the counts.
        """
        if self.data is not None:
            zero_counts = self.data.select([
                pyspark_sum((col(c) == 0).cast("int")).
                alias(c)
                for c in self.data.columns
            ])
            print("Zero Counts:")
            zero_counts.show()
        else:
            print("Data not loaded.")
[In]: def data_summary(self):
        """
        Displays a summary of the dataset, excluding certain
        columns, and filtering out rows with zero values in
        specific columns.
        """
        if self.data is not None:
            print("Data Summary (excluding 'Outcome', "
                "'SkinThickness', 'Insulin'):")
            columns_to_exclude = ['Outcome',
                                  'SkinThickness',
                                  'Insulin']
```

CHAPTER 6   DEEP LEARNING WITH PYTORCH FOR CLASSIFICATION

```
        summary_cols = [c for c in self.data.columns
                        if c not in columns_to_exclude]
        filtered_data = self.data.filter(
            (col("Glucose") != 0)
            & (col("BloodPressure") != 0)
            & (col("BMI") != 0)
        )
        filtered_data.select(summary_cols).
        describe().show()
    else:
        print("Data not loaded.")

[In]: def count_outcome(self):
    """
    Counts the occurrences of each outcome value in the
    dataset.
    """
    if self.data is not None:
        print("Outcome Counts:")
        outcome_counts = self.data.filter(
            (col("Glucose") != 0)
            & (col("BloodPressure") != 0)
            & (col("BMI") != 0)
        ).groupBy("Outcome").count()
        outcome_counts.show()
    else:
        print("Data not loaded.")
```

CHAPTER 6   DEEP LEARNING WITH PYTORCH FOR CLASSIFICATION

```
[In]: def explore_feature_distributions(self):
          """
          Plots histograms for the distribution of each
          feature in the
          dataset.
          """
          if self.data is not None:
              print("Feature Distributions:")
              for column in self.data.columns:
                  if column not in ['SkinThickness',
                  'Insulin']:
                      print(f"Feature: {column}")
                      if column in ["Glucose",
                                    "BloodPressure",
                                    "BMI"]:
                          filtered_data = self.data.filter(
                          (col(column) != 0)
                          & (col("Glucose") != 0)
                          & (col("BloodPressure") != 0)
                          & (col("BMI") != 0)
                          )
                      else:
                          filtered_data = self.data

                      plot_data=filtered_data.select(column).
                      toPandas()
                      plot_data.plot(kind='hist', title=column)
          else:
              print("Data not loaded.")

[In]: def calculate_feature_target_correlation(self):
          """
```

CHAPTER 6   DEEP LEARNING WITH PYTORCH FOR CLASSIFICATION

```
    Calculates the correlation between each
    feature and the
    target variable ('Outcome').
    """
    if self.data is not None:
        print("Feature-Target Correlation "
            "(excluding 'SkinThickness', 'Insulin'):")
        for column in self.data.columns:
            if column not in ['Outcome',
                              'SkinThickness',
                              'Insulin']:
                correlation = self.data.filter(
                    (col(column) != 0)
                    & (col("Glucose") != 0)
                    & (col("BloodPressure") != 0)
                    & (col("BMI") != 0)
                ).stat.corr(column, 'Outcome')
                print(f"{column}: {correlation}")
    else:
        print("Data not loaded.")

[In]: def copy_file_from_s3(bucket_name, file_key, local_path):
    """
    Copies a file from an AWS S3 bucket to a local path.

    Args:
        bucket_name (str): The name of the S3 bucket.
        file_key (str): The key of the file in the
        S3 bucket.
        local_path (str):
            The local path where the file will be copied.
    """
```

401

```
try:
    s3 = boto3.client('s3')
    s3.download_file(bucket_name, file_key,
    local_path)
    print(f"File {file_key} downloaded to
    {local_path}")
except Exception as e:
    print(f"An error occurred while
    downloading file",
        f"{file_key}:", str(e))
```

```
[In]: if __name__ == "__main__":
    spark = (SparkSession.builder
            .appName("PimaDatasetExplorer")
            .getOrCreate())
    local_file_path = "/home/ubuntu/airflow/dags/
    diabetes.csv"
    s3_bucket_name = "instance1bucket"
    s3_file_key = "diabetes.csv"
    copy_file_from_s3(
        s3_bucket_name,
        s3_file_key,
        local_file_path
    )
    explorer = PimaDatasetExplorer(local_file_path)
    explorer.preprocess_data()
    explorer.handle_missing_values()
    explorer.count_zeros()
    explorer.data_summary()
    explorer.count_outcome()
    explorer.explore_feature_distributions()
    explorer.calculate_feature_target_correlation()
```

Model Building, Training, and Evaluation with PyTorch

In this subsection, we consolidate the code for predicting diabetes with PyTorch with and without K-fold cross validation.

Diabetes Classification Without K-Fold Cross-Validation

```
[In]: import boto3
[In]: import logging
[In]: import numpy as np
[In]: import torch
[In]: import torch.nn as nn
[In]: import torch.optim as optim
[In]: from torch.utils.data import DataLoader, TensorDataset
[In]: from sklearn.metrics import (
          accuracy_score,
          precision_score,
          recall_score,
          f1_score,
          confusion_matrix,
          roc_auc_score
      )
[In]: from pyspark.sql import SparkSession
[In]: from pyspark.ml.feature import (
          VectorAssembler,
          StandardScaler
      )
[In]: from pyspark.sql.functions import col
[In]: class DataPreprocessor:
```

CHAPTER 6   DEEP LEARNING WITH PYTORCH FOR CLASSIFICATION

```
    """
    Preprocesses the diabetes dataset.
    """

    def __init__(self, spark, data_file_path):
        """
        Initializes the DataPreprocessor.

        Args:
            spark (pyspark.sql.SparkSession):
                The Spark session.
            data_file_path (str):
                The path to the data file.
        """
        self.spark = spark
        self.data_file_path = data_file_path

    def preprocess(self):
        """
        Performs data preprocessing.

        Returns:
            tuple:
                A tuple containing train and test
                DataFrames.
        """
        try:
            diabetes_df = self.spark.read.csv(
                self.data_file_path,
                header=True,
                inferSchema=True
            )
            diabetes_df = diabetes_df.filter(
```

404

CHAPTER 6   DEEP LEARNING WITH PYTORCH FOR CLASSIFICATION

```
        (col("Glucose") != 0)
        & (col("BloodPressure") != 0)
        & (col("BMI") != 0)
    )
    feature_cols = [
        "Pregnancies",
        "Glucose",
        "BloodPressure",
        "BMI",
        "DiabetesPedigreeFunction",
        "Age",
    ]
    assembler = VectorAssembler(
        inputCols=feature_cols,
        outputCol="features"
    )
    diabetes_df = assembler.
    transform(diabetes_df)
    scaler = StandardScaler(
        inputCol="features",
        outputCol="scaled_features"
    )
    scaler_model = scaler.fit(diabetes_df)
    diabetes_df = scaler_model.transform(
        diabetes_df)
    train_df, test_df = diabetes_df.randomSplit(
        [0.8, 0.2],
        seed=42
    )
    return train_df, test_df
except Exception as e:
```

405

CHAPTER 6   DEEP LEARNING WITH PYTORCH FOR CLASSIFICATION

```
            logging.error(
                f"Error occurred during data "
                f"preprocessing: {str(e)}"
            )
            raise e
[In]: class DiabetesClassifierModel(nn.Module):
    """
    Neural network model for diabetes classification.
    """

    def __init__(self, input_size, output_size):
        """
        Initializes the DiabetesClassifierModel.

        Args:
            input_size (int): The input size.
            output_size (int): The output size.
        """
        super(DiabetesClassifierModel, self).__init__()
        self.model = nn.Sequential(
            nn.Linear(input_size, 64),
            nn.ReLU(),
            nn.Linear(64, 32),
            nn.ReLU(),
            nn.Linear(32, 16),
            nn.ReLU(),
            nn.Linear(16, output_size),
        )

    def forward(self, x):
        """
        Defines the forward pass of the model.
```

CHAPTER 6   DEEP LEARNING WITH PYTORCH FOR CLASSIFICATION

```
        Args:
            x (torch.Tensor): The input tensor.

        Returns:
            torch.Tensor: The output tensor.
        """
        return self.model(x)

[In]: class ModelTrainer:
    """
    Trainer for the diabetes classifier model.
    """

    def __init__(
        self, model,
        criterion,
        optimizer,
        train_loader
    ):
        """
        Initializes the ModelTrainer.

        Args:
            model (torch.nn.Module):
                The neural network model.
            criterion (torch.nn.Module):
                The loss function.
            optimizer (torch.optim.Optimizer):
                The optimizer.
            train_loader (torch.utils.data.DataLoader):
                The training data loader.
        """
        self.model = model
```

```python
            self.criterion = criterion
            self.optimizer = optimizer
            self.train_loader = train_loader

    def train(self, epochs=100, lr=0.01):
        """
        Trains the model.

        Args:
            epochs (int, optional):
                The number of epochs. Defaults to 100.
            lr (float, optional):
                The learning rate. Defaults to 0.01.
        """
        try:
            for epoch in range(epochs):
                self.model.train()
                for inputs, targets in self.train_loader:
                    self.optimizer.zero_grad()
                    outputs = self.model(inputs)
                    loss = self.criterion(
                        outputs.squeeze(),
                        targets
                    )
                    loss.backward()
                    self.optimizer.step()
                if (epoch + 1) % 10 == 0:
                    logging.info(
                        f"Epoch [{epoch + 1}/{epochs}],"
                        f"Loss: {loss.item():.4f}"
                    )
```

CHAPTER 6   DEEP LEARNING WITH PYTORCH FOR CLASSIFICATION

```
        except Exception as e:
            logging.error(
                f"Error occurred during model training: "
                f"{str(e)}"
            )
            raise e
[In]: class ModelEvaluator:
        """
        Evaluator for the diabetes classifier model.
        """

        def __init__(self, model, test_loader, y_test_np):
            """
            Initializes the ModelEvaluator.

            Args:
                model (torch.nn.Module):
                    The neural network model.
                test_loader (torch.utils.data.DataLoader):
                    The test data loader.
                y_test_np (numpy.ndarray):
                    The ground truth labels for test data.
            """
            self.model = model
            self.test_loader = test_loader
            self.y_test_np = y_test_np

        def evaluate(self):
            """
            Evaluates the model.
            """
```

409

```python
try:
    with torch.no_grad():
        self.model.eval()
        predictions = []
        probabilities = []
        for inputs, _ in self.test_loader:
            outputs = self.model(inputs)
            probabilities.extend(
                torch.sigmoid(outputs)
                .squeeze().tolist()
            )
            predictions.extend(
                outputs.squeeze().tolist()
            )

    y_pred = np.array(
        [1 if pred > 0.5 else 0
          for pred in probabilities]
    )

    accuracy = accuracy_score(
                    self.y_test_np,
                    y_pred
                )
    precision = precision_score(
        self.y_test_np,
        y_pred
    )
    recall = recall_score(self.y_test_np, y_pred)
    f1 = f1_score(self.y_test_np, y_pred)

    logging.info(f"Accuracy: {accuracy:.4f}")
    logging.info(f"Precision: {precision:.4f}")
```

CHAPTER 6 DEEP LEARNING WITH PYTORCH FOR CLASSIFICATION

```
            logging.info(f"Recall: {recall:.4f}")
            logging.info(f"F1 Score: {f1:.4f}")
            conf_matrix = confusion_matrix(
                self.y_test_np,
                y_pred
            )
            logging.info(
                "Confusion Matrix:\n" + str(conf_matrix)
            )
            roc_auc = roc_auc_score(
                self.y_test_np, probabilities
            )
            logging.info(f"ROC-AUC Score: {roc_auc:.4f}")
        except Exception as e:
            logging.error(
                f"Error occurred during model "
                f"evaluation: {str(e)}"
            )
            raise e

[In]: def copy_file_from_s3(bucket_name, file_name, local_dir):
        """
        Copies a file from S3 bucket to local directory.

        Args:
            bucket_name (str):
                The name of the S3 bucket.
            file_name (str):
                The name of the file to be copied.
            local_dir (str):
```

CHAPTER 6   DEEP LEARNING WITH PYTORCH FOR CLASSIFICATION

```
            The local directory where the file will be
            copied.
        """
        try:
            s3 = boto3.client("s3")
            s3.download_file(
                bucket_name,
                file_name,
                local_dir + file_name
            )
        except Exception as e:
            logging.error(
                f"Error occurred during file copying: "
                f"{str(e)}")
            raise e
```

```
[In]: if __name__ == "__main__":
        logging.basicConfig(level=logging.INFO)
        spark = SparkSession.builder \
            .appName("DiabetesClassifier") \
            .getOrCreate()
        copy_file_from_s3(
            "instance1bucket",
            "diabetes.csv",
            "/home/ubuntu/airflow/dags/"
        )

        data_preprocessor = DataPreprocessor(
            spark,
            "/home/ubuntu/airflow/dags/diabetes.csv"
        )
        train_df, test_df = data_preprocessor.preprocess()
```

CHAPTER 6 DEEP LEARNING WITH PYTORCH FOR CLASSIFICATION

```python
X_train_np = np.array(
    train_df.select("scaled_features")
    .rdd.flatMap(lambda x: x).collect(),
    dtype=np.float32,
)
y_train_np = np.array(
    train_df.select("Outcome")
    .rdd.flatMap(lambda x: x).collect(),
    dtype=np.float32,
)

train_dataset = TensorDataset(
    torch.tensor(X_train_np),
    torch.tensor(y_train_np)
)
train_loader = DataLoader(
    train_dataset,
    batch_size=64,
    shuffle=True
)

input_size = X_train_np.shape[1]
output_size = 1
model = DiabetesClassifierModel(
    input_size,
    output_size
)
criterion = nn.BCEWithLogitsLoss()
optimizer = optim.Adam(model.parameters(), lr=0.01)
model_trainer = ModelTrainer(
    model, criterion,
    optimizer,
```

413

CHAPTER 6   DEEP LEARNING WITH PYTORCH FOR CLASSIFICATION

```
        train_loader
    )
    model_trainer.train()

    X_test_np = np.array(
        test_df.select("scaled_features")
        .rdd.flatMap(lambda x: x).collect(),
        dtype=np.float32,
    )
    y_test_np = np.array(
        test_df.select("Outcome")
        .rdd.flatMap(lambda x: x).collect()
    )

    test_dataset = TensorDataset(
        torch.tensor(X_test_np),
        torch.tensor(y_test_np, dtype=torch.float32)
    )
    test_loader = DataLoader(
        test_dataset,
        batch_size=64,
        shuffle=False
    )

    model_evaluator = ModelEvaluator(
        model,
        test_loader,
        y_test_np
    )
    model_evaluator.evaluate()
```

Diabetes Classification With K-Fold Cross-Validation

```
[In]: import boto3

[In]: import logging
[In]: import numpy as np
[In]: import torch
[In]: import torch.nn as nn
[In]: import torch.optim as optim
[In]: from torch.utils.data import DataLoader, TensorDataset
[In]: from sklearn.metrics import (
          accuracy_score,
          precision_score,
          recall_score,
          f1_score,
          confusion_matrix,
          roc_auc_score,
      )
[In]: from sklearn.model_selection import KFold
[In]: from pyspark.sql import SparkSession
[In]: from pyspark.ml.feature import VectorAssembler, StandardScaler
[In]: from pyspark.sql.functions import col

[In]: class DataPreprocessor:
          """
          Preprocesses the diabetes dataset.
          """

          def __init__(self, spark, data_file_path):
              """
```

CHAPTER 6   DEEP LEARNING WITH PYTORCH FOR CLASSIFICATION

```
        Initializes the DataPreprocessor.

        Args:
            spark (pyspark.sql.SparkSession): The Spark
            session.
            data_file_path (str): The path to the
            data file.
        """
        self.spark = spark
        self.data_file_path = data_file_path

    def preprocess(self):
        """
        Performs data preprocessing.

        Returns:
            pyspark.sql.DataFrame: The preprocessed
            DataFrame.
        """
        try:
            diabetes_df = self.spark.read.csv(
                self.data_file_path,
                header=True, inferSchema=True
            )

            diabetes_df = diabetes_df.filter(
                (col("Glucose") != 0)
                & (col("BloodPressure") != 0)
                & (col("BMI") != 0)
            )
            feature_cols = [
                "Pregnancies",
                "Glucose",
```

CHAPTER 6   DEEP LEARNING WITH PYTORCH FOR CLASSIFICATION

```
            "BloodPressure",
            "BMI",
            "DiabetesPedigreeFunction",
            "Age",
        ]
        assembler = VectorAssembler(
            inputCols=feature_cols,
            outputCol="features"
        )
        diabetes_df = assembler.transform
        (diabetes_df)
        scaler = StandardScaler(
            inputCol="features", outputCol="scaled_
            features"
        )
        scaler_model = scaler.fit(diabetes_df)
        diabetes_df = scaler_model.
        transform(diabetes_df)
        return diabetes_df
    except Exception as e:
        logging.error(
            f"Error occurred during data
            preprocessing: "
            f"{str(e)}"
        )
        raise e

[In]: class DiabetesClassifierModel(nn.Module):
        """
        Neural network model for diabetes classification.
        """
```

CHAPTER 6   DEEP LEARNING WITH PYTORCH FOR CLASSIFICATION

```
    def __init__(self, input_size, output_size):
        """
        Initializes the DiabetesClassifierModel.

        Args:
            input_size (int): The input size.
            output_size (int): The output size.
        """
        super(DiabetesClassifierModel, self).__init__()
        self.model = nn.Sequential(
            nn.Linear(input_size, 64),
            nn.ReLU(),
            nn.Linear(64, 32),
            nn.ReLU(),
            nn.Linear(32, 16),
            nn.ReLU(),
            nn.Linear(16, output_size),
        )

    def forward(self, x):
        """
        Defines the forward pass of the model.

        Args:
            x (torch.Tensor): The input tensor.

        Returns:
            torch.Tensor: The output tensor.
        """
        return self.model(x)
```

[In]:
```
class ModelTrainer:
    """
    Trainer for the diabetes classifier model.
    """
```

CHAPTER 6   DEEP LEARNING WITH PYTORCH FOR CLASSIFICATION

```
def __init__(
    self,
    model,
    criterion,
    optimizer,
    train_loader
):
    """
    Initializes the ModelTrainer.

    Args:
        model (torch.nn.Module): The neural
        network model.
        criterion (torch.nn.Module): The loss
        function.
        optimizer (torch.optim.Optimizer): The
        optimizer.
        train_loader (torch.utils.data.DataLoader):
            The training data loader.
    """
    self.model = model
    self.criterion = criterion
    self.optimizer = optimizer
    self.train_loader = train_loader

def train(self, epochs=100, lr=0.01):
    """
    Trains the model.

    Args:
        epochs (int, optional):
            The number of epochs. Defaults to 100.
```

CHAPTER 6   DEEP LEARNING WITH PYTORCH FOR CLASSIFICATION

```
            lr (float, optional):
                The learning rate. Defaults to 0.01.
        """
        try:
            for epoch in range(epochs):
                self.model.train()
                for inputs, targets in self.train_loader:
                    self.optimizer.zero_grad()
                    outputs = self.model(inputs)
                    loss = self.criterion(
                        outputs.squeeze(),
                        targets
                    )
                    loss.backward()
                    self.optimizer.step()
                if (epoch + 1) % 10 == 0:
                    logging.info(
                        f"Epoch [{epoch + 1}/"
                        f"{epochs}], Loss: "
                        f"{loss.item():.4f}"
                    )
        except Exception as e:
            logging.error(
                f"Error occurred during model training: "
                f"{str(e)}"
            )
            raise e
```

[In]: class ModelEvaluator:
    """

    Evaluator for the diabetes classifier model.
    """

CHAPTER 6 DEEP LEARNING WITH PYTORCH FOR CLASSIFICATION

```python
def __init__(self, model, test_loader, y_test_np):
    """
    Initializes the ModelEvaluator.

    Args:
        model (torch.nn.Module): The neural
        network model.
        test_loader (torch.utils.data.DataLoader):
            The test data loader.
        y_test_np (numpy.ndarray):
            The ground truth labels for test data.
    """
    self.model = model
    self.test_loader = test_loader
    self.y_test_np = y_test_np

def evaluate(self):
    """
    Evaluates the model.
    """
    try:
        with torch.no_grad():
            self.model.eval()
            predictions = []
            probabilities = []
            for inputs, _ in self.test_loader:
                outputs = self.model(inputs)
                probabilities.extend(
                    torch.sigmoid(outputs)
                    .squeeze().tolist()
                )
                predictions.extend(
```

CHAPTER 6 DEEP LEARNING WITH PYTORCH FOR CLASSIFICATION

```
            outputs.squeeze()
            .tolist()
    )

    y_pred = np.array(
        [1 if pred > 0.5 else 0
        for pred in probabilities]
    )

    accuracy = accuracy_score(
        self.y_test_np,
        y_pred
    )
    precision = precision_score(
        self.y_test_np,
        y_pred
    )
    recall = recall_score(self.y_test_
    np, y_pred)
    f1 = f1_score(self.y_test_np, y_pred)

    logging.info(f"Accuracy: {accuracy:.4f}")
    logging.info(f"Precision:
    {precision:.4f}")
    logging.info(f"Recall: {recall:.4f}")
    logging.info(f"F1 Score: {f1:.4f}")

    # Confusion Matrix
    conf_matrix = confusion_matrix(
        self.y_test_np,
        y_pred
    )
```

422

```
            logging.info(
                "Confusion Matrix:\n" +
                str(conf_matrix)
            )

            # ROC-AUC Score
            roc_auc = roc_auc_score(
                self.y_test_np,
                probabilities
            )
            logging.info(f"ROC-AUC Score: {roc_
            auc:.4f}")

            # Return the metrics as a dictionary
            return {
                'accuracy': accuracy,
                'precision': precision,
                'recall': recall,
                'f1': f1,
                'roc_auc': roc_auc
            }

        except Exception as e:
            logging.error(
                f"Error occurred during model
                evaluation: "
                f"{str(e)}")
            raise e
```

[In]:
```
def copy_file_from_s3(bucket_name, file_name, local_dir):
    """
    Copies a file from S3 bucket to local directory.
```

CHAPTER 6 DEEP LEARNING WITH PYTORCH FOR CLASSIFICATION

```
        Args:
            bucket_name (str): The name of the S3 bucket.
            file_name (str): The name of the file to
            be copied.
            local_dir (str):
                The local directory where the file will
                be copied.
        """
        try:
            s3 = boto3.client("s3")
            s3.download_file(
                bucket_name,
                file_name,
                local_dir + file_name
            )
        except Exception as e:
            logging.error(
                f"Error occurred during file copying: {str(e)}"
            )
            raise e

[In]:   if __name__ == "__main__":
            logging.basicConfig(level=logging.INFO)

            spark = SparkSession.builder.appName("DiabetesClassifier").getOrCreate()

            # Copy file from S3
            copy_file_from_s3(
                "instance1bucket",
```

```python
    "diabetes.csv",
    "/home/ubuntu/airflow/dags/"
)
data_preprocessor = DataPreprocessor(
    spark, "/home/ubuntu/airflow/dags/diabetes.csv"
)
diabetes_df = data_preprocessor.preprocess()
# Convert Spark DataFrame to numpy arrays
X_np = np.array(
    diabetes_df
    .select("scaled_features")
    .rdd.flatMap(lambda x: x).collect(),
    dtype=np.float32,
)
y_np = np.array(
    diabetes_df
    .select("Outcome")
    .rdd.flatMap(lambda x: x)
    .collect(),
    dtype=np.float32,
)

# Implementing K-fold cross-validation
k_folds = 5
kfold = KFold(
    n_splits=k_folds,
    shuffle=True,
    random_state=42
)
```

```
fold_results = []

for fold, (
    train_idx,
    test_idx) in enumerate(kfold.split(X_np)):
    logging.info(f"Training fold {fold + 1}/
    {k_folds}"
)

    X_train, X_test = X_np[train_idx], X_np[test_idx]
    y_train, y_test = y_np[train_idx], y_np[test_idx]
    # Convert to PyTorch datasets
    train_dataset = TensorDataset(
        torch.tensor(X_train),
        torch.tensor(y_train)
    )
    test_dataset = TensorDataset(
        torch.tensor(X_test),
        torch.tensor(y_test)
    )

    # Create DataLoaders
    train_loader = DataLoader(
        train_dataset,
        batch_size=64,
        shuffle=True
    )
    test_loader = DataLoader(
        test_dataset,
        batch_size=64,
        shuffle=False
    )
```

CHAPTER 6   DEEP LEARNING WITH PYTORCH FOR CLASSIFICATION

```
    # Initialize the model, criterion, and optimizer
    input_size = X_train.shape[1]
    output_size = 1
    model = DiabetesClassifierModel(input_size,
    output_size)
    criterion = nn.BCEWithLogitsLoss()
    optimizer = optim.Adam(model.parameters(),
    lr=0.01)

    # Train the model
    model_trainer = ModelTrainer(
        model,
        criterion,
        optimizer,
        train_loader
    )
    model_trainer.train()

    # Evaluate the model
    model_evaluator = ModelEvaluator(
        model,
        test_loader,
        y_test
    )
    results = model_evaluator.evaluate()

    # appended results to the fold_results list
    fold_results.append(results)

# Aggregate results across folds
avg_accuracy = np.mean(
    [result['accuracy'] for result in fold_results])
avg_precision = np.mean(
```

427

```
        [result['precision'] for result in fold_results])
    avg_recall = np.mean(
        [result['recall'] for result in fold_results])
    avg_f1 = np.mean(
        [result['f1'] for result in fold_results])
    avg_roc_auc = np.mean(
        [result['roc_auc'] for result in fold_results])

    logging.info(f"Average Accuracy: {avg_accuracy:.4f}")
    logging.info(f"Average Precision: {avg_
    precision:.4f}")
    logging.info(f"Average Recall: {avg_recall:.4f}")
    logging.info(f"Average F1 Score: {avg_f1:.4f}")
    logging.info(f"Average ROC-AUC Score: {avg_roc_
    auc:.4f}")
```

Summary

In this chapter, we examined an important task in deep learning: classification. We constructed, trained, and evaluated a PyTorch multilayer algorithm, utilizing it to predict the probability of diabetes diagnosis based on various attributes such as the number of pregnancies, glucose levels, blood pressure, Body Mass Index (BMI), age, and family history of diabetes.

We employed Spark for data preprocessing to highlight its distributed computing capabilities, crucial for managing large datasets. Despite using a relatively small dataset, we covered essential topics in machine learning, including data augmentation, transfer learning, data quality, and model complexity. Additionally, to ensure a more reliable evaluation of our model's performance, we implemented K-fold cross-validation, which allowed us to assess the model across multiple subsets of the data, reducing the risk of overfitting or underfitting.

CHAPTER 6   DEEP LEARNING WITH PYTORCH FOR CLASSIFICATION

Our code development emphasized key modeling steps such as data loading, preprocessing, defining the model architecture, training, and evaluation. Additionally, we incorporated extra functionality such as copying data from an S3 bucket to an EC2 local directory for processing, showcasing integration with AWS services, and handling errors and log messages for debugging and monitoring purposes throughout the modeling process.

In the next chapter, we will continue exploring the topic of deep learning classification, this time using TensorFlow. We will demonstrate how to replicate the same steps used with PyTorch.

CHAPTER 7

Deep Learning with TensorFlow for Classification

In the previous chapter, we constructed, trained, and evaluated a PyTorch multilayer deep learning model and used it to predict the probability of a diabetes diagnosis. Our code development emphasized key modeling steps such as data loading, preprocessing, defining the model architecture, training, and evaluation. Additionally, we incorporated extra functionality such as copying data from an S3 bucket to an EC2 local directory for processing, showcasing integration with AWS services, and handling errors and log messages for debugging and monitoring purposes throughout the modeling process. We also leveraged PySpark for data processing, showcasing its distributed computing capabilities.

In this chapter, we continue exploring the topic of deep learning classification, this time using TensorFlow. We demonstrate how to replicate the same steps used with PyTorch. To enable an apples-to-apples comparison of the two models' performance, we utilize the same Pima diabetes dataset used for PyTorch.

CHAPTER 7   DEEP LEARNING WITH TENSORFLOW FOR CLASSIFICATION

Furthermore, we introduce the use of keras_tuner for hyperparameter tuning, allowing us to automate the search for the optimal model architecture and learning rate. This approach enhances our model training process by potentially yielding better performance metrics than manually chosen hyperparameters.

The chapter begins with a review of what the diabetes dataset looked like. It then builds, trains, and evaluates a multilayer deep learning classification model in TensorFlow, comparing its performance with the results from PyTorch in the previous chapter. After evaluating the model with fixed hyperparameters and comparing its performance with PyTorch, we proceed to fine-tune it to further enhance its performance. Next, the chapter consolidates the code used throughout the chapter into a unified section. Finally, the chapter concludes with a summary of the key takeaways.

The Dataset

We explored the Pima diabetes dataset in great detail in the previous chapter. In this section, we write code to copy the CSV file from an S3 bucket to a local directory on EC2 and then print the first ten records to refresh our memory of what the features and target variable look like:

```
[In]: from pyspark.sql import SparkSession
[In]: import os

[In]: spark = SparkSession.builder \
            .appName("Read CSV from S3") \
            .getOrCreate()

[In]: bucket_name = "instance1bucket"
[In]: file_key = "diabetes.csv"
[In]: local_file_path = "/home/ubuntu/airflow/dags/
       diabetes.csv"
```

CHAPTER 7 DEEP LEARNING WITH TENSORFLOW FOR CLASSIFICATION

```
[In]: os.system(
        f"aws s3 cp s3://{bucket_name}/{file_key} "
        f"{local_file_path}"
    )
[In]: df = spark.read.csv(
            local_file_path,
            header=True,
            inferSchema=True
    )
[In]: selected_columns = [
        'Pregnancies',
        'Glucose',
        'BloodPressure',
        'BMI',
        'DiabetesPedigreeFunction',
        'Age',
        'Outcome'
    ]
[In]: df_selected = df.select(selected_columns)
[In]: df_selected.show(10)

[In]: spark.stop()
```

We will now explore the code step by step, starting by importing necessary libraries:

```
[In]: from pyspark.sql import SparkSession
[In]: import os
```

CHAPTER 7   DEEP LEARNING WITH TENSORFLOW FOR CLASSIFICATION

- from pyspark.sql import SparkSession: This imports the SparkSession class from the pyspark.sql module. SparkSession is the entry point to programming Spark with the Dataset and DataFrame API.

- import os: This imports the built-in os module, which provides a portable way to use operating system–dependent functionality.

We then create a new SparkSession with the specified app name "Read CSV from S3" or retrieve an existing one if available. The appName method sets the name of the Spark application:

```
[In]: spark = (SparkSession.builder
            .appName("Read CSV from S3")
            .getOrCreate())
```

We utilize parentheses () to facilitate code readability across multiple lines. Alternatively, one may opt for the backslash character \ at the end of each line.

In the next step, the code defines S3 bucket and file details:

```
[In]: bucket_name = "instance1bucket"
[In]: file_key = "diabetes.csv"
[In]: local_file_path = "/home/ubuntu/airflow/dags/
      diabetes.csv"
```

These variables hold the S3 bucket name, file key (path to the file within the bucket), and the local file path where the file will be copied to.

Moving forward, the code uses the os.system function to execute a shell command. It copies the file diabetes.csv from the S3 bucket specified by bucket_name and file_key to the local directory specified by local_file_path using the AWS CLI command aws s3 cp:

CHAPTER 7   DEEP LEARNING WITH TENSORFLOW FOR CLASSIFICATION

```
[In]: os.system(
        f"aws s3 cp s3://{bucket_name}/{file_key} "
        f"{local_file_path}"
    )
```

Th next line reads the CSV file located at local_file_path into a DataFrame (df) using Spark's read.csv method. It specifies that the first row of the CSV file contains headers (header=True) and infers the schema of the DataFrame from the data (inferSchema=True):

```
[In]: df = spark.read.csv(
        local_file_path,
        header=True,
        inferSchema=True
    )
```

The code then selects the appropriate columns (recall from the previous chapter that we excluded SkinThickness and Insulin because they had many invalid readings):

```
[In]: selected_columns = [
        'Pregnancies',
        'Glucose',
        'BloodPressure',
        'BMI',
        'DiabetesPedigreeFunction',
        'Age',
        'Outcome'
    ]
[In]: df_selected = df.select(selected_columns)
```

CHAPTER 7   DEEP LEARNING WITH TENSORFLOW FOR CLASSIFICATION

The next line displays the first ten records of the DataFrame df_selected using the show() method:

[In]: df_selected.show(10)

Finally, we stop the SparkSession to release the resources allocated to it. It's good practice to stop the SparkSession after finishing the Spark jobs:

[In]: spark.stop()

To execute the code, we saved it as "print_top10_records.py" and ran it in a JupyterLab terminal. After creating an environment named "myenv," we activated it using the following command:

[In]: source /home/ubuntu/myenv/bin/activate

Once the environment was activated, we executed the code using "spark-submit" with the following command:

[In]: spark-submit print_top10_ records.py

Alternatively, we could have executed the code using the Python 3 interpreter with the following command:

[In]: Python3 print_top10_ records.py

The output from the code is a table that displays the top ten rows of the Pima diabetes dataset:

[Out]:

Pregnancies	Glucose	BloodPressure	BMI	DiabetesPedigree Function	Age	Outcome
6	148	72	33.6	0.627	50	1
1	85	66	26.6	0.351	31	0
8	183	64	23.3	0.672	32	1
1	89	66	28.1	0.167	21	0
0	137	40	43.1	2.288	33	1
5	116	74	25.6	0.201	30	0
3	78	50	31.0	0.248	26	1
10	115	0	35.3	0.134	29	0
2	197	70	30.5	0.158	53	1
8	125	96	0.0	0.232	54	1

Here's an explanation of each column of the table:

- Pregnancies: Number of times pregnant.
- Glucose: Plasma glucose concentration at 2 hours in an oral glucose tolerance test.
- BloodPressure: Diastolic blood pressure (mm Hg).
- BMI (Body Mass Index): Body Mass Index, which is a measure of body fat based on height and weight.
- DiabetesPedigreeFunction: A function that scores likelihood of diabetes based on family history.
- Age: Age of the individual in years.

CHAPTER 7   DEEP LEARNING WITH TENSORFLOW FOR CLASSIFICATION

- Outcome: This is the target variable. It indicates whether the individual has diabetes or not. 1 means the person has diabetes, and 0 means the person does not have diabetes.

Notice that BloodPressure and BMI have rows with 0 values, indicating invalid readings. These will be excluded from the dataset during model training and evaluation.

Predicting Diabetes with TensorFlow

In this section, we utilize the diabetes dataset described in the previous section to train and evaluate a TensorFlow deep learning model to predict the likelihood of a Pima Indian woman having diabetes. The modeling steps largely mirror those of the previous chapter, where we developed a similar classification model using PyTorch. As in the previous chapter, we leverage PySpark for data preprocessing.

The code will map to the following modeling steps:

Step 1: Imports

- Importing necessary libraries and modules such as logging, PySpark, TensorFlow, and Scikit-Learn

Step 2: Logging Configuration

- Configuring logging settings such as the logging level and creating a logger object

Step 3: Custom Callback

- Implementing a custom Keras callback to print epoch information during model training

CHAPTER 7 DEEP LEARNING WITH TENSORFLOW FOR CLASSIFICATION

Step 4: Data Preprocessing

- Preprocessing the diabetes data using PySpark, including tasks such as reading data, filtering, feature engineering, scaling, and splitting into train and test sets

Step 5: Model Definition and Training

- Defining the architecture of a TensorFlow (Keras) sequential model and training the model using the preprocessed data

Step 6: Model Evaluation

- Evaluating the trained TensorFlow (Keras) model using test data and calculating evaluation metrics such as accuracy, precision, recall, and F1 score

Step 7: Main

- Defining the if __name__ == "__main__": block, serving as the main entry point of the script where data preprocessing, model training, and evaluation are orchestrated

Even though these steps are largely similar to PyTorch, the reader will notice that Keras often provides a higher-level interface compared with PyTorch, which can result in shorter code for certain tasks. Keras abstracts away many implementation details and provides a more intuitive and user-friendly API for building and training neural networks.

For example, the TensorFlow (Keras) code for defining the model architecture and training the model is concise and straightforward, as Keras offers pre-built layers and modules, making it easy to define neural network architectures with just a few lines of code. On the other hand, PyTorch is designed to be more flexible and customizable, which can lead to longer code snippets, especially when defining complex models

CHAPTER 7　DEEP LEARNING WITH TENSORFLOW FOR CLASSIFICATION

or implementing custom training loops. PyTorch provides a lower-level API that gives users more control over the model's behavior and training process.

When it comes to model training, Keras provides a high-level method model.fit() that automates much of the training process. This method abstracts away details such as iterating over epochs and batches, calculating loss, and performing optimization steps. This makes it easy to train models with minimal boilerplate code, which is especially beneficial for rapid prototyping and experimentation.

In contrast, PyTorch requires users to manually implement the training loop, iterating over epochs and batches, calculating loss, and performing optimization steps explicitly. While this approach may be more verbose, it offers greater flexibility and control over the training process, allowing for the implementation of custom training procedures and experimentation with different optimization techniques.

For model evaluation, Keras provides a convenient built-in method model.evaluate() for evaluating model performance. This method computes evaluation metrics such as accuracy, loss, and any other metrics specified during model compilation. This streamlined approach simplifies the evaluation process and makes it easy to obtain performance metrics directly from the model.

In PyTorch, model evaluation typically involves manual computation of evaluation metrics using external libraries such as Scikit-Learn or by implementing custom evaluation functions. While PyTorch does not have a direct equivalent to model.evaluate(), it offers the flexibility to customize the evaluation process according to specific requirements, allowing for more fine-grained control over evaluation metrics and procedures.

CHAPTER 7 DEEP LEARNING WITH TENSORFLOW FOR CLASSIFICATION

Let's now delve into the TensorFlow code step by step, starting with imports:

Step 1: Imports

```
[In]: import logging
[In]: from pyspark.sql import SparkSession
[In]: from pyspark.ml.feature import VectorAssembler, StandardScaler
[In]: from pyspark.sql.functions import col
[In]: from sklearn.metrics import (
          accuracy_score,
          precision_score,
          recall_score,
          f1_score
      )
[In]: from tensorflow.keras.models import Sequential
[In]: from tensorflow.keras.layers import Dense, ReLU
[In]: from tensorflow.keras.optimizers import SGD
[In]: from tensorflow.keras.callbacks import Callback
[In]: import numpy as np
```

Below are explanations of what these imports mean and where they will be used in the upcoming code:

- import logging: This is used for setting up logging configuration and logging messages throughout the script. It will be used in the upcoming code within the DiabetesProcessor and DiabetesModelTrainer classes for logging information about data preprocessing and model training. It will also be used within the CustomCallback class for logging epoch information during model training.

CHAPTER 7   DEEP LEARNING WITH TENSORFLOW FOR CLASSIFICATION

- from pyspark.sql import SparkSession: This is utilized for creating a SparkSession, essential for interacting with Spark and performing DataFrame operations. It is used in the main script to create a SparkSession for data preprocessing and model training.

- from pyspark.ml.feature import VectorAssembler, StandardScaler: These are required for assembling feature vectors and scaling features during data preprocessing with PySpark. They are used within the DiabetesProcessor class to assemble feature vectors and scale features before training the model.

- from pyspark.sql.functions import col: This is used for accessing DataFrame columns by name during data filtering and manipulation. It is used within the DiabetesProcessor class to filter out rows where certain columns are zero.

- from sklearn.metrics import accuracy_score, precision_score, recall_score, f1_score: These imports are utilized for computing evaluation metrics such as accuracy, precision, recall, and F1 score for the trained model on test data. They are used within the DiabetesModelEvaluator class to compute evaluation metrics for the trained TensorFlow (Keras) model.

- from tensorflow.keras.models import Sequential: This is used for defining a Sequential model architecture in TensorFlow (Keras) for training the neural network. It is used within the DiabetesModelTrainer class to define the architecture of the neural network model.

CHAPTER 7 DEEP LEARNING WITH TENSORFLOW FOR CLASSIFICATION

- from tensorflow.keras.layers import Dense, ReLU: These are required for specifying fully connected (Dense) layers and activation functions (ReLU) in the neural network architecture. They are used within the DiabetesModelTrainer class to specify the layers of the neural network model.

- from tensorflow.keras.optimizers import SGD: This is utilized for specifying the stochastic gradient descent (SGD) optimizer during the compilation of the Keras model. It's used in the code within the DiabetesModelTrainer class to specify the optimizer for training the neural network model.

- from tensorflow.keras.callbacks import Callback: This is used for defining a custom Keras callback (CustomCallback) for logging epoch information during model training. It is used within the DiabetesModelTrainer class.

- import numpy as np: This imports the NumPy library under the alias np. NumPy is a fundamental package for scientific computing in Python, providing support for multidimensional arrays and mathematical functions. It will be used in the code for converting PySpark DataFrame columns to NumPy arrays for model training and evaluation.

Step 2: Logging Configuration

The two lines of code below set up logging configuration and create a logger object in Python:

[In]: logging.basicConfig(level=logging.INFO)
[In]: logger = logging.getLogger(__name__)

CHAPTER 7 DEEP LEARNING WITH TENSORFLOW FOR CLASSIFICATION

The first line, logging.basicConfig(level=logging.INFO), initializes the logging system and configures it to log messages with a level of INFO or higher. This means that only messages with a severity level of INFO, WARNING, ERROR, or CRITICAL will be logged, while messages with a DEBUG level will be ignored. The basicConfig() function allows us to configure the root logger, which is the main logger used by the Python logging module.

The second line, logger = logging.getLogger(__name__), creates a logger object named logger. The getLogger() function is used to obtain a logger instance, and __name__ is a special variable in Python that represents the name of the current module. By using __name__, the logger will have the same name as the current module in which it is defined. This helps in identifying the source of log messages, as the logger name will correspond to the module name.

Step 3: Custom Callback

The code below defines a custom Keras callback named CustomCallback in Python:

```
[In]: class CustomCallback(Callback):
          """Custom Keras callback for printing epoch
          information."""
          def on_epoch_end(self, epoch, logs=None):
              """Prints epoch information every 10 epochs."""
              if (epoch + 1) % 10 == 0:
                  logger.info(
                      f"Epoch [{epoch + 1}/{self.
                      params['epochs']}], "
                      f"Loss: {logs['loss']:.4f}"
                  )
```

444

CHAPTER 7   DEEP LEARNING WITH TENSORFLOW FOR CLASSIFICATION

The CustomCallback class inherits from the Callback class provided by the Keras library, allowing for custom actions to be performed during the training of a neural network model:

- The docstring """Custom Keras callback for printing epoch information.""" provides a brief description of the purpose of this callback.

- Inside the class, there is a method on_epoch_end(self, epoch, logs=None), which is called by Keras at the end of each epoch during model training. This method prints epoch information every ten epochs.

- Within the on_epoch_end method, there's a conditional statement that checks if the current epoch number plus one is divisible by 10. If it is, it logs information about the current epoch using the logger. info() function. The logged information includes the epoch number, total number of epochs, and the loss value from the logs dictionary.

Step 4: Data Preprocessing

In this step, we preprocess the Pima diabetes data before we feed it into the algorithm. The code below defines a Python class called DiabetesProcessor, which is responsible for preprocessing this data using PySpark:

```
[In]: class DiabetesProcessor:
          """Class for preprocessing diabetes data."""
          @staticmethod
          def preprocess_data(spark, data_file_path,
              train_parquet_path, test_parquet_path):
              """Preprocesses the diabetes data using
              PySpark."""
              try:
```

CHAPTER 7 DEEP LEARNING WITH TENSORFLOW FOR CLASSIFICATION

```
diabetes_df = spark.read.csv(
    data_file_path,
    header=True,
    inferSchema=True
)
diabetes_df = diabetes_df.filter(
    (col("Glucose") != 0)
    & (col("BloodPressure") != 0)
    & (col("BMI") != 0)
)
feature_cols = [
    "Pregnancies",
    "Glucose",
    "BloodPressure",
    "BMI",
    "DiabetesPedigreeFunction",
    "Age"
]
assembler = VectorAssembler(
    inputCols=feature_cols,
    outputCol="features"
)
diabetes_df = assembler.transform(diabetes_df)
scaler = StandardScaler(
    inputCol="features",
    outputCol="scaled_features"
)
scaler_model = scaler.fit(diabetes_df)
diabetes_df = scaler_model.transform(diabetes_df)
```

CHAPTER 7 DEEP LEARNING WITH TENSORFLOW FOR CLASSIFICATION

```
        train_df, test_df = diabetes_df.randomSplit(
            [0.8, 0.2],
            seed=42
        )
        train_df.write.parquet(
            train_parquet_path,
            mode="overwrite"
        )
        test_df.write.parquet(
            test_parquet_path,
            mode="overwrite"
        )
        logger.info("Data preprocessing completed.")
    except Exception as e:
        logger.error(
            f"Error occurred during data "
            f"preprocessing: {str(e)}"
        )
```

Below is more information about the DiabetesProcessor class and the preprocess_data method inside it:

- The docstring """"Class for preprocessing diabetes data."""" provides a brief description of the purpose of this class.

- Inside the class, there's a static method preprocess_data(spark, data_file_path, train_parquet_path, test_parquet_path), which performs the actual data preprocessing. This method takes four arguments: spark (the SparkSession object), data_file_path (the path to the diabetes data file), train_parquet_path (the path to save the preprocessed training data), and test_

CHAPTER 7   DEEP LEARNING WITH TENSORFLOW FOR CLASSIFICATION

parquet_path (the path to save the preprocessed test data). The @staticmethod decorator is used to declare this method as a static method, indicating that it does not require access to the instance or its attributes and can be called directly on the class without needing to create an instance of DiabetesProcessor.

- Within the method, the diabetes data is read from a CSV file using Spark's read.csv() method. Subsequently, it is filtered to remove rows where the Glucose, BloodPressure, and BMI columns have a value of zero, indicating invalid readings.

- Feature engineering is performed next, where the selected feature columns (including "Pregnancies", "Glucose", "BloodPressure", "BMI", "DiabetesPedigreeFunction", and "Age") are assembled into a single feature vector using VectorAssembler. This vectorization process prepares the data for further analysis. The inputCols parameter of VectorAssembler specifies the columns to be assembled, and the outputCol parameter specifies the name of the output column containing the assembled feature vector.

- The assembled feature vector is then scaled using StandardScaler. The inputCol parameter specifies the column containing the feature vectors to be scaled, and the outputCol parameter specifies the name of the output column containing the scaled feature vectors.

- The transform function is called on the VectorAssembler and StandardScaler objects to apply the transformations to the diabetes data DataFrame. This function applies the transformations defined by

448

CHAPTER 7   DEEP LEARNING WITH TENSORFLOW FOR CLASSIFICATION

the VectorAssembler and StandardScaler to the input DataFrame and returns a new DataFrame with the transformed features.

- The preprocessed data is then split into training and testing sets using PySpark's randomSplit() method. This ensures that the model is trained on a portion of the data (80% in this case) and evaluated on another (20%), facilitating unbiased model assessment.

- Finally, the preprocessed data is written to Parquet files for later use. Saving data as Parquet in PySpark offers several benefits, including efficient storage, columnar storage format, built-in compression, and compatibility with big data processing frameworks.

- Error handling is implemented using a try–except block to catch any exceptions that may occur during the preprocessing process. If an exception occurs, an error message is logged using logger.error(). This ensures that any issues encountered during data preprocessing are appropriately handled and logged for debugging purposes.

Note Saving data to Parquet format is a widely adopted method for handling large datasets in big data environments. Parquet's columnar storage format stores data by columns rather than rows, making it highly efficient for queries that target specific columns, thereby reducing disk reads and speeding up query performance. It also offers better compression rates than formats like CSV or JSON, resulting in smaller file sizes and faster I/O operations—key benefits for large-scale data processing.

449

Parquet supports schema evolution, allowing you to add or remove columns without rewriting the entire dataset, which is invaluable in dynamic production environments. Its compatibility with big data frameworks like Apache Spark, Hadoop, and AWS Glue ensures efficient processing across distributed systems. The columnar nature of Parquet, combined with features like predicate pushdown—where filtering occurs as early as possible during the data reading phase to minimize the amount of data that needs to be processed—optimizes data scans by skipping irrelevant blocks, further enhancing performance. Additionally, Parquet's support for complex data types, such as nested structures, arrays, and maps, adds versatility in storing and querying sophisticated data models.

While Parquet is excellent for storing and querying large datasets, it's worth noting that alternative methods like AWS Glue and Spark SQL can further enhance data management and processing. Although these tools fall outside the scope of this book, AWS Glue's ability to automatically discover, catalog, and transform data and Spark SQL's capability to perform complex transformations and queries directly within Spark applications are powerful options to consider.

Step 5: Model Definition and Training

The code in this step defines the DiabetesModelTrainer class for training the TensorFlow (Keras) model:

```
[In]: class DiabetesModelTrainer:
          """Class for training the TensorFlow (Keras)
          model."""
          @staticmethod
          def train_tensorflow_model(
              X_train,
```

CHAPTER 7   DEEP LEARNING WITH TENSORFLOW FOR CLASSIFICATION

```
        y_train,
        epochs=100,
        lr=0.01
):
    """Trains the TensorFlow (Keras) model."""
    try:
        model = Sequential([
            Dense(64, input_dim=X_train.shape[1]),
            ReLU(),
            Dense(32),
            ReLU(),
            Dense(16),
            ReLU(),
            Dense(1, activation='sigmoid')
        ])
        optimizer = SGD(learning_rate=lr)
        model.compile(loss='binary_crossentropy',
                      optimizer=optimizer,
                      metrics=['accuracy'])
        custom_callback = CustomCallback()
        model.fit(
            X_train,
            y_train,
            epochs=epochs,
            verbose=0,
            callbacks=[custom_callback]
        )
        logger.info("Model training completed.")
        return model
```

```
            except Exception as e:
                logger.error(
                    f"Error occurred during "
                    f"model training: {str(e)}"
                )
```

In this code

- The docstring """Class for training the TensorFlow (Keras) model.""" provides a brief description of the purpose of this class.

- Inside the class, there's a static method train_tensorflow_model(X_train, y_train, epochs=100, lr=0.01), which is responsible for training the TensorFlow (Keras) model. This method takes four arguments: X_train (the input features for training), y_train (the target labels for training), epochs (the number of epochs for training, with a default value of 100), and lr (the learning rate for the optimizer, with a default value of 0.01). The @staticmethod decorator is used to declare this method as a static method, indicating that it does not require access to the instance or its attributes and can be called directly on the class without needing to create an instance of DiabetesModelTrainer.

- Within the method, a neural network model is defined using Keras's Sequential API. This model consists of several fully connected (Dense) layers with rectified linear unit (ReLU) activation functions, ending with an output layer with a sigmoid activation function.

CHAPTER 7 DEEP LEARNING WITH TENSORFLOW FOR CLASSIFICATION

- An optimizer (stochastic gradient descent with a specified learning rate) is configured, and the model is compiled with a binary cross-entropy loss function and accuracy as the evaluation metric.

- A custom callback (CustomCallback) is instantiated to log information during model training.

- The fit method is called on the model object with the input features (X_train) and target labels (y_train) to train the model for the specified number of epochs. The custom_callback is passed to the callbacks argument to enable logging of epoch information during training.

- Upon successful completion of model training, a log message is generated indicating that the model training has been completed, and the trained model is returned.

- Error handling is implemented using a try–except block to catch any exceptions that may occur during the model training process. If an exception occurs, an error message is logged using logger.error(). This ensures that any issues encountered during model training are appropriately handled and logged for debugging purposes.

Step 6: Model Evaluation

In this step, we define the DiabetesModelEvaluator class, which is responsible for evaluating the TensorFlow (Keras) model trained for diabetes prediction. It includes methods for assessing the performance of the model using various evaluation metrics such as accuracy, precision, recall, and F1 score:

CHAPTER 7   DEEP LEARNING WITH TENSORFLOW FOR CLASSIFICATION

```
[In]: class DiabetesModelEvaluator:
          """Class for evaluating the TensorFlow (Keras)
          model."""
          @staticmethod
          def evaluate_model(model, X_test, y_test):
              """Evaluates the TensorFlow (Keras) model."""
              try:
                  y_pred = (model.predict(X_test) > 0.5).
                  astype(int)
                  accuracy = accuracy_score(y_test, y_pred)
                  precision = precision_score(y_test, y_pred)
                  recall = recall_score(y_test, y_pred)
                  f1 = f1_score(y_test, y_pred)
                  logger.info(f"Accuracy: {accuracy:.4f}")
                  logger.info(f"Precision: {precision:.4f}")
                  logger.info(f"Recall: {recall:.4f}")
                  logger.info(f"F1 Score: {f1:.4f}")
              except Exception as e:
                  logger.error(
                      f"Error occurred during model "
                      f"evaluation: {str(e)}"
                  )
```

Let's take a closer look at the code:

- The docstring """Class for evaluating the TensorFlow (Keras) model.""" provides a brief description of the purpose of this class.

- Inside the class, there's a static method evaluate_model(model, X_test, y_test), which evaluates the performance of the TensorFlow (Keras) model. This method takes three arguments: model (the

454

CHAPTER 7 DEEP LEARNING WITH TENSORFLOW FOR CLASSIFICATION

trained Keras model), X_test (the input features for testing), and y_test (the true labels for testing). The @staticmethod decorator is used to declare this method as a static method, indicating that it does not require access to the instance or its attributes and can be called directly on the class without needing to create an instance of DiabetesModelEvaluator.

- Within the method, predictions are generated using the trained model (model.predict(X_test)) and thresholded to obtain binary predictions. Several evaluation metrics, including accuracy, precision, recall, and F1 score, are computed using Scikit-Learn's metric functions (accuracy_score, precision_score, recall_score, and f1_score, respectively).

- Evaluation results, including accuracy, precision, recall, and F1 score, are logged using the logger at an information level to provide insights into the model's performance.

- Error handling is implemented using a try-except block to catch any exceptions that may occur during the evaluation process. If an exception occurs, an error message is logged using logger.error(). This ensures that any issues encountered during model evaluation are appropriately handled and logged for debugging purposes.

CHAPTER 7   DEEP LEARNING WITH TENSORFLOW FOR CLASSIFICATION

Step 7: Main

In this final step, we define the main function, which ensures that the following code is only executed when the script is run directly, not when it's imported as a module into another script:

```
[In]: if __name__ == "__main__":
          try:
              spark = (SparkSession.builder
                      .appName("DiabetesClassification")
                      .config("spark.executor.instances", "4")
                      .config("spark.executor.memory", "4g")
                      .config("spark.executor.cores", "2")
                      .config("spark.driver.memory", "2g")
                      .getOrCreate())

              data_file_path = "/home/ubuntu/airflow/dags/diabetes.csv"
              train_parquet_path = \
                  "/home/ubuntu/airflow/dags/diabetes_train.parquet"
              test_parquet_path = \
                  "/home/ubuntu/airflow/dags/diabetes_test.parquet"

              DiabetesProcessor.preprocess_data(
                  spark,
                  data_file_path,
                  train_parquet_path,
                  test_parquet_path
              )

              train_diabetes_df = spark.read.parquet(
                  train_parquet_path
              )
```

```
    X_train = np.array(train_diabetes_df.select(
                "scaled_features").rdd.flatMap(
                lambda x: x).collect()
    )
    y_train = np.array(
        train_diabetes_df.select("Outcome")
        .rdd.flatMap(lambda x: x).collect()
    )

    trained_model =
        DiabetesModelTrainer.train_tensorflow_model(
        X_train,
        y_train
    )

    test_diabetes_df = spark.read.parquet(test_
    parquet_path)
    X_test = np.array(
        test_diabetes_df.select("scaled_features")
        .rdd.flatMap(lambda x: x).collect()
    )
    y_test = np.array(
        test_diabetes_df.select("Outcome")
        .rdd.flatMap(lambda x: x).collect()
    )

    DiabetesModelEvaluator.evaluate_model(
        trained_model,
        X_test,
        y_test
    )
except Exception as e:
    logger.error(f"An error occurred: {str(e)}")
```

CHAPTER 7   DEEP LEARNING WITH TENSORFLOW FOR CLASSIFICATION

Below is an explanation of this code:

- The if __name__ == "__main__": block is the entry point of the Python script. It ensures that the following code is only executed when the script is run directly, not when it's imported as a module into another script.

- Inside the try block, several steps are executed to preprocess diabetes data, train a TensorFlow (Keras) model, and evaluate its performance. These steps are enclosed in a try-except block to handle any potential exceptions.

- First, a SparkSession is created using the SparkSession.builder API. The SparkSession serves as the entry point to Spark functionality, allowing the script to interact with Spark and perform distributed data processing.

- The SparkSession is configured with various properties such as the application name (appName), number of executor instances (spark.executor.instances), executor memory (spark.executor.memory), executor cores (spark.executor.cores), and driver memory (spark.driver.memory). These configurations optimize resource allocation for data processing tasks.

- Paths to the input diabetes data file (data_file_path), preprocessed training data Parquet file (train_parquet_path), and preprocessed test data Parquet file (test_parquet_path) are defined.

- The DiabetesProcessor.preprocess_data() method is called to preprocess the diabetes data using PySpark. This method reads the raw diabetes data from the CSV file, applies data cleaning and feature engineering

CHAPTER 7   DEEP LEARNING WITH TENSORFLOW FOR CLASSIFICATION

transformations, and writes the preprocessed data to Parquet files for efficient storage and retrieval.

- The preprocessed training data is then loaded into a Spark DataFrame (train_diabetes_df) from the Parquet file, and the input features (X_train) and target labels (y_train) are extracted from the DataFrame. These features and labels are converted to NumPy arrays for training the TensorFlow (Keras) model.

- The DiabetesModelTrainer.train_tensorflow_model() method is called to train the TensorFlow (Keras) model using the preprocessed training data. This method constructs, compiles, and trains a neural network model using the Sequential API provided by Keras.

- After training the model, the preprocessed test data is loaded into a Spark DataFrame (test_diabetes_df) from the Parquet file, and the input features (X_test) and target labels (y_test) are extracted from the DataFrame. Similarly, these features and labels are converted to NumPy arrays for model evaluation.

- Finally, the DiabetesModelEvaluator.evaluate_model() method is called to evaluate the performance of the trained TensorFlow (Keras) model using the test data. This method computes evaluation metrics such as accuracy, precision, recall, and F1 score and logs the results using the logger.

- If any exceptions occur during the execution of the try block, they are caught and logged using logger.error(). This ensures that any errors or issues encountered during data preprocessing, model training, or evaluation are appropriately handled and logged for debugging purposes.

CHAPTER 7 DEEP LEARNING WITH TENSORFLOW FOR CLASSIFICATION

To run the code, we saved it as "diabetes_classification_tensorflow.py" and executed it in a JupyterLab terminal. After creating an environment named "myenv," we activated it with the following command:

[In]: source /home/ubuntu/myenv/bin/activate

Once the environment was activated, we executed the code using Python 3 with the following command:

[In]: python3 diabetes_classification_tensorflow.py

The first output from the code is the Epoch and Loss table below. This shows the loss value at every tenth epoch during the training process of the TensorFlow (Keras) model:

Epoch	Loss
10/100	0.6353
20/100	0.5995
30/100	0.5723
40/100	0.5581
50/100	0.5513
60/100	0.5472
70/100	0.5316
80/100	0.5275
90/100	0.5181
100/100	0.5123

CHAPTER 7   DEEP LEARNING WITH TENSORFLOW FOR CLASSIFICATION

In this table

- The epoch number indicates the current iteration of the training process out of the total number of epochs specified.

- The loss value represents the error or discrepancy between the predicted outcomes and the actual outcomes. It is a measure of how well the model is performing, with lower values indicating better performance.

The second output is the metrics table. This presents various evaluation metrics calculated after testing the trained TensorFlow (Keras) model on the test dataset:

Metric	Value
Accuracy	0.7692
Precision	0.7059
Recall	0.5854
F1 Score	0.6400

Below are the definitions of the metrics in the table:

- Accuracy: The accuracy metric measures the proportion of correct predictions out of all predictions made by the model.

- Precision: Precision indicates the proportion of true positive predictions out of all positive predictions made by the model. It measures the model's ability to avoid false positive predictions.

CHAPTER 7   DEEP LEARNING WITH TENSORFLOW FOR CLASSIFICATION

- Recall: Recall, also known as sensitivity or true positive rate, measures the proportion of true positive predictions out of all actual positive instances in the dataset.

- F1 Score: The F1 score is the weighted mean of precision and recall. It provides a balance between precision and recall, giving a single metric to evaluate the model's performance.

Based on the Epoch/Loss table and the metrics table, we can draw several conclusions regarding the performance of the TensorFlow (Keras) model:

Epoch/Loss Table:

- The loss value decreases consistently as the number of epochs increases. This indicates that the model is learning and improving its performance over successive epochs.

- The decrease in loss over epochs suggests that the model is effectively minimizing errors and becoming more accurate in its predictions.

Metrics Table:

- The accuracy metric provides an overall measure of the model's correctness in predicting both positive and negative instances. An accuracy of 0.77 indicates that the model correctly predicts around 77% of the outcomes on the test dataset.

- Precision, recall, and F1 score provide insights into the model's performance in predicting positive outcomes (i.e., cases of diabetes).

- The precision of 0.71 suggests that around 71% of the predicted positive outcomes are correct, minimizing false positive predictions.

- The recall of 0.59 indicates that around 59% of actual positive instances are correctly identified by the model, minimizing false negative predictions.

- The F1 score of 0.64, which is the weighted mean of precision and recall, provides a balance between precision and recall. It suggests that the model achieves a reasonable balance between minimizing false positives and false negatives.

Overall, the model's performance is reasonable given the known issues discussed in the previous chapter: class imbalance, skewed features, and potential data quality issues in the Pima dataset. Key areas for improvement would include addressing data anomalies, resampling to balance class distribution, data augmentation to diversify the dataset, and considering transfer learning for leveraging pre-trained models. Balancing model complexity, optimizing data representation, and fine-tuning hyperparameters are also crucial for enhancing performance, especially with small datasets like Pima.

TensorFlow vs. PyTorch

Building upon the results from the previous chapter where we utilized PyTorch and the current chapter where TensorFlow was employed, we are now positioned to conduct a comparative analysis of the performance of these two models. This allows for an apples-to-apples comparison since we utilized the same Pima diabetes dataset. By examining training loss, precision, recall, and F1 score, we can glean insights into the predictive capabilities of both algorithms in identifying cases of diabetes. Such

CHAPTER 7   DEEP LEARNING WITH TENSORFLOW FOR CLASSIFICATION

a comparison is invaluable in guiding decisions related to framework selection and model deployment in real-world production settings.

Starting with the loss value comparison, the table below compares the loss values of the two models generated during training:

Epoch	TensorFlow	PyTorch
10/100	0.6353	0.5899
20/100	0.5995	0.4330
30/100	0.5723	0.4608
40/100	0.5581	0.4612
50/100	0.5513	0.5155
60/100	0.5472	0.4025
70/100	0.5316	0.3406
80/100	0.5275	0.4323
90/100	0.5181	0.4822
100/100	0.5123	0.3203

Based on the data in the table above, we can observe the following:

- TensorFlow demonstrates a consistent decline in loss values over the training process, indicating effective optimization and learning. PyTorch, on the other hand, exhibits fluctuations in loss values, suggesting potential variations in optimization dynamics or learning behavior.

- The loss values for TensorFlow and PyTorch models vary at each epoch, indicating potential differences in the convergence behavior of the models. The patterns of loss values for TensorFlow and PyTorch models do

CHAPTER 7   DEEP LEARNING WITH TENSORFLOW FOR CLASSIFICATION

not follow identical trends, suggesting differences in optimization or learning dynamics between the two frameworks.

Moving on to evaluation metric comparison, the table below compares accuracy, precision, recall, and F1 score generated on test data:

Metric	TensorFlow	PyTorch
Accuracy	0.7692	0.7607
Precision	0.7059	0.6585
Recall	0.5854	0.6585
F1 Score	0.6400	0.6585

Based on this data, we can make the following remarks:

- Accuracy: The accuracy of the PyTorch model (0.76) is slightly lower than that of the TensorFlow model (0.77). Both models exhibit high accuracy, suggesting they are effective in making correct predictions on the test dataset. The slight difference in accuracy may be attributed to variations in model optimization or data handling.

- Precision: The precision of the PyTorch model (0.66) is lower than that of the TensorFlow model (0.71). This indicates that the TensorFlow model has a lower rate of false positives, making it slightly better at identifying only the true positive instances.

- Recall: The recall of the PyTorch model (0.66) is higher compared with the TensorFlow model (0.59). This suggests that the PyTorch model captures a larger proportion of actual positive instances, which may be beneficial for tasks where identifying as many positives as possible is crucial.

- F1 Score: The F1 score of the PyTorch model (0.66) is slightly higher than that of the TensorFlow model (0.64). The F1 score balances precision and recall, indicating that the PyTorch model provides a slightly better overall balance between precision and recall.

Overall, both TensorFlow and PyTorch models demonstrate comparable performance in terms of accuracy, precision, recall, and F1 score. Although there are minor differences between the models, they fall within the margin of error or may result from variations in optimization strategies or random initialization. Despite their high performance, there is room for improvement. Enhancing data quality, employing resampling and data augmentation techniques, exploring transfer learning, adjusting model complexity, improving data representation, and fine-tuning hyperparameters can further refine model performance.

Optimizing Model Performance with Hyperparameter Tuning

The results so far were obtained from a model with fixed hyperparameters. While these results are reasonable given the smallness of the input sample, TensorFlow offers a powerful tool for improving model performance: hyperparameter tuning. By leveraging keras_tuner, we can systematically search for the best combination of hyperparameters, optimizing the model beyond what manual selection can achieve.

CHAPTER 7   DEEP LEARNING WITH TENSORFLOW FOR CLASSIFICATION

To enhance our model, we replace the fixed hyperparameters with a hyperparameter tuning process using keras_tuner. This approach allows us to explore a wider range of model configurations and identify the one that performs best.

Step 1: Required Imports

Before we proceed with hyperparameter tuning, we need to ensure that the necessary modules are imported. The HyperModel class from keras_tuner will help us define the model architecture with tunable hyperparameters, while RandomSearch will facilitate the search for the optimal combination of these hyperparameters:

```
[In]: from keras_tuner import HyperModel, RandomSearch
```

Step 2: Define the DiabetesHyperModel Class

Next, we create the DiabetesHyperModel class, which extends the HyperModel class from keras_tuner. This class allows us to define the model architecture, specifically indicating which parameters are tunable. In this case, we specify that the number of units in each dense layer and the learning rate are the hyperparameters to be tuned.

Within the build method, we define a basic sequential model with three dense layers. For each dense layer, the number of units is determined by hp.Int, which allows keras_tuner to explore different values within a specified range. Similarly, the learning rate is set using hp.Float, providing a range for the tuner to explore. The model is compiled with a standard binary cross-entropy loss function and accuracy as a metric:

```
[In]: class DiabetesHyperModel(HyperModel):
        """HyperModel for tuning the Keras model."""
        def build(self, hp):
            model = Sequential()
            model.add(
                Dense(
                    units=hp.Int('units_1',
```

467

CHAPTER 7   DEEP LEARNING WITH TENSORFLOW FOR CLASSIFICATION

```
                min_value=32,
                max_value=128,
                step=32
            ),
            input_dim=X_train.shape[1]
        )
    )
    model.add(ReLU())
    model.add(
        Dense(
            units=hp.Int(
                'units_2',
                min_value=16,
                max_value=64,
                step=16
            )
        )
    )
    model.add(ReLU())
    model.add(
        Dense(
            units=hp.Int(
                'units_3',
                min_value=8,
                max_value=32,
                step=8
            )
        )
    )
    model.add(ReLU())
    model.add(Dense(1, activation='sigmoid'))
```

CHAPTER 7 DEEP LEARNING WITH TENSORFLOW FOR CLASSIFICATION

```
            optimizer = SGD(
                learning_rate=hp.Float(
                    'lr',
                    1e-4,
                    1e-2,
                    sampling='log'
                )
            )
            model.compile(loss='binary_crossentropy',
optimizer=optimizer, metrics=['accuracy'])
            return model
```

Step 3: Integrate the Tuner into DiabetesModelTrainer

With the DiabetesHyperModel class defined, we now integrate keras_tuner into the model training process. We modify the train_tensorflow_model method in the DiabetesModelTrainer class to use RandomSearch, a tuner that systematically explores the hyperparameter space defined in the DiabetesHyperModel class.

Within this method, we instantiate the tuner, specifying parameters like the number of trials (max_trials) and the number of executions per trial (executions_per_trial). The tuner then searches for the best model configuration by iterating over different combinations of hyperparameters. Once the best model is found, it is returned and can be further trained and evaluated:

```
[In]: class DiabetesModelTrainer:
          """Class for training the TensorFlow (Keras)
          model."""
          @staticmethod
          def train_tensorflow_model(
              X_train,
              y_train,
```

```python
        X_test,
        y_test,
        epochs=50
    ):
        """Trains the TensorFlow (Keras) model using
        keras_tuner.
        """
        try:
            tuner = RandomSearch(
                DiabetesHyperModel(),
                objective='val_accuracy',
                max_trials=5,
                executions_per_trial=3,
                directory='tuner_dir',
                project_name='diabetes_tuning'
            )
            tuner.search(
                X_train,
                y_train,
                epochs=epochs,
                validation_data=(X_test, y_test),
                verbose=1
            )
            best_model = tuner.get_best_models(num_models=1)[0]
            logger.info("Best model training completed.")
            return best_model
        except Exception as e:
            logger.error(
                f"Error occurred during model training: "
                f"{str(e)}"
            )
```

CHAPTER 7   DEEP LEARNING WITH TENSORFLOW FOR CLASSIFICATION

Note   In this code, we opted to use Keras Tuner for hyperparameter tuning because it is a native tool within the TensorFlow ecosystem, designed specifically to work seamlessly with TensorFlow models. Keras Tuner provides a straightforward and flexible interface for defining and searching through hyperparameters, making it an ideal choice for tuning deep learning models built with TensorFlow.

Alternatively, we could have used GridSearchCV from the Scikit-Learn library, which is a widely used method for hyperparameter tuning across various machine learning models. While GridSearchCV is powerful and versatile, we chose Keras Tuner for this example due to its native integration with TensorFlow, which simplifies the tuning process and ensures compatibility with the specific features and functions of TensorFlow. For those interested, we covered GridSearchCV in detail in *Distributed Machine Learning with PySpark: Migrating Effortlessly from Pandas and Scikit-Learn* (Apress, November 2023).

The table below compares the results from the tuned model with those from the fixed hyperparameter model. As demonstrated, the model with tuned hyperparameters shows improved performance across multiple metrics, confirming the effectiveness of hyperparameter tuning in optimizing model performance:

Metric	Tuned Model	Fixed Hyperparameter Model
Accuracy	0.7863	0.7692
Precision	0.7500	0.7059
Recall	0.5854	0.5854
F1 Score	0.6575	0.6400

Hyperparameter tuning is just one way to improve model performance. In Chapter 9, we will further dig into tuning and explore additional methods to improve model performance, providing a deeper understanding of how to optimize deep learning models effectively.

Bringing It All Together

In this section, we consolidate the individual code snippets for building, training, and evaluating the TensorFlow (Keras) model for diabetes prediction with and without Keras Tuner.

Building and Training a TensorFlow Model with Fixed Hyperparameters

```
[In]: import logging
[In]: from pyspark.sql import SparkSession
[In]: from pyspark.ml.feature import VectorAssembler,
      StandardScaler
[In]: from pyspark.sql.functions import col
[In]: from sklearn.metrics import (
          accuracy_score,
          precision_score,
          recall_score,
          f1_score
      )
[In]: from tensorflow.keras.models import Sequential
[In]: from tensorflow.keras.layers import Dense, ReLU
[In]: from tensorflow.keras.optimizers import SGD
[In]: from tensorflow.keras.callbacks import Callback
[In]: import numpy as np
```

CHAPTER 7 DEEP LEARNING WITH TENSORFLOW FOR CLASSIFICATION

```
[In]: logging.basicConfig(level=logging.INFO)
[In]: logger = logging.getLogger(__name__)

[In]: class CustomCallback(Callback):
          """Custom Keras callback for printing epoch
          information."""
          def on_epoch_end(self, epoch, logs=None):
              """Prints epoch information every 10 epochs."""
              if (epoch + 1) % 10 == 0:
                  logger.info(
                      f"Epoch [{epoch + 1}/{self.
                      params['epochs']}],"
                      f"Loss: {logs['loss']:.4f}"
                  )

[In]: class DiabetesProcessor:
          """Class for preprocessing diabetes data."""
          @staticmethod
          def preprocess_data(spark, data_file_path,
              train_parquet_path, test_parquet_path):
              """Preprocesses the diabetes data using
              PySpark."""
              try:
                  diabetes_df = spark.read.csv(
                      data_file_path,
                      header=True,
                      inferSchema=True
                  )
                  diabetes_df = diabetes_df.filter(
                      (col("Glucose") != 0)
                      & (col("BloodPressure") != 0)
                      & (col("BMI") != 0)
                  )
```

```python
feature_cols = [
    "Pregnancies",
    "Glucose",
    "BloodPressure",
    "BMI",
    "DiabetesPedigreeFunction",
    "Age"
]
assembler = VectorAssembler(
    inputCols=feature_cols,
    outputCol="features")
diabetes_df = assembler.
transform(diabetes_df),
scaler = StandardScaler(
    inputCol="features",
    outputCol="scaled_features"
)
scaler_model = scaler.fit(diabetes_df)
diabetes_df = scaler_model.
transform(diabetes_df)
train_df, test_df = diabetes_df.randomSplit(
    [0.8, 0.2],
    seed=42
)
train_df.write.parquet(
    train_parquet_path
    mode="overwrite"
)
test_df.write.parquet(
    test_parquet_path,
    mode="overwrite"
)
```

CHAPTER 7 DEEP LEARNING WITH TENSORFLOW FOR CLASSIFICATION

```
            logger.info("Data preprocessing completed.")
        except Exception as e:
            logger.error(
                f"Error occurred during data "
                f"preprocessing: {str(e)}"
            )
```

```
[In]: class DiabetesModelTrainer:
    """Class for training the TensorFlow (Keras) model."""
    
    @staticmethod
    def train_tensorflow_model(
        X_train,
        y_train,
        epochs=100,
        lr=0.01
    ):
        """Trains the TensorFlow (Keras) model."""
        try:
            model = Sequential([
                Dense(64, input_dim=X_train.shape[1]),
                ReLU(),
                Dense(32),
                ReLU(),
                Dense(16),
                ReLU(),
                Dense(1, activation='sigmoid')
            ])
            optimizer = SGD(learning_rate=lr)
            model.compile(loss='binary_crossentropy',
                          optimizer=optimizer,
                          metrics=['accuracy'])
```

```
            custom_callback = CustomCallback()
            model.fit(X_train, y_train, epochs=epochs,
                verbose=0, callbacks=[custom_callback])
            logger.info("Model training completed.")
            return model

    except Exception as e:
        logger.error(
            f"Error occurred during model training: "
            f"{str(e)}"
        )

[In]: class DiabetesModelEvaluator:
    """Class for evaluating the TensorFlow (Keras)
    model."""
    @staticmethod
    def evaluate_model(model, X_test, y_test):
        """Evaluates the TensorFlow (Keras) model."""
        try:
            y_pred = (model.predict(X_test) > 0.5).
            astype(int)
            accuracy = accuracy_score(y_test, y_pred)
            precision = precision_score(y_test, y_pred)
            recall = recall_score(y_test, y_pred)
            f1 = f1_score(y_test, y_pred)
            logger.info(f"Accuracy: {accuracy:.4f}")
            logger.info(f"Precision: {precision:.4f}")
            logger.info(f"Recall: {recall:.4f}")
            logger.info(f"F1 Score: {f1:.4f}")
        except Exception as e:
```

CHAPTER 7   DEEP LEARNING WITH TENSORFLOW FOR CLASSIFICATION

```
        logger.error(
            f"Error occurred during model "
            f"evaluation: {str(e)}"
        )
In]: if __name__ == "__main__":
        try:
            spark = (SparkSession.builder
                .appName("DiabetesClassification")
                .config("spark.executor.instances", "4")
                .config("spark.executor.memory", "4g")
                .config("spark.executor.cores", "2")
                .config("spark.driver.memory", "2g")
                .getOrCreate())

            data_file_path = "/home/ubuntu/airflow/dags/
                            diabetes.csv"
            train_parquet_path = \
                "/home/ubuntu/airflow/dags/diabetes_train.
                parquet"
            test_parquet_path = \
                "/home/ubuntu/airflow/dags/diabetes_test.
                parquet"

            DiabetesProcessor.preprocess_data(
                spark,
                data_file_path,
                train_parquet_path,
                test_parquet_path
            )

            train_diabetes_df = spark.read.parquet(
                train_parquet_path
            )
```

CHAPTER 7   DEEP LEARNING WITH TENSORFLOW FOR CLASSIFICATION

```
            X_train = np.array(train_diabetes_df.select(
                    "scaled_features")
                    .rdd.flatMap(lambda x: x).collect()
            )
            y_train = np.array(
                train_diabetes_df.select("Outcome")
                .rdd.flatMap(lambda x: x).collect())

            trained_model =
                DiabetesModelTrainer.train_tensorflow_model(
                X_train,
                y_train
            )

            test_diabetes_df = spark.read.parquet(test_
            parquet_path)
            X_test = np.array(
                test_diabetes_df.select("scaled_features")
                .rdd.flatMap(lambda x: x).collect())
            y_test = np.array(
                test_diabetes_df.select("Outcome")
                .rdd.flatMap(lambda x: x).collect()
            )

            DiabetesModelEvaluator.evaluate_model(
                trained_model,
                X_test,
                y_test
            )
        except Exception as e:
            logger.error(f"An error occurred: {str(e)}")
```

Optimizing the TensorFlow Model with Hyperparameter Tuning

```
[In]: import logging
[In]: from pyspark.sql import SparkSession
[In]: from pyspark.ml.feature import VectorAssembler,
      StandardScaler
[In]: from pyspark.sql.functions import col
[In]: from sklearn.metrics import (
          accuracy_score,
          precision_score,
          recall_score,
          f1_score
      )
[In]: from tensorflow.keras.models import Sequential
[In]: from tensorflow.keras.layers import Dense, ReLU
[In]: from tensorflow.keras.optimizers import SGD
[In]: from tensorflow.keras.callbacks import Callback
[In]: from keras_tuner import HyperModel, RandomSearch
[In]: import numpy as np

[In]: logging.basicConfig(level=logging.INFO)
[In]: logger = logging.getLogger(__name__)

[In]: class CustomCallback(Callback):
          """Custom Keras callback for printing epoch
          information."""
          def on_epoch_end(self, epoch, logs=None):
              """Prints epoch information every 10 epochs."""
              if (epoch + 1) % 10 == 0:
                  logger.info(
```

CHAPTER 7 DEEP LEARNING WITH TENSORFLOW FOR CLASSIFICATION

```python
                f"Epoch [{epoch + 1}/{self.
                params['epochs']}], "
                f"Loss: {logs['loss']:.4f}"
            )
[In]: class DiabetesProcessor:
    """Class for preprocessing diabetes data."""
    @staticmethod
    def preprocess_data(
        spark,
        data_file_path,
        train_parquet_path,
        test_parquet_path
    ):
        """Preprocesses the diabetes data using
        PySpark."""
        try:
            diabetes_df = spark.read.csv(
                data_file_path,
                header=True,
                inferSchema=True
            )
            diabetes_df = diabetes_df.filter(
                (col("Glucose") != 0)
                & (col("BloodPressure") != 0)
                & (col("BMI") != 0)
            )
            feature_cols = [
                "Pregnancies",
                "Glucose",
                "BloodPressure",
                "BMI",
```

CHAPTER 7   DEEP LEARNING WITH TENSORFLOW FOR CLASSIFICATION

```
        "DiabetesPedigreeFunction",
        "Age"
    ]
    assembler = VectorAssembler(
        inputCols=feature_cols,
        outputCol="features"
    )
    diabetes_df = assembler.
    transform(diabetes_df)
    scaler = StandardScaler(
        inputCol="features",
        outputCol="scaled_features"
    )
    scaler_model = scaler.fit(diabetes_df)
    diabetes_df = scaler_model.
    transform(diabetes_df)
    train_df, test_df = diabetes_df.randomSplit(
        [0.8, 0.2],
        seed=42
    )
    train_df.write.parquet(
        train_parquet_path,
        mode="overwrite"
    )
    test_df.write.parquet(
        test_parquet_path,
        mode="overwrite"
    )
    logger.info("Data preprocessing completed.")
except Exception as e:
```

CHAPTER 7   DEEP LEARNING WITH TENSORFLOW FOR CLASSIFICATION

```
            logger.error(
                f"Error occurred during data
                preprocessing: "
                f"{str(e)}"
            )
[In]: class DiabetesHyperModel(HyperModel):
        """HyperModel for tuning the Keras model."""
        def build(self, hp):
            model = Sequential()
            model.add(
                Dense(
                    units=hp.Int('units_1',
                    min_value=32,
                    max_value=128,
                    step=32
                ),
                input_dim=X_train.shape[1]
                )
            )
            model.add(ReLU())
            model.add(
                Dense(
                    units=hp.Int(
                        'units_2',
                        min_value=16,
                        max_value=64,
                        step=16
                    )
                )
            )
```

```
            model.add(ReLU())
            model.add(
                Dense(
                    units=hp.Int(
                        'units_3',
                        min_value=8,
                        max_value=32,
                        step=8
                    )
                )
            )
            model.add(ReLU())
            model.add(Dense(1, activation='sigmoid'))
            optimizer = SGD(
                learning_rate=hp.Float(
                    'lr',
                    1e-4,
                    1e-2,
                    sampling='log'
                )
            )
            model.compile(loss='binary_crossentropy',
            optimizer=optimizer, metrics=['accuracy'])
            return model
```

```
[In]: class DiabetesModelTrainer:
          """Class for training the TensorFlow (Keras)
          model."""
          @staticmethod
          def train_tensorflow_model(
              X_train,
              y_train,
```

```
        X_test,
        y_test,
        epochs=50
    ):
        """Trains the TensorFlow (Keras) model using
        keras_tuner.
        """
        try:
            tuner = RandomSearch(
                DiabetesHyperModel(),
                objective='val_accuracy',
                max_trials=5,
                executions_per_trial=3,
                directory='tuner_dir',
                project_name='diabetes_tuning'
            )
            tuner.search(
                X_train,
                y_train,
                epochs=epochs,
                validation_data=(X_test, y_test),
                verbose=1
            )
            best_model = tuner.get_best_models(num_
            models=1)[0]
            logger.info("Best model training completed.")
            return best_model
        except Exception as e:
            logger.error(
                f"Error occurred during model training: "
                f"{str(e)}"
            )
```

```
[In]: class DiabetesModelEvaluator:
          """Class for evaluating the TensorFlow (Keras)
          model."""
          @staticmethod
          def evaluate_model(model, X_test, y_test):
              """Evaluates the TensorFlow (Keras) model."""
              try:
                  y_pred = (model.predict(X_test) > 0.5).
                  astype(int)
                  accuracy = accuracy_score(y_test, y_pred)
                  precision = precision_score(y_test, y_pred)
                  recall = recall_score(y_test, y_pred)
                  f1 = f1_score(y_test, y_pred)
                  logger.info(f"Accuracy: {accuracy:.4f}")
                  logger.info(f"Precision: {precision:.4f}")
                  logger.info(f"Recall: {recall:.4f}")
                  logger.info(f"F1 Score: {f1:.4f}")
              except Exception as e:
                  logger.error(
                      f"Error occurred during model
                      evaluation: "
                      f"{str(e)}"
                  )

[In]: if __name__ == "__main__":
          try:
              spark = (
                  SparkSession.builder
                      .appName(
                  "DiabetesProcessingAndTensorFlowTraining"
                      )
                      .config("spark.executor.instances", "4")
```

CHAPTER 7 DEEP LEARNING WITH TENSORFLOW FOR CLASSIFICATION

```
            .config("spark.executor.memory", "4g")
            .config("spark.executor.cores", "2")
            .config("spark.driver.memory", "2g")
            .getOrCreate()
)

data_file_path = "/home/ubuntu/airflow/dags/
diabetes.csv"
train_parquet_path = ("/home/ubuntu/airflow/dags/"
                      "diabetes_train.parquet")
test_parquet_path = ("/home/ubuntu/airflow/dags/"
                     "diabetes_test.parquet")

DiabetesProcessor.preprocess_data(
    spark,
    data_file_path,
    train_parquet_path,
    test_parquet_path
)

train_diabetes_df = spark.read.parquet(
    train_parquet_path
)
X_train = np.array(
    train_diabetes_df
    .select("scaled_features")
    .rdd.flatMap(lambda x: x)
    .collect()
)
y_train = np.array(
    train_diabetes_df
    .select("Outcome")
```

CHAPTER 7   DEEP LEARNING WITH TENSORFLOW FOR CLASSIFICATION

```
    .rdd.flatMap(lambda x: x)
    .collect()
)
test_diabetes_df = spark.read.parquet(
    test_parquet_path
)
X_test = np.array(
    test_diabetes_df
    .select("scaled_features")
    .rdd.flatMap(lambda x: x)
    .collect()
)
y_test = np.array(
    test_diabetes_df
    .select("Outcome")
    .rdd.flatMap(lambda x: x)
    .collect()
)
best_model = DiabetesModelTrainer.train_
tensorflow_model(
    X_train,
    y_train,
    X_test,
    y_test
)

DiabetesModelEvaluator.evaluate_model(
    best_model,
    X_test,
    y_test
)
```

```
        except Exception as e:
            logger.error(f"An error occurred: {str(e)}")
```

Summary

In this chapter, we continued exploring the topic of deep learning classification, this time using TensorFlow after using PyTorch in the previous chapter. We demonstrated how to replicate the same steps used with PyTorch, including data loading, preprocessing, defining the model architecture, training, and evaluation. Additionally, we incorporated extra functionality such as copying data from an S3 bucket to an EC2 local directory for processing, showcasing integration with AWS services, and handling errors and log messages for debugging and monitoring purposes throughout the modeling process.

To enable an apples-to-apples comparison of the two models' performance, we utilized the same Pima diabetes dataset that was used for PyTorch. The main conclusion was that the performance of both models (PyTorch and TensorFlow) is comparable.

We also introduced hyperparameter tuning using keras_tuner, which helped improve the performance of the TensorFlow model. By automatically searching for the optimal configuration, we were able to achieve better results than with manually chosen hyperparameters.

In the next chapter, we pivot to a different yet equally important topic: how to streamline the pipeline on AWS using Apache Airflow.

CHAPTER 8

Scalable Deep Learning Pipelines with Apache Airflow

In the preceding chapters, we built scalable deep learning pipelines and executed them manually on AWS EC2 instance. While this non-automated approach is common in the industry, it may not always be the most efficient. On the one hand, it lacks visibility into the status and progress of workflows, necessitating separate implementation of monitoring and logging. On the other hand, it can also result in challenges when modifying or extending the workflow. Indeed, scaling workflows to manage large volumes of data or computational resources can be daunting without a dedicated orchestration framework. Additionally, ensuring reproducibility of experiments becomes challenging without a standardized approach, relying heavily on version control and documentation.

Adopting Apache Airflow to run workflows addresses these issues effectively. It facilitates automated execution based on predefined schedules or triggers, minimizing the need for manual intervention. Furthermore, it offers comprehensive monitoring and logging capabilities out of the box, enabling users to easily track workflow progress and troubleshoot any issues that arise. When you run DAGs (Directed Acyclic Graphs) in Airflow, you ensure tasks are executed in the correct order based on their dependencies. Airflow is purposefully designed to scale with the size and complexity of workflows, efficiently managing large volumes of data and computational resources. Moreover, it promotes reproducibility by encapsulating workflows as code, facilitating version control, and enabling sharing among team members for experiment reproduction and change tracking over time.

In this chapter, we demonstrate how to build scalable data processing pipelines with Apache Airflow. We will create entire AWS workflows that include preprocessing data using PySpark, developing deep learning models using PyTorch and TensorFlow, and running the code using Airflow. More specifically, we will use the Tesla stock and diabetes datasets from Chapters 4 and 7, respectively, as well as the standalone regression and classification scripts from those chapters. This will allow us to demonstrate how a Python script is normally run and compare it with the more efficient execution as an Apache Airflow DAG.

We will begin by creating an Apache Airflow pipeline for Tesla stock price prediction using PySpark and PyTorch and then proceed to do the same for diabetes classification using PySpark and TensorFlow.

An Airflow Pipeline for Tesla Stock Price Prediction

In this section, we demonstrate how to build a scalable data processing workflow with Apache Airflow. We will create an AWS pipeline that includes preprocessing data using PySpark, developing a deep learning model using PyTorch, and running the code using Airflow.

To accomplish this, we will utilize the same Tesla dataset as in Chapter 4. The code will also remain consistent with that chapter, with the main difference being a more modular approach in this chapter. A modular

CHAPTER 8 SCALABLE DEEP LEARNING PIPELINES WITH APACHE AIRFLOW

approach can streamline development and enhance system flexibility. Each component, such as PySpark for big data processing, PyTorch for machine learning tasks, and Airflow for workflow orchestration, can be treated as a distinct module. This modular design allows for easier integration, customization, and reuse across projects. By encapsulating tasks within Airflow DAGs, each module can be orchestrated and scheduled independently, enabling efficient workflow management. Moreover, designing modules for reusability ensures consistency and reduces development overhead across different projects. Scalability and performance considerations are essential, particularly for PySpark jobs handling large datasets and PyTorch models optimized for efficient training and inference.

In Chapter 4, we executed the code manually through a main() function. The main difference between this approach and running the code as an Airflow DAG is that the main() function orchestrates the execution of various tasks, whereas, when integrating code into an Airflow DAG, a separate main() function is typically not needed to orchestrate tasks in the same way. Instead, tasks are defined explicitly as part of the Airflow DAG using operators such as PythonOperator or BashOperator. Each operator corresponds to a specific task, and dependencies between tasks are specified using the >> operator within the DAG definition.

To illustrate this difference, we will first examine how we would run the Tesla stock price prediction code without an Apache Airflow DAG using the main() function. After that, we will demonstrate how to run the exact same code (without the main function) as a DAG.

CHAPTER 8   SCALABLE DEEP LEARNING PIPELINES WITH APACHE AIRFLOW

Tesla Stock Price Prediction Without Airflow DAG

In this subsection, we illustrate how a modular standalone Python script is run without an Airflow DAG, using Tesla stock prediction with PySpark and PyTorch as an example. We have developed five modules for this project:

- data_processing.py: This module is responsible for loading and preprocessing Tesla stock data with PySpark.

- model_training.py: This module handles the training of a PyTorch deep learning regression model.

- model_evaluation.py: This module is used for evaluating the PyTorch deep learning model.

- utils.py: This module provides utility functions for converting Spark DataFrame columns to PyTorch tensors.

- main.py: This final module orchestrates the overall workflow.

Let's go through the five modules one by one, starting with the data_processing.py module:

Module 1: data_processing.py

This module is designed to preprocess the Tesla stock data:

```
[In]: from pyspark.sql import SparkSession
[In]: from pyspark.ml.feature import VectorAssembler, StandardScaler
[In]: from pyspark.sql import DataFrame
```

```
[In]: def load_data(file_path: str) -> DataFrame:
        """
        Load stock price data from a CSV file using
        SparkSession.
        """
        spark = SparkSession.builder \
                .appName("StockPricePrediction") \
                .getOrCreate()
        df = spark.read.csv(
                file_path,
                header=True,
                inferSchema=True
        )
        return df

[In]: def preprocess_data(df: DataFrame) -> DataFrame:
        """
        Preprocess the data by assembling feature
        vectors using
        VectorAssembler and scaling them using
        StandardScaler.
        """
        assembler = VectorAssembler(
            inputCols=['Open', 'High', 'Low', 'Volume'],
            outputCol='features'
        )
        df = assembler.transform(df)

        scaler = StandardScaler(
            inputCol="features",
            outputCol="scaled_features",
```

CHAPTER 8   SCALABLE DEEP LEARNING PIPELINES WITH APACHE AIRFLOW

```
        withStd=True,
        withMean=True
    )
    scaler_model = scaler.fit(df)
    df = scaler_model.transform(df)
    df = df.select('scaled_features', 'Close')
    return df
```

Let's delve into the code step by step:

- Firstly, the code imports necessary modules from PySpark, including SparkSession for initializing Spark functionality and VectorAssembler and StandardScaler from the machine learning (ML) feature module. Additionally, it imports DataFrame from pyspark.sql, which is a representation of structured data.

- The load_data function takes a file path as input, pointing to a CSV file containing stock price data. It utilizes SparkSession.builder to create a Spark session named "StockPricePrediction". Then, it reads the CSV file into a DataFrame df, enabling Spark to handle distributed data manipulation efficiently. The header=True argument indicates that the first row contains column names, while inferSchema=True instructs Spark to infer column types from the data.

- The preprocess_data function takes a DataFrame df as input and performs preprocessing steps required for the machine learning task. Firstly, it utilizes VectorAssembler to assemble feature vectors. Here, it selects columns 'Open', 'High', 'Low', and 'Volume' from the DataFrame to construct a new column named 'features' containing vectors of these features.

CHAPTER 8   SCALABLE DEEP LEARNING PIPELINES WITH APACHE AIRFLOW

- Even though the subsequent machine learning tasks (i.e., training and evaluation) are conducted using PyTorch instead of PySpark, assembling features with VectorAssembler remains beneficial especially when dealing with large datasets. By consolidating relevant features into a single vector column, it streamlines the data preparation process and ensures consistency between the data fed into PyTorch and the data it expects. This consolidation simplifies the interface between data preprocessing and the machine learning model, facilitating a smoother transition from data preparation to model training and enhancing overall workflow efficiency.

- Next, the module employs StandardScaler to scale the feature vectors. Scaling is essential to normalize features and ensure that they have similar ranges, which can improve the performance and convergence of the machine learning algorithm. The withStd=True argument indicates that the data should be scaled to have unit standard deviation, while withMean=True centers the data before scaling.

- After scaling, the function selects the 'scaled_features' column, containing the scaled feature vectors, and the 'Close' column, which represents the target variable (closing stock prices). The processed DataFrame df is returned, ready for model training and evaluation.

CHAPTER 8　SCALABLE DEEP LEARNING PIPELINES WITH APACHE AIRFLOW

Module 2: model_training.py

This module utilizes the PyTorch library to create and train a neural network regression model for a supervised learning task (Tesla stock price prediction):

```
[In]: import torch
[In]: import torch.nn as nn
[In]: import torch.optim as optim
[In]: from torch.utils.data import DataLoader, TensorDataset
[In]: def create_data_loader(
        features,
        labels,
        batch_size=32
    ) -> DataLoader:
        """
        Convert the preprocessed data into PyTorch
        tensors and
        Create DataLoader objects for both the training
        and test
        sets.
        """
        dataset = TensorDataset(features, labels)
        return DataLoader(
            dataset,
            batch_size=batch_size,
            shuffle=True
        )

[In]: def train_model(
        model, train_loader, criterion, optimizer, num_epochs
    ):
        """
```

Train the model on the training data using the
DataLoader
and the defined loss function and optimizer.
Iterate over
the data for a specified number of epochs,
calculate the
loss, and update the model parameters.
"""
for epoch in range(num_epochs):
    for inputs, labels in train_loader:
        optimizer.zero_grad()
        outputs = model(inputs)
        loss = criterion(outputs, labels.
        unsqueeze(1))
        loss.backward()
        optimizer.step()
print(
    f"Epoch [{epoch + 1}/{num_epochs}], "
    f"Loss: {loss.item():.4f}"
)
```

Let's break the code down for better understanding:

- The code begins by importing necessary modules from PyTorch, including torch for the core functionality, torch.nn for building neural network components, torch.optim for optimization algorithms, and DataLoader and TensorDataset from torch.utils.data for handling data loading and batching. DataLoader is particularly important from a PySpark perspective because it efficiently manages the loading and batching of data during training. In distributed environments like PySpark, where large datasets are often processed,
```

CHAPTER 8   SCALABLE DEEP LEARNING PIPELINES WITH APACHE AIRFLOW

DataLoader ensures that data is fed into the model in manageable batches, which helps optimize memory usage and computational efficiency. It also handles shuffling and parallel data loading, which are crucial for improving model performance and reducing training time.

- The create_data_loader function is defined to convert preprocessed data into PyTorch tensors and create DataLoader objects for both the training and test sets. It takes features and labels as input, which are preprocessed data from the previous module. These are converted into a TensorDataset, which combines the features and labels into a single dataset. Then, a DataLoader is created from this dataset, specifying the desired batch_size and enabling shuffling of the data for each epoch.

- Next, the train_model function is defined to train the neural network regression model on the training data using the DataLoader, a defined loss function (criterion), and an optimizer. This function iterates over the data for a specified number of epochs, each time calculating the loss and updating the model parameters based on the gradients computed during backpropagation.

- Within the training loop, for each batch of inputs and labels loaded from the train_loader, the optimizer's gradients are zeroed (optimizer.zero_grad()), ensuring that gradients from previous iterations do not accumulate. Then, the model is applied to the inputs

CHAPTER 8 SCALABLE DEEP LEARNING PIPELINES WITH APACHE AIRFLOW

to generate predictions (outputs = model(inputs)). The loss between these predictions and the actual labels is calculated using the specified criterion.

- After computing the loss, the gradients are propagated backward through the network (loss.backward()), enabling PyTorch to compute the gradients of the loss with respect to the model parameters. Finally, the optimizer updates the model parameters based on these gradients (optimizer.step()).

- Throughout the training process, information such as the current epoch and loss is printed for monitoring the model's progress. This iterative training loop continues for the specified number of epochs, gradually improving the model's performance over time.

Module 3: model_evaluation.py

This module is designed to evaluate the performance of the trained PyTorch model on a test dataset:

```
[In]: import torch
```

```
[In]: def evaluate_model(model, test_loader, criterion):
          """
          Evaluate the trained model on the test data to
          assess its
          performance.
          Calculate the test loss and additional
          evaluation metrics
          such as the R-squared score.
          """
          with torch.no_grad():
              model.eval()
```

CHAPTER 8   SCALABLE DEEP LEARNING PIPELINES WITH APACHE AIRFLOW

```python
predictions = []
targets = []
test_loss = 0.0
for inputs, labels in test_loader:
    outputs = model(inputs)
    loss = criterion(outputs, labels.
    unsqueeze(1))
    test_loss += loss.item() * inputs.size(0)
    predictions.extend(outputs.squeeze().
    tolist())
    targets.extend(labels.tolist())

test_loss /= len(test_loader.dataset)
predictions = torch.tensor(predictions)
targets = torch.tensor(targets)
ss_res = torch.sum((targets - predictions) ** 2)
ss_tot = torch.sum((targets - torch.
mean(targets)) ** 2)
r_squared = 1 - ss_res / ss_tot
print(f"Test Loss: {test_loss:.4f}")
print(f"R-squared Score: {r_squared:.4f}")
return test_loss, r_squared.item()
```

Let's break the code down to understand its functionality:

- The code begins by importing the torch library, which is the core library for PyTorch.

- The evaluate_model function is defined to assess the performance of the model on a test dataset. It takes three main inputs: model, the trained PyTorch model to be evaluated; test_loader, the DataLoader object containing the test data; and criterion, the loss function used for evaluation, typically the same as the one used during training.

CHAPTER 8   SCALABLE DEEP LEARNING PIPELINES WITH APACHE AIRFLOW

- Within the function, the evaluation process is encapsulated in a with torch.no_grad() block (which disables gradient calculation since gradients are not needed during evaluation), and the model is switched to evaluation mode using model.eval().

- Next, the function initializes empty lists predictions and targets to store the model's predictions and actual target values, respectively. Additionally, test_loss is initialized to accumulate the total loss over the entire test dataset.

- The function iterates over batches of data from the test_loader, computing predictions using the model for each batch. The loss between the predicted outputs and the actual labels is calculated using the specified criterion. The test loss is accumulated by multiplying the loss by the batch size and adding it to test_loss.

- The predicted outputs are then appended to the predictions list after removing the extra dimension using outputs.squeeze().tolist(). Similarly, the actual labels are appended to the targets list.

- After iterating through all batches, the average test loss across the entire dataset is computed by dividing test_loss by the total number of samples in the test dataset.

- The predictions and targets lists are converted into PyTorch tensors. The script then calculates the residual sum of squares (ss_res) and the total sum of squares (ss_tot) to compute the R-squared score, a metric indicating how well the regression model fits the data.

CHAPTER 8 SCALABLE DEEP LEARNING PIPELINES WITH APACHE AIRFLOW

- Finally, the test loss and R-squared score are printed for evaluation purposes. The function returns both the test loss and the R-squared score as a tuple.

Module 4: utils.py

Utility modules typically serve as a collection of functions that are used across different parts of the project. In our case, there is only one function, spark_df_to_tensor, which is responsible for converting Spark DataFrame columns to PyTorch tensors:

```
[In]: import numpy as np
[In]: import torch
[In]: from typing import Tuple
[In]: from pyspark.sql import DataFrame

[In]: def spark_df_to_tensor(
        df: DataFrame
    ) -> Tuple[torch.Tensor, torch.Tensor]:
        """
        Converts Spark DataFrame columns to PyTorch tensors.
        """
        features = torch.tensor(
            np.array(
                df.rdd.map(lambda x: x.scaled_features.
                toArray())
                .collect()
            ),
            dtype=torch.float32
        )
        labels = torch.tensor(
            np.array(
                df.rdd.map(lambda x: x.Close).collect()
            ),
```

CHAPTER 8   SCALABLE DEEP LEARNING PIPELINES WITH APACHE AIRFLOW

```
        dtype=torch.float32
    )
    return features, labels
```

The spark_df_to_tensor function takes a DataFrame as input and returns a tuple containing features and labels as PyTorch tensors. Inside the function, it uses NumPy and PyTorch to perform the necessary conversions. This function is crucial for preprocessing data before training and evaluating the deep learning regression model.

Module 5: main.py

We now come to the main module, which orchestrates the entire workflow—data preprocessing, model training, model evaluation, and utility function:

```
[In]: import logging
[In]: import torch
[In]: import torch.optim as optim
[In]: import torch.nn as nn
[In]: from torch.utils.data import DataLoader
[In]: from typing import Tuple
[In]: from pyspark.sql import DataFrame
[In]: from data_processing import load_data, preprocess_data
[In]: from model_training import create_data_loader,
      train_model
[In]: from model_evaluation import evaluate_model
[In]: import utils

[In]: logging.basicConfig(level=logging.INFO)
[In]: logger = logging.getLogger(__name__)

[In]: def main(
        data_file_path: str,
        num_epochs: int = 100,
```

503

CHAPTER 8 SCALABLE DEEP LEARNING PIPELINES WITH APACHE AIRFLOW

```
    batch_size: int = 32,
    learning_rate: float = 0.001
):
    """
    Main function for training and evaluating a deep
    learning
    model for Tesla stock price prediction.

    Args:
        data_file_path (str):
            The path to the CSV file containing
            stock price
            data.
        num_epochs (int):
            The number of epochs for training the model.
            Defaults to 100.
        batch_size (int):
            The batch size for data loading during
            training.
            Defaults to 32.
        learning_rate (float):
            The learning rate for the optimizer.
            Defaults to
            0.001.

    Raises:
        FileNotFoundError: If the specified data
        file is not
        found.
    """

    try:
        df = load_data(data_file_path)
        df = preprocess_data(df)
```

CHAPTER 8 SCALABLE DEEP LEARNING PIPELINES WITH APACHE AIRFLOW

```python
train_df, test_df = df.randomSplit([0.8, 0.2],
seed=42)

train_features, train_labels = utils.spark_df_
to_tensor(
    train_df
)
test_features, test_labels = utils.spark_df_
to_tensor(
    test_df
)
train_loader = create_data_loader(
    train_features,
    train_labels,
    batch_size=batch_size
)
test_loader = create_data_loader(
    test_features,
    test_labels,
    batch_size=batch_size
)

input_size = train_features.shape[1]
output_size = 1
model = nn.Sequential(
    nn.Linear(input_size, 64),
    nn.ReLU(),
    nn.Linear(64, 32),
    nn.ReLU(),
    nn.Linear(32, 16),
```

```
            nn.ReLU(),
            nn.Linear(16, output_size)
        )

        criterion = nn.MSELoss()
        optimizer = optim.Adam(
            model.parameters(),
            lr=learning_rate
        )

        train_model(
            model,
            train_loader,
            criterion,
            optimizer,
            num_epochs
        )

        evaluate_model(model, test_loader, criterion)

        torch.save(model.state_dict(), 'trained_model.pth')

    except FileNotFoundError:
        logger.error(f"Data file not found at {data_file_path}")
```
```
[In]: if __name__ == "__main__":
        data_file_path = "/home/ubuntu/airflow/dags/TSLA_stock.csv"
        main(data_file_path)
```

The code starts by importing PyTorch modules (torch.optim, torch.nn), PySpark DataFrame (pyspark.sql.DataFrame), and custom modules (data_processing, model_training, model_evaluation, and utils). These imports are important for data processing, model training, and evaluation. It then configures the logging to display INFO-level messages and initializes a logger named name.

In the next step, the module defines the main function. This takes several arguments such as data_file_path, num_epochs, batch_size, and learning_rate. Inside this function, the following steps are performed:

- Data Loading and Preprocessing: The function loads data from a CSV file specified by data_file_path, preprocesses it, and then splits it into training and testing sets.

- Data Conversion: It converts Spark DataFrame features and labels into PyTorch tensors using a custom function utils.spark_df_to_tensor.

- Model Definition: A feedforward neural network model is defined using nn.Sequential. The model consists of several linear layers with ReLU activation functions.

- Loss Function and Optimizer: Mean Squared Error (MSE) loss is used as the criterion for training the model, and Adam optimizer is employed to optimize the model parameters.

- Model Training: The train_model function is called to train the model using the training data loader, criterion, optimizer, and specified number of epochs.

- Model Evaluation: After training, the trained model is evaluated using the evaluate_model function and the testing data loader.

CHAPTER 8 SCALABLE DEEP LEARNING PIPELINES WITH APACHE AIRFLOW

- Model Saving: Finally, the trained model's state dictionary is saved to a file named 'trained_model.pth'.
- Finally, the name == "main" block: ensures that the main function is executed only when the script is run directly, not when it's imported as a module into another script. It sets the data_file_path variable to the path of the Tesla stock CSV file and calls the main function with this path as an argument.

To run the code, we first activated the "myenv" environment, then navigated to the directory "home/ubuntu/airflow/dags" on EC2 instance where the code is stored, and executed the "main.py" module to orchestrate the Tesla stock price prediction workflow:

```
[In]: source activate myenv
[In]: cd home/ubuntu/airflow/dags
[In]: python3 main.py
```

The first output from the code is the loss for each epoch during the training process of the Tesla stock price prediction model:

Epoch	Loss
1	49250.46
2	55661.49
3	24399.93
4	10579.51
5	2365.65
6	1972.29
7	1864.98

(continued)

CHAPTER 8   SCALABLE DEEP LEARNING PIPELINES WITH APACHE AIRFLOW

Epoch	Loss
8	1288.25
9	736.96
10	625.62
...	...
100	9.70

Here's an explanation for the two columns in the table:

- Epoch: This column indicates the number of training epochs, which are iterations over the entire training dataset.

- Loss: This column represents the training loss for each epoch. The training loss is a measure of how well the model's predictions match the actual values in the training data. Higher loss values at the beginning of training epochs are expected, as the model starts with random weights and biases and gradually adjusts them to minimize the loss.

Here's how the training loss evolves over the epochs:

- In the initial epochs (e.g., epochs 1 and 2), the loss values are extremely high (i.e., 49,250 and 55,661), indicating significant errors in the model's predictions. This is expected at the beginning of training when the model's parameters are randomly initialized.

- As training progresses, the loss values steadily decrease. For instance, by epoch 10, the loss has decreased to 625.6, indicating that the model's predictions are getting closer to the actual Tesla stock prices.

CHAPTER 8   SCALABLE DEEP LEARNING PIPELINES WITH APACHE AIRFLOW

- By epoch 100, the training loss has significantly decreased to 9.7, which suggests that the model has learned the patterns in the training data well. This low training loss indicates that the model's predictions closely match the actual Tesla stock prices on the training data.

The second output is the test loss and R-squared, as shown below:

Metric	Value
Test Loss	25.3015
R-squared	0.9976

In this table, we provide key metrics for evaluating the performance of the Tesla stock price prediction model:

- Test Loss: This metric represents the error of the model when making predictions on a separate test dataset. In the context of a Tesla stock price prediction model, a test loss of 25.3 means that, on average, the model's predictions deviated from the actual Tesla stock prices by approximately $25.3. Lower test loss values generally indicate better performance, as they signify that the model's predictions are closer to the actual prices.

- R-squared Score: The R-squared score, also known as the coefficient of determination, measures how well the model's predictions explain the variance in the actual Tesla stock prices. An R-squared score of 0.9976 indicates that approximately 99.76% of the variability in the Tesla stock prices is explained by the model. In other words, the model is highly accurate in predicting Tesla stock prices, as it can explain most of the variability observed in the actual prices.

The decreasing trend in the training loss indicates that the model has learned and improved its predictions over the course of the training epochs. The test loss and R-squared metrics suggest that the Tesla stock price prediction model performs very well, with low test loss and a high R-squared score, indicating accurate predictions and a strong ability to explain the variability in the Tesla stock prices.

Tesla Stock Price Prediction with Airflow DAG

Now, we arrive at the essence of this chapter: orchestrating workflows as Airflow DAGs instead of standalone scripts with a main() function. In the previous subsection, we executed the Tesla stock price prediction code manually through a main() function, which orchestrated the execution of various tasks (preprocessing, model training, and model evaluation). When integrating code into an Airflow DAG, a separate main() function is not needed to orchestrate tasks in the same way. Instead, tasks are explicitly defined as part of the Airflow DAG using operators such as PythonOperator or BashOperator. Each operator corresponds to a specific task, and dependencies between tasks are specified using the >> operator within the DAG definition.

CHAPTER 8   SCALABLE DEEP LEARNING PIPELINES WITH APACHE AIRFLOW

To demonstrate how to run the exact same code as in the previous subsection as a DAG, we will follow the steps below to define and orchestrate the tesla_stock_prediction DAG:

- Imports: Necessary libraries and modules are imported.

- Default Arguments: Configuration settings for the DAG, such as owner, start date, and retries, are defined.

- DAG Definition: The DAG object is created with a unique identifier, description, and other settings.

- Preprocessing Data Task: Defines a PythonOperator (preprocess_data_task) that executes the preprocess_data_task function to load and preprocess the Tesla stock data.

- Training Model Task: Defines a PythonOperator (train_model_task) that executes the train_model_task function to train a deep learning regression model using the preprocessed data.

- Evaluating Model Task: Defines a PythonOperator (evaluate_model_task) that executes the evaluate_model_task function to evaluate the trained model.

- Task Dependencies: Sets dependencies between tasks using the >> operator, indicating the order in which tasks should be executed.

Let's walk through the code step by step, starting with importing the necessary modules:

Step 1: Imports

In this section, we import various libraries and modules required for the workflow:

CHAPTER 8   SCALABLE DEEP LEARNING PIPELINES WITH APACHE AIRFLOW

```
from datetime import datetime, timedelta
from airflow import DAG
from airflow.operators.python_operator import PythonOperator
import logging
import torch
import torch.optim as optim
import torch.nn as nn
from torch.utils.data import DataLoader
import numpy as np
from pyspark.sql import DataFrame
from typing import Tuple
from utils import spark_df_to_tensor
from data_processing import load_data, preprocess_data
from model_training import create_data_loader, train_model
from model_evaluation import evaluate_model
```

Each import statement brings in functionality required for specific tasks within the DAG, such as defining the workflow, handling data, performing deep learning tasks, and logging messages for monitoring and debugging purposes. More specifically

- datetime, timedelta: These are imported to work with dates and time intervals. They are used in the default_args dictionary to set the start date and retry delay.

- DAG, PythonOperator: These classes are imported from the Airflow library for defining the workflow and tasks. They are used to create the DAG (DAG class) and define tasks (PythonOperator class) such as preprocess_data_task, train_model_task, and evaluate_model_task.

CHAPTER 8 SCALABLE DEEP LEARNING PIPELINES WITH APACHE AIRFLOW

- logging: This module is imported to configure logging settings and log messages during the execution of the DAG. It is used to configure the logger (logging.basicConfig) and log messages (logger.error).

- torch, optim, nn, DataLoader: These are imported from the PyTorch library for deep learning tasks. They are used in the train_model_task and evaluate_model_task to define the neural network regression model, loss function, optimizer, and data loader.

- numpy: This module is imported for numerical operations. It is used in the train_model_task for array operations, as PyTorch interoperates with NumPy arrays.

- DataFrame: This class is imported from the PySpark library to represent data. It is used in the load_data and preprocess_data functions from the data_processing module, which are called in preprocess_data_task.

- Tuple: This class is imported for representing tuples. It is used for type annotations in data processing tasks.

- utils, data_processing, model_training, model_evaluation: These are custom modules imported from the project directory. They contain functions and utilities used for various tasks:

  - utils: Contains spark_df_to_tensor function

  - data_processing: Contains load_data and preprocess_data functions

- model_training: Contains create_data_loader and train_model functions

- model_evaluation: Contains evaluate_model function

Step 2: Default Arguments

In this section, we define the default arguments for the Airflow DAG. We also set the data file path and logging:

```
[In]: data_file_path = "/home/ubuntu/airflow/dags/TSLA_
      stock.csv"
```

```
[In]: logging.basicConfig(level=logging.INFO)
[In]: logger = logging.getLogger(__name__)
```

```
[In]: default_args = {
          'owner': 'airflow',
          'depends_on_past': False,
          'start_date': datetime(2024, 5, 1),
          'email_on_failure': False,
          'email_on_retry': False,
          'retries': 1,
          'retry_delay': timedelta(minutes=5),
      }
```

The default arguments are as follows:

- owner: Specifies the owner of the DAG

- depends_on_past: Determines whether a task instance should depend on the success of the previous task's instance

- start_date: Specifies the start date of the DAG

- email_on_failure: Determines whether to send an email on failure

CHAPTER 8 SCALABLE DEEP LEARNING PIPELINES WITH APACHE AIRFLOW

- email_on_retry: Determines whether to send an email on retry

- retries: Specifies the number of retries for failed task instances

- retry_delay: Specifies the delay between retries

Step 3: DAG Definition

In this section, we define the Airflow DAG itself:

```
[In]: dag = DAG(
        'tesla_stock_prediction',
        default_args=default_args,
        description='DAG for Tesla stock price prediction',
        schedule_interval=None,
    )
```

Here's what each parameter represents:

- 'tesla_stock_prediction': This is the unique identifier for our DAG. It's used to refer to the DAG throughout Airflow.

- default_args=default_args: This parameter specifies the default arguments for the DAG, which we defined earlier. These default arguments determine various settings and behaviors for the DAG execution.

- description='DAG for Tesla stock price prediction': This provides a description of our DAG, explaining its purpose or function.

- schedule_interval=None: This parameter determines the schedule interval for the DAG. In this case, it's set to None, indicating that the DAG is not scheduled to run at regular intervals, but rather triggered manually or externally.

CHAPTER 8   SCALABLE DEEP LEARNING PIPELINES WITH APACHE AIRFLOW

The scheduling functionality of the DAG is an important attribute compared with running the code without a DAG. Setting the schedule_interval to a specific value using a cron expression, such as '0 0 * * *', would mean that the DAG would run at 00:00 (midnight) every day of every month, regardless of the day of the week. Other examples of cron expressions include

- '0 12 * * MON-FRI': The DAG would run at 12:00 p.m. (noon) every Monday to Friday.
- '0 0 1 * *': The DAG would run at 00:00 (midnight) on the first day of every month.
- '0 0 * JAN,FEB,MAR *': The DAG would run at 00:00 (midnight) every day in January, February, and March.
- '0 0 1,15 * *': The DAG would run at 00:00 (midnight) on the 1st and 15th day of every month.

For example, the configuration below would ensure that the DAG runs at 00:00 (midnight) every day:

```
[In]: dag = DAG(
        'tesla_stock_prediction',
        default_args=default_args,
        description=('An example DAG with a specific schedule
                    interval'),
        schedule_interval='0 0 * * *'
    )
```

We can also schedule DAGs to run daily (every day at midnight), weekly (every Monday at midnight), or monthly (first day of every month at midnight), as follows:

CHAPTER 8   SCALABLE DEEP LEARNING PIPELINES WITH APACHE AIRFLOW

```
[In]: dag = DAG(
          'tesla_stock_prediction',
          default_args=default_args,
          description='A DAG scheduled to run daily',
          schedule_interval='@daily'
      )

[In]: dag = DAG(
          'tesla_stock_prediction',
          default_args=default_args,
          description='A DAG scheduled to run daily',
          schedule_interval='@weekly'
      )

[In]: dag = DAG(
          'tesla_stock_prediction',
          default_args=default_args,
          description='A DAG scheduled to run daily',
          schedule_interval='@monthly'
      )
```

Step 4: Preprocessing Data Task

In this section, we define a task called "preprocess_data_task" that performs the preprocessing of the Tesla stock data:

```
[In]: def preprocess_data_task():
          try:
              df = load_data(data_file_path)
              df = preprocess_data(df)
          except Exception as e:
              logger.error(f"Error in preprocessing data: {str(e)}")
              raise
```

```
[In]: preprocess_data_task = PythonOperator(
          task_id='preprocess_data_task',
          python_callable=preprocess_data_task,
          dag=dag,
      )
```

This task is responsible for preprocessing the Tesla stock data before it's used for model training and evaluation. It's an essential step in our workflow to ensure that the data is in the appropriate format for further processing.

Here's what each part of the code does:

- We define a function preprocess_data_task() that encapsulates the preprocessing logic. Inside the function, we load the data using the load_data() function and then preprocess it using the preprocess_data() function.

- We use a try-except block to handle any exceptions that may occur during the preprocessing process. If an error occurs, we log the error message using the logger and raise the exception.

- We create a PythonOperator named preprocess_data_task that executes the preprocess_data_task() function when triggered. We specify the task ID as "preprocess_data_task," the function to be called as python_callable=preprocess_data_task, and the DAG to which the task belongs as dag=dag.

CHAPTER 8 SCALABLE DEEP LEARNING PIPELINES WITH APACHE AIRFLOW

Step 5: Training Model Task

In this section, we define a task called "train_model_task" that trains the deep learning regression model using the preprocessed data:

```
[In]: def train_model_task():
          try:
              df = load_data(data_file_path)
              df = preprocess_data(df)
              train_df, test_df = df.randomSplit([0.8, 0.2],
              seed=42)
              train_features, train_labels = spark_df_
              to_tensor(
                  train_df
              )
              train_loader = create_data_loader(
                  train_features,
                  train_labels,
                  batch_size=32
              )

              input_size = train_features.shape[1]
              output_size = 1
              model = nn.Sequential(
                  nn.Linear(input_size, 64),
                  nn.ReLU(),
                  nn.Linear(64, 32),
                  nn.ReLU(),
                  nn.Linear(32, 16),
                  nn.ReLU(),
                  nn.Linear(16, output_size)
              )
```

```
        criterion = nn.MSELoss()
        optimizer = optim.Adam(model.parameters(),
        lr=0.001)
        train_model(
            model,
            train_loader,
            criterion,
            optimizer,
            num_epochs=100
        )
    except Exception as e:
        logger.error(f"Error in training model:
        {str(e)}")
        raise
[In]: train_model_task = PythonOperator(
        task_id='train_model_task',
        python_callable=train_model_task,
        dag=dag,
    )
```

This task is responsible for training the deep learning regression model using the preprocessed data. It's a crucial step in our workflow to create a model that can predict Tesla stock prices.

Here's what each part of the code does:

- We define a function train_model_task() that encapsulates the training logic. Inside the function, we load the preprocessed data; split it into training and testing sets; convert the data to PyTorch tensors; create a data loader; define the model architecture, loss function, and optimizer; and, finally, train the model using the train_model() function.

CHAPTER 8   SCALABLE DEEP LEARNING PIPELINES WITH APACHE AIRFLOW

- We use a try-except block to handle any exceptions that may occur during the training process. If an error occurs, we log the error message using the logger and raise the exception.

- We create a PythonOperator named train_model_task that executes the train_model_task() function when triggered. We specify the task ID as "train_model_task," the function to be called as python_callable=train_model_task, and the DAG to which the task belongs as dag=dag.

Step 6: Evaluating Model Task

In this section, we define a task called "evaluate_model_task" that evaluates the trained deep learning regression model using the test data:

```
[In]: def evaluate_model_task():
        try:
            df = load_data(data_file_path)
            df = preprocess_data(df)
            train_df, test_df = df.randomSplit([0.8, 0.2],
            seed=42)
            test_features, test_labels = spark_df_to_
            tensor(test_df)
            test_loader = create_data_loader(
                test_features,
                test_labels,
                batch_size=32
            )

            input_size = test_features.shape[1]
            output_size = 1
            model = nn.Sequential(
                nn.Linear(input_size, 64),
```

```
            nn.ReLU(),
            nn.Linear(64, 32),
            nn.ReLU(),
            nn.Linear(32, 16),
            nn.ReLU(),
            nn.Linear(16, output_size)
        )
        model.load_state_dict(torch.load('trained_
        model.pth'))

        criterion = nn.MSELoss()
        evaluate_model(model, test_loader, criterion)
    except Exception as e:
        logger.error(f"Error in evaluating model:
        {str(e)}")
        raise

[In]: evaluate_model_task = PythonOperator(
        task_id='evaluate_model_task',
        python_callable=evaluate_model_task,
        dag=dag,
    )
```

This task is responsible for evaluating the performance of the trained deep learning regression model using the test data. It's a crucial step in our workflow to assess how well the model generalizes to unseen data.

Here's what each part of the code does:

- We define a function evaluate_model_task() that encapsulates the evaluation logic. Inside the function, we load the preprocessed data, split it into training and testing sets, convert the test data to PyTorch tensors, create a data loader, load the trained model, and, finally, evaluate the model using the evaluate_model() function.

- We use a try-except block to handle any exceptions that may occur during the evaluation process. If an error occurs, we log the error message using the logger and raise the exception.

- We create a PythonOperator named evaluate_model_task that executes the evaluate_model_task() function when triggered. We specify the task ID as "evaluate_model_task," the function to be called as python_callable=evaluate_model_task, and the DAG to which the task belongs as dag=dag.

Step 7: Task Dependencies

In this section, we define the dependencies between the tasks in our DAG:

```
[In]: preprocess_data_task >> train_model_task >> evaluate_model_task
```

These task dependencies ensure that our workflow executes in the correct sequence, with each task depending on the successful completion of its prerequisite tasks. In this case, data preprocessing must be completed before model training can begin, and model training must be completed before model evaluation can take place.

Here's what each part of the code does:

- preprocess_data_task >> train_model_task: This statement sets the dependency relationship between the "preprocess_data_task" and "train_model_task" tasks. It means that the "train_model_task" task should only start execution after the "preprocess_data_task" task has successfully completed.

CHAPTER 8   SCALABLE DEEP LEARNING PIPELINES WITH APACHE AIRFLOW

- train_model_task >> evaluate_model_task: This statement sets the dependency relationship between the "train_model_task" and "evaluate_model_task" tasks. It means that the "evaluate_model_task" task should only start execution after the "train_model_task" task has successfully completed.

To execute the code on the AWS EC2 instance, we first activated the virtual environment named "myenv" to ensure that the necessary dependencies are available. Next, we navigated to the directory where the DAG for stock price prediction is located. Finally, we triggered the DAG named "stock_price_prediction" in Apache Airflow, using the commands below:

```
[In]: source /home/ubuntu/myenv/bin/activate
[In]: cd /home/ubuntu/airflow/dags
[In]: airflow dags trigger stock_price_prediction
```

The output from the DAG closely mirrors that of the main function without the DAG, albeit with minor discrepancies primarily stemming from variations in train loss, test loss, and R-squared values. These differences can be attributed to variances in the randomized selection of train and test samples.

The first output is the training loss values recorded at each epoch during the training process of the PyTorch model for predicting the Tesla stock prices:

CHAPTER 8   SCALABLE DEEP LEARNING PIPELINES WITH APACHE AIRFLOW

Epoch	Loss
1	30359.38
2	25931.21
3	34454.88
4	19568.01
5	4835.51
6	2016.92
7	1076.98
8	2502.23
9	1652.30
10	1186.91
...	...
100	8.50

In this table, we have two columns:

- Epoch: This column represents the number of training epochs, which are iterations over the entire training dataset. Each epoch consists of one forward pass and one backward pass through the entire dataset.

- Loss: This column indicates the loss or error value computed by the model during each epoch. In the context of stock price prediction, this loss value represents the discrepancy between the predicted stock prices and the actual observed prices in the training data. Lower values indicate better performance, as they imply that the model's predictions are closer to the actual prices.

CHAPTER 8 SCALABLE DEEP LEARNING PIPELINES WITH APACHE AIRFLOW

Here's an interpretation of that data:

- In the initial epochs, the loss values are relatively high (e.g., 30,359 at epoch 1), indicating that the model's predictions are far from the actual stock prices in the training dataset.

- As training progresses, the loss values gradually decrease (e.g., 1,187 at epoch 10). This suggests that the model is learning from the data and making better predictions over time. Toward the later epochs, the loss values become very low (e.g., 8.5 at epoch 100), indicating that the model has converged and is making highly accurate predictions.

The second output is the performance metrics of the trained model on a separate test dataset:

Metric	Value
Test Loss	25.3016
R-squared Score	0.9976

In this table, the meaning of each of the two performance metrics is as follows:

- Test Loss: This metric measures the error or loss of the model's predictions specifically on a test dataset. It represents how well the model generalizes to unseen data. A lower test loss indicates better performance, as it suggests that the model's predictions are closer to the actual values in the test dataset. In this case, the test loss is reported as 25.3.

CHAPTER 8   SCALABLE DEEP LEARNING PIPELINES WITH APACHE AIRFLOW

- R-squared Score (Coefficient of Determination): This metric evaluates the goodness of fit of the model's predictions to the actual values in the test dataset. It ranges from 0 to 1, where 1 indicates a perfect fit. An R-squared score of 0.9976 suggests that the model explains approximately 99.76% of the variance in the test data around its mean. In other words, the model's predictions align very closely with the actual values in the test dataset.

A low test loss coupled with a high R-squared score indicates that the model performs well in making accurate predictions on new data.

An Airflow Pipeline for Diabetes Prediction

In the previous section, we created an Airflow pipeline for a regression task—predicting Tesla stock prices. In this section, we will build a similar Airflow pipeline for a classification task—predicting diabetes. More precisely, we will create an AWS pipeline that includes preprocessing data using PySpark, developing a deep learning classification model using TensorFlow, and running the code using Airflow.

Similar to our approach in the regression task, we will first demonstrate how to execute the diabetes classification code without an Airflow DAG. This will involve using a main() function to orchestrate the tasks in the pipeline. We will utilize the same diabetes dataset as introduced in Chapter 7. The code structure will largely mirror that of the aforementioned chapter, with the primary distinction being a more modular approach presented in this chapter. Following the demonstration of running the code without the Airflow DAG, we will illustrate how to integrate it with the DAG.

CHAPTER 8   SCALABLE DEEP LEARNING PIPELINES WITH APACHE AIRFLOW

Diabetes Prediction Without Airflow DAG

In this subsection, we illustrate how a diabetes prediction pipeline is run without an Airflow DAG using a main() function. The code has been organized into the following four modules:

- preprocessing.py: This module is responsible for loading and preprocessing the diabetes data with PySpark.
- model_training.py: This module handles the training of a TensorFlow deep learning classification model.
- model_evaluation.py: This module is used for evaluating the TensorFlow deep learning classification model.
- main.py: This module contains the main() function, which orchestrates the overall workflow.

Let's go through these four modules one by one, starting with the preprocessing.py module:

Module 1: preprocessing.py

This module contains the code for preprocessing diabetes data using PySpark before feeding it into a TensorFlow model:

```
[In]: from pyspark.sql import SparkSession
[In]: from pyspark.ml.feature import VectorAssembler, 
      StandardScaler
[In]: import logging
[In]: from pyspark.sql.functions import col

[In]: logger = logging.getLogger(__name__)

[In]: def preprocess_data(
          data_file_path,
```

CHAPTER 8 SCALABLE DEEP LEARNING PIPELINES WITH APACHE AIRFLOW

```
    train_parquet_file_path,
    test_parquet_file_path
):
    """
    Preprocess diabetes data.

    Args:
        data_file_path (str):
            Path to the diabetes data CSV file.
        train_parquet_file_path (str):
            Path to save the preprocessed training data.
        test_parquet_file_path (str):
            Path to save the preprocessed testing data.
    """
    try:
        spark = (SparkSession.builder
                 .appName("DiabetesProcessing")
                 .getOrCreate())

        diabetes_df = spark.read.csv(
            data_file_path,
            header=True,
            inferSchema=True.
        )

        diabetes_df = diabetes_df.filter(
            (col("Glucose") != 0)
            & (col("BloodPressure") != 0)
            & (col("BMI") != 0)
        )

        feature_cols = ["Pregnancies",
                        "Glucose",
```

CHAPTER 8   SCALABLE DEEP LEARNING PIPELINES WITH APACHE AIRFLOW

```python
                "BloodPressure",
                "BMI",
                "DiabetesPedigreeFunction",
                "Age"]

assembler = VectorAssembler(
    inputCols=feature_cols,
    outputCol="features"
)
diabetes_df = assembler.transform(diabetes_df)

scaler = StandardScaler(
    inputCol="features",
    outputCol="scaled_features"
)
scaler_model = scaler.fit(diabetes_df)
diabetes_df = scaler_model.transform(diabetes_df)

train_df, test_df = diabetes_df.randomSplit(
    [0.8, 0.2],
    seed=42
)

train_df.write.parquet(
    train_parquet_file_path,
    mode="overwrite"
)
test_df.write.parquet(
    test_parquet_file_path,
    mode="overwrite"
)
```

```
            logger.info(
                "Data preprocessing completed successfully."
            )
        except Exception as e:
            logger.error(
                f"Error occurred during preprocessing:
                {str(e)}"
            )
```

This module encapsulates the preprocessing steps required for the diabetes dataset in a PySpark environment, making it ready for consumption by the TensorFlow model. Below is a brief explanation of the code:

- Importing Required Libraries: The module imports necessary libraries such as SparkSession, VectorAssembler, StandardScaler, logging, and functions from pyspark.sql.functions.

- Logger Setup: It sets up a logger named after the module.

- preprocess_data Function: This function is the main entry point for preprocessing. It takes three arguments:

  - data_file_path: Path to the diabetes data CSV file

  - train_parquet_file_path: Path to save the preprocessed training data

  - test_parquet_file_path: Path to save the preprocessed testing data

- SparkSession Setup: It creates a SparkSession named "DiabetesProcessing".

CHAPTER 8   SCALABLE DEEP LEARNING PIPELINES WITH APACHE AIRFLOW

- Reading Data: The CSV file containing the diabetes data is read into a DataFrame using Spark's read.csv method.

- Data Filtering: Rows where Glucose, BloodPressure, and BMI are equal to 0 are filtered out.

- Feature Selection: The feature columns are selected—"Pregnancies", "Glucose", "BloodPressure", "BMI", "DiabetesPedigreeFunction", and "Age".

- Feature Vectorization: A feature vector is created using VectorAssembler, which combines the selected feature columns into a single vector column named "features".

- Feature Scaling: Features are scaled using StandardScaler to standardize them.

- Data Splitting: The preprocessed data is split into training and testing sets using a 80–20 split ratio.

- Data Saving: The preprocessed training and testing data are saved as Parquet files.

- Logging: Logs a message indicating successful completion of data preprocessing. If any exception occurs during preprocessing, it logs an error message with details of the exception.

Module 2: model_training.py

This module is responsible for training the TensorFlow classification model using the preprocessed data:

```
[In]: import numpy as np
[In]: from pyspark.sql import SparkSession
[In]: from tensorflow.keras.models import Sequential
[In]: from tensorflow.keras.layers import Dense
```

CHAPTER 8 SCALABLE DEEP LEARNING PIPELINES WITH APACHE AIRFLOW

```python
[In]: from tensorflow.keras.optimizers import SGD
[In]: from tensorflow.keras.callbacks import Callback
[In]: import logging

[In]: logger = logging.getLogger(__name__)

[In]: class CustomCallback(Callback):
        """Custom callback for printing epoch information."""
        def on_epoch_end(self, epoch, logs=None):
            """Print epoch information at the end of each
            epoch."""
            if (epoch + 1) % 10 == 0:
                logger.info(
                    f"Epoch [{epoch + 1}/{self.
                    params['epochs']}], "
                    f"Loss: {logs['loss']:.4f}"
                )

[In]: def train_tensorflow_model(
        train_parquet_file_path,
        model_save_path
    ):
        """Train a TensorFlow model.

        Args:
            train_parquet_file_path (str):
                Path to the preprocessed training data.
            model_save_path (str):
                Path to save the trained model.
        """
        try:
            spark = (SparkSession.builder
                    .appName("TensorFlowTraining")
                    .getOrCreate())
```

534

CHAPTER 8   SCALABLE DEEP LEARNING PIPELINES WITH APACHE AIRFLOW

```python
train_diabetes_df = spark.read.parquet(
    train_parquet_file_path
)

X_train = np.array(
    train_diabetes_df.select("scaled_features")
    .rdd.flatMap(lambda x: x)
    .collect()
)
y_train = np.array(
    train_diabetes_df.select("Outcome")
    .rdd.flatMap(lambda x: x)
    .collect()
)

model = Sequential()
model.add(Dense(
    64,
    input_dim=X_train.shape[1],
    activation='relu'
))
model.add(Dense(32, activation='relu'))
model.add(Dense(1, activation='sigmoid'))

optimizer = SGD(learning_rate=0.01)
model.compile(
    loss='binary_crossentropy',
    optimizer=optimizer,
    metrics=['accuracy']
)
```

```
            custom_callback = CustomCallback()
            model.fit(
                X_train,
                y_train,
                epochs=100,
                verbose=0,
                callbacks=[custom_callback]
            )
            model.save(model_save_path)

            logger.info("Model training completed
            successfully.")
        except Exception as e:
            logger.error(
                f"Error occurred during model training:
                {str(e)}"
            )
```

Let's break down the code:

- Importing Required Libraries: The module imports necessary libraries such as NumPy, SparkSession from PySpark, and various components from TensorFlow, including Sequential for creating sequential models, Dense for adding densely connected layers, SGD for stochastic gradient descent optimization, Callback for creating custom callbacks, and logging for logging messages.

- Custom Callback: Defines a custom callback named CustomCallback, which inherits from TensorFlow's Callback class. This callback is used to print epoch information at the end of each epoch during training.

- train_tensorflow_model Function: This function is the main entry point for training the TensorFlow model. It takes two arguments:
  - train_parquet_file_path: Path to the preprocessed training data in Parquet format
  - model_save_path: Path to save the trained TensorFlow model
- SparkSession Setup: Creates a SparkSession named "TensorFlowTraining".
- Loading Data: The Parquet training data is loaded into a PySpark DataFrame named train_diabetes_df.
- Data Conversion: The PySpark DataFrame is converted to NumPy arrays (X_train and y_train) to be used as input and output for the TensorFlow model.
- Model Definition: Defines a sequential model with three layers—two dense layers with ReLU activation functions and one dense layer with a sigmoid activation function.
- Model Compilation: Compiles the model using binary cross-entropy loss function, SGD optimizer with a specified learning rate, and accuracy as the evaluation metric.
- Callback Setup: Creates an instance of the custom callback (custom_callback) defined earlier.
- Model Training: Fits the model to the training data for 100 epochs, using the custom callback for printing epoch information.

CHAPTER 8　SCALABLE DEEP LEARNING PIPELINES WITH APACHE AIRFLOW

- Model Saving: Saves the trained model to the specified path.

- Logging: Logs a message indicating successful completion of model training. If any exception occurs during training, it logs an error message with details of the exception.

Module 3: model_evaluation.py

This module is responsible for evaluating the performance of the TensorFlow classification model using the preprocessed test data:

```
[In]: import numpy as np
[In]: from pyspark.sql import SparkSession
[In]: from sklearn.metrics import accuracy_score, precision_
       score, [In]: recall_score, f1_score
[In]: from tensorflow.keras.models import load_model
[In]: import logging

[In]: logger = logging.getLogger(__name__)

[In]: def evaluate_model(
         test_parquet_file_path,
         model_load_path
     ):
         """
         Evaluate a TensorFlow model.

         Args:
             test_parquet_file_path (str):
                 Path to the preprocessed testing data.
             model_load_path (str):
                 Path to load the trained model.
         """
         try:
```

CHAPTER 8   SCALABLE DEEP LEARNING PIPELINES WITH APACHE AIRFLOW

```
(spark = SparkSession.builder
 .appName("ModelEvaluation")
 .getOrCreate())

model = load_model(model_load_path)

test_diabetes_df = spark.read.parquet(
    test_parquet_file_path
)

X_test = np.array(
    test_diabetes_df.select("scaled_features")
    .rdd.flatMap(lambda x: x)
    .collect()
)
y_test = np.array(
    test_diabetes_df.select("Outcome")
    .rdd.flatMap(lambda x: x)
    .collect()
)

y_pred = (model.predict(X_test) > 0.5).astype(int)

accuracy = accuracy_score(y_test, y_pred)
precision = precision_score(y_test, y_pred)
recall = recall_score(y_test, y_pred)
f1 = f1_score(y_test, y_pred)

logger.info(f"Accuracy: {accuracy:.4f}")
logger.info(f"Precision: {precision:.4f}")
logger.info(f"Recall: {recall:.4f}")
logger.info(f"F1 Score: {f1:.4f}")
```

```
        except Exception as e:
            logger.error(
                f"Error occurred during model evaluation: 
                {str(e)}"
            )
```

Let's break down the code:

- Importing Required Libraries: The module imports necessary libraries such as NumPy, SparkSession from PySpark, various evaluation metrics from sklearn. metrics, load_model from TensorFlow, and logging.

- evaluate_model Function: This function is the main entry point for evaluating the TensorFlow model. It takes two arguments:

  - test_parquet_file_path: Path to the preprocessed testing data in Parquet format

  - model_load_path: Path to load the trained TensorFlow model

- SparkSession Setup: Creates a SparkSession named "ModelEvaluation".

- Model Loading: Loads the trained TensorFlow model from the specified path using load_model.

- Loading Test Data: The Parquet test data is loaded into a PySpark DataFrame named test_diabetes_df.

- Data Conversion: The PySpark DataFrame is converted to NumPy arrays (X_test and y_test) to be used for model evaluation.

CHAPTER 8 SCALABLE DEEP LEARNING PIPELINES WITH APACHE AIRFLOW

- Model Prediction: Uses the loaded model to predict the outcomes (y_pred) for the test data. Thresholding at 0.5 is applied to convert probabilities to binary predictions.

- Evaluation Metrics Calculation: Calculates evaluation metrics such as accuracy, precision, recall, and F1 score using accuracy_score, precision_score, recall_score, and f1_score from sklearn.metrics.

- Logging: Logs the evaluation metrics, including accuracy, precision, recall, and F1 score.

- Error Handling: If any exception occurs during model evaluation, it logs an error message with details of the exception.

Module 4: main.py

We have now arrived at the main module. This script is the main entry point (main.py) for orchestrating the workflow of preprocessing, training, and evaluating the TensorFlow classification model for diabetes prediction:

```
[In]: import logging
[In]: from preprocessing import preprocess_data
[In]: from model_training import train_tensorflow_model
[In]: from model_evaluation import evaluate_model

[In]: logging.basicConfig(level=logging.INFO)
[In]: logger = logging.getLogger(__name__)

[In]: def main():
          preprocess_data(
              "/home/ubuntu/airflow/dags/diabetes.csv",
```

CHAPTER 8 SCALABLE DEEP LEARNING PIPELINES WITH APACHE AIRFLOW

```
            "/home/ubuntu/airflow/dags/diabetes_train.
            parquet",
            "/home/ubuntu/airflow/dags/diabetes_test.parquet"
    )
    logger.info("Data preprocessing completed.")
    train_tensorflow_model(
        "/home/ubuntu/airflow/dags/diabetes_train.parquet",
        "/home/ubuntu/airflow/dags/diabetes_model.h5"
    )
    logger.info("Model training completed.")
    evaluate_model(
        "/home/ubuntu/airflow/dags/diabetes_test.parquet",
        "/home/ubuntu/airflow/dags/diabetes_model.h5"
    )
    logger.info("Model evaluation completed.")
```

```
[In]: if __name__ == "__main__":
        main()
```

Let's break down the code:

- Importing Required Libraries: The script imports logging for logging messages and the functions preprocess_data, train_tensorflow_model, and evaluate_model from the respective modules.

- Logging Setup: It configures logging to display messages with a level of INFO.

- Main Function: This function is the main entry point for orchestrating the workflow. It performs the following steps:

CHAPTER 8   SCALABLE DEEP LEARNING PIPELINES WITH APACHE AIRFLOW

- Calls the preprocess_data function to preprocess the diabetes data, saving the preprocessed training and testing data as Parquet files.

- Logs a message indicating completion of data preprocessing.

- Calls the train_tensorflow_model function to train the TensorFlow model using the preprocessed training data, saving the trained model.

- Logs a message indicating completion of model training.

- Calls the evaluate_model function to evaluate the trained model using the preprocessed testing data.

- Logs a message indicating completion of model evaluation.

- __name__ == "__main__" Check: This check ensures that the main function is only executed if the script is run directly, not if it is imported as a module into another script.

This script orchestrates the entire workflow, from data preprocessing to model training and evaluation, logging messages to indicate the completion of each step. The paths provided to the preprocessing, training, and evaluation functions indicate where the input data and trained model will be stored or retrieved.

To run the code, we first activated the "myenv" environment, then navigated to the directory "home/ubuntu/airflow/dags" on EC2 instance where the code is stored, and executed the "main.py" module to orchestrate the diabetes prediction workflow:

CHAPTER 8 SCALABLE DEEP LEARNING PIPELINES WITH APACHE AIRFLOW

```
[In]: source activate myenv
[In]: cd home/ubuntu/airflow/dags
[In]: python3 main.py
```

The first output from the code provides the loss value for every tenth epoch during the training process of the TensorFlow diabetes classification model:

Epoch	Loss
10	0.5857
20	0.5705
30	0.5663
40	0.5611
50	0.5543
60	0.5484
70	0.5410
80	0.5359
90	0.5296
100	0.5273

Each row in the output table corresponds to a specific epoch, with the loss value recorded at the end of each epoch. The meaning of the columns is as follows:

- Epoch: This column indicates the epoch number, representing each complete pass through the entire training dataset during the training process.

CHAPTER 8 SCALABLE DEEP LEARNING PIPELINES WITH APACHE AIRFLOW

- Loss: This column represents the loss value obtained after each epoch. The loss value is a measure of how well the model's predictions match the actual labels in the training dataset. Lower loss values indicate better performance, as they suggest that the model's predictions are closer to the actual labels.

For example, in the first epoch (epoch 10), the loss value was about 0.59, indicating that after the first pass through the training dataset, the model's predictions had an average deviation from the actual labels of approximately 0.59. As training progresses through subsequent epochs, the loss generally decreases, indicating that the model is improving its ability to make accurate predictions. By the end of training (epoch 100), the loss has decreased to 0.53, suggesting that the model has achieved a lower average deviation from the actual labels compared with the earlier epochs.

The second output represents the evaluation metrics obtained after evaluating the diabetes classification model on the test dataset:

Metric	Value
Accuracy	0.7692
Precision	0.7059
Recall	0.5854
F1 Score	0.6400

The meaning of each metric is as follows:

- Accuracy: This metric represents the overall accuracy of the model, indicating the proportion of correctly classified instances among all instances. In this case, the model achieved an accuracy of 0.7692, meaning approximately 76.92% of the instances were correctly classified.

- Precision: Precision is a measure of the model's ability to correctly identify positive instances among all instances predicted as positive. A precision of 0.7059 indicates that out of all instances predicted as positive by the model, approximately 70.59% were true positive instances.

- Recall: Recall, also known as sensitivity, measures the proportion of actual positive instances that were correctly classified by the model. A recall of 0.5854 means that the model correctly identified approximately 58.54% of all actual positive instances.

- F1 Score: The F1 score is the harmonic or weighted mean of precision and recall, providing a single metric that balances both precision and recall. A higher F1 score indicates better overall performance. In this case, the F1 score is 0.64, suggesting a reasonable balance between precision and recall.

As we already know from Chapter 7, the results are less than optimal and suggest that there is room for improvement in the performance of the model. This is mainly due to a limited and an unbalanced sample with potential data quality issues.

Diabetes Prediction with Airflow DAG

In the previous subsection, we manually executed the diabetes prediction code through a main() function, orchestrating the execution of various tasks (preprocessing, model training, and model evaluation). In the current subsection, we'll convert this main() function into an Apache Airflow DAG. When integrating code into a DAG, a separate main() function isn't necessary to orchestrate tasks. Instead, tasks are explicitly defined within the Airflow DAG using operators such as PythonOperator

CHAPTER 8 SCALABLE DEEP LEARNING PIPELINES WITH APACHE AIRFLOW

or BashOperator. Each operator corresponds to a specific task, and dependencies between tasks are specified using the >> operator within the DAG definition.

Let's demonstrate how to run the same code as in the previous subsection as a DAG named diabetes_processing_and_training. This DAG replaces the main() function, with the three modules (preprocessing.py, model_training.py, and model_evaluation.py) remaining unchanged.

We start with the import statements.

Step 1: Imports

The below import statements bring in the necessary classes and functions required for defining and executing tasks within an Apache Airflow DAG for data preprocessing, model training, and model evaluation:

```
[In]: from datetime import datetime, timedelta
[In]: from airflow import DAG
[In]: from airflow.operators.python_operator import
      PythonOperator
[In]: from preprocessing import preprocess_data
[In]: from model_training import train_tensorflow_model
[In]: from model_evaluation import evaluate_model
```

Here's what each line does:

- from datetime import datetime, timedelta: This line imports the datetime and timedelta classes from the datetime module. These classes are used for working with dates and times, and timedelta is particularly useful for adding or subtracting time intervals.

- from airflow import DAG: This line imports the DAG class from the airflow module. DAG is a fundamental concept in Apache Airflow, representing a collection of tasks with dependencies.

CHAPTER 8 SCALABLE DEEP LEARNING PIPELINES WITH APACHE AIRFLOW

- from airflow.operators.python_operator import PythonOperator: This line imports the PythonOperator class from the python_operator module within the operators package in Airflow. PythonOperator is used to execute Python functions as tasks within an Airflow DAG.

- from preprocessing import preprocess_data: This line imports the preprocess_data function from the preprocessing module for preprocessing diabetes data.

- from model_training import train_tensorflow_model: This line imports the train_tensorflow_model function from the model_training module for training the TensorFlow machine learning classification model.

- from model_evaluation import evaluate_model: This line imports the evaluate_model function from the model_evaluation module for evaluating the trained TensorFlow machine learning model.

Step 2: Define Default Arguments

In this step, we define the default_args dictionary, which specifies various default arguments such as owner, start date, email settings, retries, and retry delay:

```
[In]: default_args = {
          'owner': 'airflow',
          'depends_on_past': False,
          'start_date': datetime(2024, 2, 25),
          'email_on_failure': False,
          'email_on_retry': False,
          'retries': 1,
          'retry_delay': timedelta(minutes=5),
      }
```

CHAPTER 8   SCALABLE DEEP LEARNING PIPELINES WITH APACHE AIRFLOW

This block of code defines a dictionary named default_args containing various configuration parameters for an Apache Airflow DAG. Each keyvalue pair in the dictionary represents a specific configuration option:

- 'owner': 'airflow': Specifies the owner of the DAG. This is typically the username or identifier of the person or team responsible for maintaining the DAG.

- 'depends_on_past': False: Indicates whether the execution of a task depends on the success of the previous run of the same task. Setting it to False means tasks can run independently of their previous execution status.

- 'start_date': datetime(2024, 2, 25): Sets the start date for the DAG's execution. Here, datetime(2024, 2, 25) creates a datetime object representing February 25, 2024, which specifies when the DAG should start running tasks.

- 'email_on_failure': False: Determines whether email notifications should be sent when a task within the DAG fails. Setting it to False disables failure email notifications.

- 'email_on_retry': False: Controls whether email notifications should be sent when a task is retried after failure. Similar to 'email_on_failure', setting it to False disables retry email notifications.

- 'retries': 1: Defines the number of times a task should be retried in case of failure. Here, tasks will be retried once after the initial failure.

CHAPTER 8 SCALABLE DEEP LEARNING PIPELINES WITH APACHE AIRFLOW

- 'retry_delay': timedelta(minutes=5): Specifies the time interval between task retries. In this case, tasks will be retried after a delay of 5 minutes if they fail.

Step 3: Create the DAG Object

In this step, we instantiate the DAG object using the DAG class. This creates a DAG named 'diabetes_processing_and_training' with the specified default arguments:

```
[In]: dag = DAG(
          'diabetes_processing_and_training',
          default_args=default_args,
          description=('Process diabetes data with PySpark '
                       'and train TensorFlow model'),
          schedule_interval=None,
      )
```

This code block initializes a DAG object named dag with specific attributes and configurations:

- 'diabetes_processing_and_training': Specifies the name of the DAG. This name is used to identify the DAG within Airflow.

- default_args=default_args: Assigns the default_args dictionary defined earlier to the default_args parameter of the DAG constructor. This ensures that the default configuration parameters specified in default_args are applied to the DAG.

- description=('Process diabetes data with PySpark and train TensorFlow model'): Provides a description for the DAG. This description serves as a brief summary of what the DAG is designed to accomplish.

CHAPTER 8   SCALABLE DEEP LEARNING PIPELINES WITH APACHE AIRFLOW

- schedule_interval=None: Sets the schedule interval for the DAG's execution. Here, None indicates that the DAG is not scheduled to run on a regular interval. Instead, it can be triggered manually or by external events.

Step 4: Define Tasks

In this step, we define three tasks corresponding to the three modules of the previous subsection: preprocessing data, training the TensorFlow model, and evaluating the trained model.

We begin by defining the preprocess_task that will execute the preprocess_data function with the specified arguments as part of the Airflow DAG:

```
[In]: preprocess_task = PythonOperator(
          task_id='preprocess_data',
          python_callable=preprocess_data,
          op_kwargs={
              'data_file_path': ("/home/ubuntu/airflow/dags"
                                 "/diabetes.csv"),
              'train_parquet_file_path': ("/home/ubuntu/
              airflow/dags"
                                          "/diabetes_train.
                                          parquet"),
              'test_parquet_file_path': ("/home/ubuntu/
              airflow/dags"
                                         "/diabetes_test.
                                         parquet")
          },
          dag=dag,
      )
```

551

This code creates an instance of the PythonOperator class named preprocess_task. Let's break down each part of this instantiation:

- PythonOperator: This is a class from the Airflow library used to execute a Python function as a task in an Airflow DAG.

- task_id='preprocess_data': This specifies the unique identifier for the task. In this case, it's named 'preprocess_data'.

- python_callable=preprocess_data: This parameter specifies the Python function that the operator will execute. Here, it's set to preprocess_data, indicating that the preprocess_data function will be executed as the task.

- op_kwargs: This parameter specifies any keyword arguments (kwargs) that need to be passed to the Python function (preprocess_data). It's a dictionary containing the arguments required by the preprocess_data function. These arguments include the file paths for the data file, training Parquet file, and testing Parquet file:

  - data_file_path: The path to the diabetes data CSV file

  - train_parquet_file_path: The path to save the preprocessed training data as a Parquet file

  - test_parquet_file_path: The path to save the preprocessed testing data as a Parquet file

- dag=dag: This parameter specifies the Airflow DAG to which the task belongs. It assigns the task to the DAG object created earlier in the script (dag).

CHAPTER 8   SCALABLE DEEP LEARNING PIPELINES WITH APACHE AIRFLOW

The next code is the same as the previous one, but it defines a different task named train_task:

```
[In]: train_task = PythonOperator(
          task_id='train_tensorflow_model',
          python_callable=train_tensorflow_model,
          op_kwargs={
              'train_parquet_file_path': ("/home/ubuntu/
              airflow/dags"
                                          "/diabetes_train.
                                          parquet"),
              'model_save_path': ("/home/ubuntu/airflow/dags"
                                  "/diabetes_model.h5")
          },
          dag=dag,
      )
```

This task will execute the train_tensorflow_model function with the specified arguments as part of the Airflow DAG. Let's break it down:

- PythonOperator: This is a class from the Airflow library used to execute a Python function as a task in an Airflow DAG.

- task_id='train_tensorflow_model': This specifies the unique identifier for the task. In this case, it's named 'train_tensorflow_model'.

- python_callable=train_tensorflow_model: This parameter specifies the Python function that the operator will execute. Here, it's set to train_tensorflow_model, indicating that the train_tensorflow_model function will be executed as the task.

CHAPTER 8 SCALABLE DEEP LEARNING PIPELINES WITH APACHE AIRFLOW

- op_kwargs: This parameter specifies any keyword arguments (kwargs) that need to be passed to the Python function (train_tensorflow_model). It's a dictionary containing the arguments required by the train_tensorflow_model function:
    - train_parquet_file_path: The path to the preprocessed training data Parquet file
    - model_save_path: The path to save the trained TensorFlow model file
- dag=dag: This parameter specifies the Airflow DAG to which the task belongs. It assigns the task to the DAG object created earlier in the script (dag).

The next code is also similar to the previous ones, but it defines yet another task named evaluate_task:

```
[In]: evaluate_task = PythonOperator(
        task_id='evaluate_model',
        python_callable=evaluate_model,
        op_kwargs={
            'test_parquet_file_path': ("/home/ubuntu/
            airflow/dags"
                                        "/diabetes_test.
                                        parquet"),
            'model_load_path': ("/home/ubuntu/airflow/dags"
                                        "/diabetes_model.h5")
        },
        dag=dag,
    )
```

This code defines a task (evaluate_task) that will execute the evaluate_model function with the specified arguments as part of the Airflow DAG. Let's break down each part of the code:

- PythonOperator: This class is from the Airflow library and is used to execute a Python function as a task in an Airflow DAG.
- task_id='evaluate_model': This specifies the unique identifier for the task. In this case, it's named 'evaluate_model'.
- python_callable=evaluate_model: This parameter specifies the Python function that the operator will execute. Here, it's set to evaluate_model, indicating that the evaluate_model function will be executed as the task.
- op_kwargs: This parameter specifies any keyword arguments (kwargs) that need to be passed to the Python function (evaluate_model). It's a dictionary containing the arguments required by the evaluate_model function:
    - test_parquet_file_path: The path to the preprocessed testing data Parquet file
    - model_load_path: The path from which to load the trained TensorFlow model file
- dag=dag: This parameter specifies the Airflow DAG to which the task belongs. It assigns the task to the DAG object created earlier in the script (dag).

CHAPTER 8 SCALABLE DEEP LEARNING PIPELINES WITH APACHE AIRFLOW

Step 5: Define Task Dependencies

In this final step, we define dependencies between tasks:

[In]: preprocess_task >> train_task >> evaluate_task

This line of code defines the task dependencies within the Airflow DAG:

- preprocess_task >> train_task >> evaluate_task: This line specifies the order in which tasks should be executed. It uses the >> operator to define dependencies between tasks.

- preprocess_task: This task represents the preprocessing step, which must be completed before proceeding to the next step.

- train_task: This task represents the training of the TensorFlow model. It depends on the completion of the preprocessing task (preprocess_task). Once the preprocessing is done, the training can begin.

- evaluate_task: This task represents the evaluation of the TensorFlow model. It depends on the completion of the training task (train_task). Once the training is completed, the model can be evaluated.

This step ensures that the tasks are executed in the correct order: preprocessing data, then training the TensorFlow model, and finally evaluating its performance. It establishes the workflow within the Airflow DAG.

To execute the code on the AWS EC2 instance, we first activated the virtual environment named "myenv" to ensure that the necessary dependencies are available. Next, we navigated to the directory where the DAG for diabetes prediction is located. Finally, we triggered the DAG named "diabetes_processing_and_training" in Apache Airflow, using the commands below:

```
[In]: source /home/ubuntu/myenv/bin/activate
[In]: cd /home/ubuntu/airflow/dags
[In]: airflow dags trigger diabetes_processing_and_training
```

The output generated by the DAG closely resembles that of the main function operating without the DAG, albeit with slight discrepancies attributable to variations in the random selection process for training and testing samples.

The first output from the code provides the loss value for every tenth epoch during the training process of the TensorFlow diabetes classification model:

Epoch	Loss
10	0.6083
20	0.5830
30	0.5728
40	0.5623
50	0.5605
60	0.5589
70	0.5485
80	0.5469
90	0.5413
100	0.5332

CHAPTER 8   SCALABLE DEEP LEARNING PIPELINES WITH APACHE AIRFLOW

The second output provides metrics that are based on test data. They collectively provide insight into the overall performance and effectiveness of the trained model in predicting diabetes outcomes. A higher accuracy, precision, recall, and F1 score indicate better model performance and predictive capability:

Metric	Value
Accuracy	0.7436
Precision	0.6571
Recall	0.5610
F1 Score	0.6053

As previously discussed in Chapter 7 and the preceding subsection, the outcomes indicate suboptimal performance, highlighting potential areas for enhancing the model's effectiveness. This can primarily be attributed to the constrained and imbalanced nature of the dataset, along with potential data quality concerns.

Bringing It All Together

In this section, we consolidate the individual codes for predicting Tesla stock price and diabetes with and without Apache Airflow DAGs.

Tesla Stock Price Prediction Without Airflow DAG

In this subsection, we present the code for executing the Tesla stock price prediction algorithm using a main() function, i.e., without an Airflow DAG:

Module 1: data_processing.py

CHAPTER 8   SCALABLE DEEP LEARNING PIPELINES WITH APACHE AIRFLOW

```
[In]: from pyspark.sql import SparkSession
[In]: from pyspark.ml.feature import VectorAssembler,
      StandardScaler
[In]: from pyspark.sql import DataFrame

[In]: def load_data(file_path: str) -> DataFrame:
          """
          Load stock price data from a CSV file using
          SparkSession.
          """
          spark = SparkSession.builder \
                  .appName("StockPricePrediction") \
                  .getOrCreate()
          df = spark.read.csv(
                  file_path,
                  header=True,
                  inferSchema=True
          )
          return df

[In]: def preprocess_data(df: DataFrame) -> DataFrame:
          """
          Preprocess the data by assembling feature
          vectors using
          VectorAssembler and scaling them using
          StandardScaler.
          """
          assembler = VectorAssembler(
              inputCols=['Open', 'High', 'Low', 'Volume'],
              outputCol='features'
          )
          df = assembler.transform(df)
```

559

CHAPTER 8   SCALABLE DEEP LEARNING PIPELINES WITH APACHE AIRFLOW

```
            scaler = StandardScaler(
                inputCol="features",
                outputCol="scaled_features",
                withStd=True,
                withMean=True
            )
            scaler_model = scaler.fit(df)
            df = scaler_model.transform(df)
            df = df.select('scaled_features', 'Close')
            return df
```

   Module 2: model_training.py

```
[In]: import torch
[In]: import torch.nn as nn
[In]: import torch.optim as optim
[In]: from torch.utils.data import DataLoader, TensorDataset

[In]: def create_data_loader(
          features,
          labels,
          batch_size=32
      ) -> DataLoader:
          """
          Convert the preprocessed data into PyTorch
          tensors and
          Create DataLoader objects for both the training
          and test
          sets.
          """
          dataset = TensorDataset(features, labels)
          return DataLoader(
              dataset,
```

```
        batch_size=batch_size,
        shuffle=True
    )

[In]: def train_model(
        model, train_loader, criterion, optimizer, num_epochs
    ):
        """
        Train the model on the training data using the
        DataLoader
        and the defined loss function and optimizer.
        Iterate over
        the data for a specified number of epochs,
        calculate the
        loss, and update the model parameters.
        """
        for epoch in range(num_epochs):
            for inputs, labels in train_loader:
                optimizer.zero_grad()
                outputs = model(inputs)
                loss = criterion(outputs, labels.
                unsqueeze(1))
                loss.backward()
                optimizer.step()
            print(
                f"Epoch [{epoch + 1}/{num_epochs}], "
                f"Loss: {loss.item():.4f}"
            )
```

CHAPTER 8 SCALABLE DEEP LEARNING PIPELINES WITH APACHE AIRFLOW

Module 3: model_evaluation.py

```
[In]: import torch
```

```
[In]: def evaluate_model(model, test_loader, criterion):
          """
          Evaluate the trained model on the test data to
          assess its
          performance.
          Calculate the test loss and additional
          evaluation metrics
          such as the R-squared score.
          """
          with torch.no_grad():
              model.eval()
              predictions = []
              targets = []
              test_loss = 0.0
              for inputs, labels in test_loader:
                  outputs = model(inputs)
                  loss = criterion(outputs, labels.
                  unsqueeze(1))
                  test_loss += loss.item() * inputs.size(0)
                  predictions.extend(outputs.squeeze().
                  tolist())
                  targets.extend(labels.tolist())

              test_loss /= len(test_loader.dataset)
              predictions = torch.tensor(predictions)
              targets = torch.tensor(targets)
              ss_res = torch.sum((targets - predictions) ** 2)
              ss_tot = torch.sum((targets - torch.
              mean(targets)) ** 2)
```

CHAPTER 8   SCALABLE DEEP LEARNING PIPELINES WITH APACHE AIRFLOW

```
        r_squared = 1 - ss_res / ss_tot
        print(f"Test Loss: {test_loss:.4f}")
        print(f"R-squared Score: {r_squared:.4f}")
        return test_loss, r_squared.item()
```

Module 4: utils.py

```
[In]: import numpy as np
[In]: import torch
[In]: from typing import Tuple
[In]: from pyspark.sql import DataFrame

[In]: def spark_df_to_tensor(
        df: DataFrame
    ) -> Tuple[torch.Tensor, torch.Tensor]:
        """
        Converts Spark DataFrame columns to PyTorch tensors.
        """
        features = torch.tensor(
            np.array(
                df.rdd.map(lambda x: x.scaled_features.
                toArray())
                .collect()
            ),
            dtype=torch.float32
        )
        labels = torch.tensor(
            np.array(
                df.rdd.map(lambda x: x.Close).collect()
            ),
            dtype=torch.float32
        )
        return features, labels
```

CHAPTER 8 SCALABLE DEEP LEARNING PIPELINES WITH APACHE AIRFLOW

Module 5: main.py

```python
[In]: import logging
[In]: import torch
[In]: import torch.optim as optim
[In]: import torch.nn as nn
[In]: from torch.utils.data import DataLoader
[In]: from typing import Tuple
[In]: from pyspark.sql import DataFrame
[In]: from data_processing import load_data, preprocess_data
[In]: from model_training import create_data_loader,
      train_model
[In]: from model_evaluation import evaluate_model
[In]: import utils

[In]: logging.basicConfig(level=logging.INFO)
[In]: logger = logging.getLogger(__name__)

[In]: def main(
          data_file_path: str,
          num_epochs: int = 100,
          batch_size: int = 32,
          learning_rate: float = 0.001
      ):
          """
          Main function for training and evaluating a deep
          learning
          model for Tesla stock price prediction.

          Args:
              data_file_path (str):
                  The path to the CSV file containing
                  stock price
                  data.
```

CHAPTER 8 SCALABLE DEEP LEARNING PIPELINES WITH APACHE AIRFLOW

```
    num_epochs (int):
        The number of epochs for training the model.
        Defaults to 100.
    batch_size (int):
        The batch size for data loading during
        training.
        Defaults to 32.
    learning_rate (float):
        The learning rate for the optimizer.
        Defaults to
        0.001.

Raises:
    FileNotFoundError: If the specified data
    file is not
    found.
"""
try:
    df = load_data(data_file_path)
    df = preprocess_data(df)

    train_df, test_df = df.randomSplit([0.8, 0.2],
    seed=42)

    train_features, train_labels = utils.spark_df_
    to_tensor(
        train_df
    )
    test_features, test_labels = utils.spark_df_
    to_tensor(
        test_df
    )
```

565

```python
    train_loader = create_data_loader(
        train_features,
        train_labels,
        batch_size=batch_size
    )
    test_loader = create_data_loader(
        test_features,
        test_labels,
        batch_size=batch_size
    )

    input_size = train_features.shape[1]
    output_size = 1
    model = nn.Sequential(
        nn.Linear(input_size, 64),
        nn.ReLU(),
        nn.Linear(64, 32),
        nn.ReLU(),
        nn.Linear(32, 16),
        nn.ReLU(),
        nn.Linear(16, output_size)
    )

    criterion = nn.MSELoss()
    optimizer = optim.Adam(
        model.parameters(),
        lr=learning_rate
    )

    train_model(
        model,
        train_loader,
        criterion,
```

```
            optimizer,
            num_epochs
        )
        evaluate_model(model, test_loader, criterion)

        torch.save(model.state_dict(), 'trained_
        model.pth')

    except FileNotFoundError:
        logger.error(f"Data file not found at
        {data_file_path}")
```

```
[In]: if __name__ == "__main__":
         data_file_path = "/home/ubuntu/airflow/dags/TSLA_
         stock.csv"
         main(data_file_path)
```

Tesla Stock Price Prediction with Airflow DAG

In this subsection, we present the code for executing the Tesla stock price prediction algorithm using an Airflow DAG. This replaces the main() function discussed in the previous subsection. Before sharing the code, we would like to add a few comments that can improve tasks when working with various datasets. While these techniques are not implemented in the current code for simplicity and clarity, they are valuable considerations for optimizing and scaling workflows in production environments.

First, when working with small datasets shared across multiple tasks, broadcast variables can improve efficiency. Broadcasting the dataset allows for caching a read-only copy on each node, reducing network traffic and speeding up task execution. This is particularly useful if the same dataset will be used in multiple tasks, such as training and evaluation.

CHAPTER 8　SCALABLE DEEP LEARNING PIPELINES WITH APACHE AIRFLOW

You can implement this with a single line of code:

```
[In]: broadcast_df = sc.broadcast(df.collect())
```

Second, to make DAGs more flexible and scalable, you can generate dynamic DAGs based on configuration files or parameters. This approach allows the DAG to adapt to different datasets or models by defining dataset paths, model configurations, and other parameters in a YAML or JSON file. These configurations can then be loaded within the DAG definition to dynamically create tasks and workflows.

Here's a load configuration example:

```
[In]: import yaml
```

```
[In]: with open('/home/ubuntu/airflow/dags/config.yaml', 'r')
      as file:
          config = yaml.safe_load(file)
```

Finally, to safeguard against interruptions during model training, you can implement model checkpoints. This involves saving the model's state at regular intervals so that training can resume from the last saved checkpoint in case of failure.

Here are basic steps:

- Save checkpoints during training:

  ```
  [In]: torch.save({
            'model_state_dict': model.state_dict(),
            'optimizer_state_dict': optimizer.state_dict(),
            'epoch': epoch,
            'loss': loss
        }, checkpoint_path)
  ```

CHAPTER 8 SCALABLE DEEP LEARNING PIPELINES WITH APACHE AIRFLOW

- Load from a checkpoint to resume training:

    ```
    [In]: checkpoint = torch.load(checkpoint_path)
    [In]: model.load_state_dict(checkpoint['model_
         state_dict'])
    [In]: optimizer.load_state_dict(checkpoint['optimizer_
         state_dict'])
    ```

This approach ensures that long-running training processes can continue from the last checkpoint if interrupted, preventing loss of progress.

Below is the code we have used in this chapter to execute the Tesla stock price prediction algorithm using an Airflow DAG:

```
from datetime import datetime, timedelta
from airflow import DAG
from airflow.operators.python_operator import PythonOperator
import logging
import torch
import torch.optim as optim
import torch.nn as nn
from torch.utils.data import DataLoader
import numpy as np
from pyspark.sql import DataFrame
from typing import Tuple
from utils import spark_df_to_tensor
from data_processing import load_data, preprocess_data
from model_training import create_data_loader, train_model
from model_evaluation import evaluate_model
[In]: data_file_path = "/home/ubuntu/airflow/dags/TSLA_
stock.csv"

[In]: logging.basicConfig(level=logging.INFO)
[In]: logger = logging.getLogger(__name__)
```

CHAPTER 8   SCALABLE DEEP LEARNING PIPELINES WITH APACHE AIRFLOW

```
[In]: default_args = {
          'owner': 'airflow',
          'depends_on_past': False,
          'start_date': datetime(2024, 5, 1),
          'email_on_failure': False,
          'email_on_retry': False,
          'retries': 1,
          'retry_delay': timedelta(minutes=5),
      }
```

```
[In]: dag = DAG(
          'tesla_stock_prediction',
          default_args=default_args,
          description='DAG for Tesla stock price prediction',
          schedule_interval=None,
      )
```

```
[In]: def preprocess_data_task():
          try:
              df = load_data(data_file_path)
              df = preprocess_data(df)
          except Exception as e:
              logger.error(f"Error in preprocessing data: {str(e)}")
              raise
```

```
[In]: preprocess_data_task = PythonOperator(
          task_id='preprocess_data_task',
          python_callable=preprocess_data_task,
          dag=dag,
      )
```

```
[In]: def train_model_task():
          try:
              df = load_data(data_file_path)
              df = preprocess_data(df)
              train_df, test_df = df.randomSplit([0.8, 0.2],
              seed=42)
              train_features, train_labels = spark_df_
              to_tensor(
                  train_df
              )
              train_loader = create_data_loader(
                  train_features,
                  train_labels,
                  batch_size=32
              )
              input_size = train_features.shape[1]
              output_size = 1
              model = nn.Sequential(
                  nn.Linear(input_size, 64),
                  nn.ReLU(),
                  nn.Linear(64, 32),
                  nn.ReLU(),
                  nn.Linear(32, 16),
                  nn.ReLU(),
                  nn.Linear(16, output_size)
              )
              criterion = nn.MSELoss()
              optimizer = optim.Adam(model.parameters(),
              lr=0.001)
              train_model(
                  model,
```

CHAPTER 8 SCALABLE DEEP LEARNING PIPELINES WITH APACHE AIRFLOW

```
            train_loader,
            criterion,
            optimizer,
            num_epochs=100
        )
    except Exception as e:
        logger.error(f"Error in training model:
        {str(e)}")
        raise
```

```
[In]: train_model_task = PythonOperator(
        task_id='train_model_task',
        python_callable=train_model_task,
        dag=dag,
    )
```

```
[In]: def evaluate_model_task():
        try:
            df = load_data(data_file_path)
            df = preprocess_data(df)
            train_df, test_df = df.randomSplit([0.8, 0.2],
            seed=42)
            test_features, test_labels = spark_df_to_
            tensor(test_df)
            test_loader = create_data_loader(
                test_features,
                test_labels,
                batch_size=32
            )

            input_size = test_features.shape[1]
            output_size = 1
            model = nn.Sequential(
```

```
            nn.Linear(input_size, 64),
            nn.ReLU(),
            nn.Linear(64, 32),
            nn.ReLU(),
            nn.Linear(32, 16),
            nn.ReLU(),
            nn.Linear(16, output_size)
        )
        model.load_state_dict(torch.load('trained_
        model.pth'))

        criterion = nn.MSELoss()
        evaluate_model(model, test_loader, criterion)
    except Exception as e:
        logger.error(f"Error in evaluating model:
        {str(e)}")
        raise
```

```
[In]: evaluate_model_task = PythonOperator(
        task_id='evaluate_model_task',
        python_callable=evaluate_model_task,
        dag=dag,
    )
[In]: preprocess_data_task >> train_model_task >> evaluate_
      model_task
```

Diabetes Prediction Without Airflow DAG

In this subsection, we present the code for executing the diabetes prediction algorithm using a main() function:

CHAPTER 8   SCALABLE DEEP LEARNING PIPELINES WITH APACHE AIRFLOW

Module 1: preprocessing.py

```
[In]: from pyspark.sql import SparkSession
[In]: from pyspark.ml.feature import VectorAssembler,
      StandardScaler
[In]: import logging
[In]: from pyspark.sql.functions import col

[In]: logger = logging.getLogger(__name__)

[In]: def preprocess_data(
          data_file_path,
          train_parquet_file_path,
          test_parquet_file_path
      ):
          """
          Preprocess diabetes data.

          Args:
              data_file_path (str):
                  Path to the diabetes data CSV file.
              train_parquet_file_path (str):
                  Path to save the preprocessed training data.
              test_parquet_file_path (str):
                  Path to save the preprocessed testing data.
          """
          try:
              spark = (SparkSession.builder
                      .appName("DiabetesProcessing")
                      .getOrCreate())

              diabetes_df = spark.read.csv(
                  data_file_path,
                  header=True,
```

CHAPTER 8   SCALABLE DEEP LEARNING PIPELINES WITH APACHE AIRFLOW

```python
        inferSchema=True.
    )
    diabetes_df = diabetes_df.filter(
        (col("Glucose") != 0)
        & (col("BloodPressure") != 0)
        & (col("BMI") != 0)
    )
    feature_cols = ["Pregnancies",
                    "Glucose",
                    "BloodPressure",
                    "BMI",
                    "DiabetesPedigreeFunction",
                    "Age"]
    assembler = VectorAssembler(
        inputCols=feature_cols,
        outputCol="features"
    )
    diabetes_df = assembler.transform(diabetes_df)
    scaler = StandardScaler(
        inputCol="features",
        outputCol="scaled_features"
    )
    scaler_model = scaler.fit(diabetes_df)
    diabetes_df = scaler_model.transform(diabetes_df)
    train_df, test_df = diabetes_df.randomSplit(
        [0.8, 0.2],
        seed=42
    )
```

CHAPTER 8   SCALABLE DEEP LEARNING PIPELINES WITH APACHE AIRFLOW

```
            train_df.write.parquet(
                train_parquet_file_path,
                mode="overwrite"
            )
            test_df.write.parquet(
                test_parquet_file_path,
                mode="overwrite"
            )
            logger.info(
                "Data preprocessing completed successfully."
            )
        except Exception as e:
            logger.error(
                f"Error occurred during preprocessing:
                {str(e)}"
            )
```

Module 2: model_training.py

```
[In]: import numpy as np
[In]: from pyspark.sql import SparkSession
[In]: from tensorflow.keras.models import Sequential
[In]: from tensorflow.keras.layers import Dense
[In]: from tensorflow.keras.optimizers import SGD
[In]: from tensorflow.keras.callbacks import Callback
[In]: import logging

[In]: logger = logging.getLogger(__name__)

[In]: class CustomCallback(Callback):
          """Custom callback for printing epoch information."""
          def on_epoch_end(self, epoch, logs=None):
```

CHAPTER 8   SCALABLE DEEP LEARNING PIPELINES WITH APACHE AIRFLOW

```
        """Print epoch information at the end of each
        epoch."""
        if (epoch + 1) % 10 == 0:
            logger.info(
                f"Epoch [{epoch + 1}/{self.
                params['epochs']}], "
                f"Loss: {logs['loss']:.4f}"
            )
```

```
[In]: def train_tensorflow_model(
    train_parquet_file_path,
    model_save_path
):
    """Train a TensorFlow model.

    Args:
        train_parquet_file_path (str):
            Path to the preprocessed training data.
        model_save_path (str):
            Path to save the trained model.
    """
    try:
        spark = (SparkSession.builder
                 .appName("TensorFlowTraining")
                 .getOrCreate())

        train_diabetes_df = spark.read.parquet(
            train_parquet_file_path
        )

        X_train = np.array(
            train_diabetes_df.select("scaled_features")
            .rdd.flatMap(lambda x: x)
```

CHAPTER 8 SCALABLE DEEP LEARNING PIPELINES WITH APACHE AIRFLOW

```python
        .collect()
    )
    y_train = np.array(
        train_diabetes_df.select("Outcome")
        .rdd.flatMap(lambda x: x)
        .collect()
    )

    model = Sequential()
    model.add(Dense(
        64,
        input_dim=X_train.shape[1],
        activation='relu'
    ))
    model.add(Dense(32, activation='relu'))
    model.add(Dense(1, activation='sigmoid'))

    optimizer = SGD(learning_rate=0.01)
    model.compile(
        loss='binary_crossentropy',
        optimizer=optimizer,
        metrics=['accuracy']
    )

    custom_callback = CustomCallback()
    model.fit(
        X_train,
        y_train,
        epochs=100,
        verbose=0,
        callbacks=[custom_callback]
    )
```

CHAPTER 8 SCALABLE DEEP LEARNING PIPELINES WITH APACHE AIRFLOW

```
        model.save(model_save_path)

        logger.info("Model training completed
        successfully.")
    except Exception as e:
        logger.error(
            f"Error occurred during model training:
            {str(e)}"
        )
```

Module 3: model_evaluation.py

```
[In]: import numpy as np
[In]: from pyspark.sql import SparkSession
[In]: from sklearn.metrics import accuracy_score, precision_
      score, [In]: recall_score, f1_score
[In]: from tensorflow.keras.models import load_model
[In]: import logging

[In]: logger = logging.getLogger(__name__)

[In]: def evaluate_model(
        test_parquet_file_path,
        model_load_path
    ):
        """
        Evaluate a TensorFlow model.

        Args:
            test_parquet_file_path (str):
                Path to the preprocessed testing data.
            model_load_path (str):
                Path to load the trained model.
        """
```

CHAPTER 8   SCALABLE DEEP LEARNING PIPELINES WITH APACHE AIRFLOW

```python
    try:
        (spark = SparkSession.builder
         .appName("ModelEvaluation")
         .getOrCreate())

        model = load_model(model_load_path)

        test_diabetes_df = spark.read.parquet(
            test_parquet_file_path
        )

        X_test = np.array(
            test_diabetes_df.select("scaled_features")
            .rdd.flatMap(lambda x: x)
            .collect()
        )
        y_test = np.array(
            test_diabetes_df.select("Outcome")
            .rdd.flatMap(lambda x: x)
            .collect()
        )

        y_pred = (model.predict(X_test) > 0.5).
        astype(int)

        accuracy = accuracy_score(y_test, y_pred)
        precision = precision_score(y_test, y_pred)
        recall = recall_score(y_test, y_pred)
        f1 = f1_score(y_test, y_pred)

        logger.info(f"Accuracy: {accuracy:.4f}")
        logger.info(f"Precision: {precision:.4f}")
        logger.info(f"Recall: {recall:.4f}")
```

CHAPTER 8 SCALABLE DEEP LEARNING PIPELINES WITH APACHE AIRFLOW

```
            logger.info(f"F1 Score: {f1:.4f}")
        except Exception as e:
            logger.error(
                f"Error occurred during model evaluation:
                {str(e)}"
            )
```

Module 4: main.py

```
[In]: import logging
[In]: from preprocessing import preprocess_data
[In]: from model_training import train_tensorflow_model
[In]: from model_evaluation import evaluate_model

[In]: logging.basicConfig(level=logging.INFO)
[In]: logger = logging.getLogger(__name__)

[In]: def main():
          preprocess_data(
              "/home/ubuntu/airflow/dags/diabetes.csv",
              "/home/ubuntu/airflow/dags/diabetes_train.
              parquet",
              "/home/ubuntu/airflow/dags/diabetes_test.parquet"
      )
      logger.info("Data preprocessing completed.")
      train_tensorflow_model(
          "/home/ubuntu/airflow/dags/diabetes_train.parquet",
          "/home/ubuntu/airflow/dags/diabetes_model.h5"
      )
      logger.info("Model training completed.")
      evaluate_model(
          "/home/ubuntu/airflow/dags/diabetes_test.parquet",
          "/home/ubuntu/airflow/dags/diabetes_model.h5"
      )
```

581

CHAPTER 8   SCALABLE DEEP LEARNING PIPELINES WITH APACHE AIRFLOW

```
        logger.info("Model evaluation completed.")
[In]: if __name__ == "__main__":
        main()
```

Diabetes Prediction with Airflow DAG

In this subsection, we present the code for executing the diabetes prediction algorithm using an Airflow DAG. This replaces the main() function discussed in the previous subsection:

```
[In]: from datetime import datetime, timedelta
[In]: from airflow import DAG
[In]: from airflow.operators.python_operator import
        PythonOperator
[In]: from preprocessing import preprocess_data
[In]: from model_training import train_tensorflow_model
[In]: from model_evaluation import evaluate_model
[In]: default_args = {
          'owner': 'airflow',
          'depends_on_past': False,
          'start_date': datetime(2024, 2, 25),
          'email_on_failure': False,
          'email_on_retry': False,
          'retries': 1,
          'retry_delay': timedelta(minutes=5),
      }
[In]: dag = DAG(
          'diabetes_processing_and_training',
          default_args=default_args,
          description=('Process diabetes data with PySpark '
                       'and train TensorFlow model'),
```

```
        schedule_interval=None,
    )

[In]: preprocess_task = PythonOperator(
        task_id='preprocess_data',
        python_callable=preprocess_data,
        op_kwargs={
            'data_file_path': ("/home/ubuntu/airflow/dags"
                                "/diabetes.csv"),
            'train_parquet_file_path': ("/home/ubuntu/
                                        airflow/dags"
                                        "/diabetes_train.
                                        parquet"),
            'test_parquet_file_path': ("/home/ubuntu/
                                        airflow/dags"
                                        "/diabetes_test.
                                        parquet")
        },
        dag=dag,
    )

[In]: train_task = PythonOperator(
        task_id='train_tensorflow_model',
        python_callable=train_tensorflow_model,
        op_kwargs={
            'train_parquet_file_path': ("/home/ubuntu/
            airflow/dags"
                                        "/diabetes_train.
                                        parquet"),
            'model_save_path': ("/home/ubuntu/airflow/dags"
                                "/diabetes_model.h5")
        },
        dag=dag,
    )
```

CHAPTER 8   SCALABLE DEEP LEARNING PIPELINES WITH APACHE AIRFLOW

```
[In]: evaluate_task = PythonOperator(
        task_id='evaluate_model',
        python_callable=evaluate_model,
        op_kwargs={
            'test_parquet_file_path': ("/home/ubuntu/
                                        airflow/dags"
                                        "/diabetes_test.
                                        parquet"),
            'model_load_path': ("/home/ubuntu/airflow/dags"
                                "/diabetes_model.h5")
        },
        dag=dag,
    )
[In]: preprocess_task >> train_task >> evaluate_task
```

Summary

In this chapter, we demonstrated how to build scalable data processing pipelines with Apache Airflow. We created entire AWS workflows that included preprocessing data using PySpark, developing deep learning models using PyTorch and TensorFlow, and running the code on AWS using Airflow for both regression and classification tasks. Additionally, we enhanced the code by making it modular and showcased how to run it both with and without Apache Airflow DAGs.

In the next chapter, we will delve into advanced techniques for enhancing the performance of deep learning models. Topics will include handling overfitting, implementing regularization techniques (such as dropout), employing early stopping strategies, and fine-tuning models for improved accuracy and efficiency.

CHAPTER 9

Techniques for Improving Model Performance

In the preceding chapters, we trained and evaluated models with default parameters and had no attempt to improve their performance. In this chapter, we examine various techniques to improve the performance of deep learning models. Achieving a model that generalizes well to unseen data is a critical goal in machine learning. Overfitting, a common issue, occurs when a model performs well on training data but poorly on validation or test datasets.

We will explore techniques to address overfitting, including dropout, early stopping, and L1 and L2 regularization, which impose constraints on the model's parameters to prevent overfitting. Dropout involves randomly "dropping out" (i.e., setting to zero) a subset of neurons during the training process, which helps prevent the network from becoming too reliant on specific neurons and encourages the network to generalize better. Early stopping involves monitoring the model's performance on a validation set during training and stopping the training process once the performance stops improving.

L1 and L2 regularization add a penalty term to the loss function used to train the model. L1 regularization (known in linear models as Lasso) adds the absolute value of the weights to the loss function, encouraging sparsity. In neural networks, this encourages sparsity in the weights, leading to fewer connections and potentially simpler models. L2 regularization (known in linear models as Ridge) adds the squared value of the weights to the loss function, encouraging smaller weights overall. In neural networks, L2 regularization helps prevent overfitting by penalizing large weights, although its effect on sparsity may be less pronounced compared with L1 regularization.

Additionally, we will cover hyperparameter optimization, including fine-tuning the learning rate and network architecture, to identify the best-performing model on the validation set. This will be done both manually and automatically using Keras Tuner.

The chapter begins with an examination of the baseline model, which will serve as a benchmark for comparing the results of regularization and hyperparameter fine-tuning. We will then delve into each technique in detail: dropout, early stopping, L1 and L2 regularization, learning rate adjustments, and modifying model capacity by varying the number of layers and neurons. Before combining all the code used throughout the chapter into a unified section, we will demonstrate how to use Keras Tuner for automatic hyperparameter optimization. The chapter concludes with a final section highlighting key takeaways.

The Baseline Model

We start with a baseline neural network model architecture, consistent with Chapter 5 for predicting Tesla stock price with TensorFlow, but with some modifications to align with the objectives of the current chapter. The model consists of several densely connected layers with ReLU activation

CHAPTER 9 TECHNIQUES FOR IMPROVING MODEL PERFORMANCE

functions, followed by an output layer. This model serves as our starting point for implementing regularization techniques (dropout, early stopping, L1 and L2 regularization) and fine-tuning enhancements.

The script starts with importing necessary libraries:

- logging: For logging information and errors
- numpy: For numerical operations
- tensorflow: For building and training the neural network
- pyspark.sql: For handling Spark DataFrames
- pyspark.ml.feature: For feature transformation
- sklearn.metrics: For calculating the R2 score
- matplotlib.pyplot: For plotting the training loss

```
[In]: import logging
[In]: import numpy as np
[In]: import tensorflow as tf
[In]: from pyspark.sql import SparkSession
[In]: from pyspark.ml.feature import VectorAssembler, StandardScaler
[In]: from sklearn.metrics import r2_score
[In]: from tensorflow.keras.models import Sequential
[In]: from tensorflow.keras.layers import Dense
[In]: import matplotlib.pyplot as plt
```

Following this, we define the TSLARegressor class, which encapsulates the entire process of data preprocessing, model training, and evaluation. The class starts with an init method, which initializes the TSLARegressor object. This method sets up a logger, initializes a Spark session, and defines input_shape and model attributes for later use:

CHAPTER 9   TECHNIQUES FOR IMPROVING MODEL PERFORMANCE

```
[In]: class TSLARegressor:
          """
          Class for TSLA stock price regression.
          """
          def __init__(self):
              """
              Initialize TSLARegressor object.
              """
              self.logger = logging.getLogger(__name__)
              self.spark = SparkSession.builder \
                          .appName("TSLA_Regression") \
                          .getOrCreate()
              self.input_shape = None
              self.model = None
```

Moving on to what's inside the class, we have the preprocess_data method. This method preprocesses the input data by reading the CSV file into a Spark DataFrame, assembling the feature columns into a single vector column, scaling the features using StandardScaler, and splitting the data into training and testing sets based on the train_ratio:

```
[In]: def preprocess_data(
          self,
          file_path,
          feature_cols=['Open', 'High', 'Low', 'Volume'],
          train_ratio=0.8,
          seed=42
      ):
          """
          Preprocesses the data.
```

```
Parameters:
    - file_path (str): Path to the CSV file.
    - feature_cols (list): List of feature columns.
    - train_ratio (float): Ratio of training data.
    - seed (int): Random seed for splitting data.

Returns:
    - tuple: Tuple containing train and test
      DataFrames.
"""
df = self.spark.read.csv(
    file_path,
    header=True,
    inferSchema=True
)
assembler = VectorAssembler(
    inputCols=feature_cols, outputCol='features')
df = assembler.transform(df)
scaler = StandardScaler(
    inputCol='features',
    outputCol='scaled_features',
    withMean=True,
    withStd=True
)
scaler_model = scaler.fit(df)
df = scaler_model.transform(df)
train_df, test_df = df.randomSplit(
    [train_ratio, 1-train_ratio],
    seed=seed
)
return train_df, test_df
```

CHAPTER 9   TECHNIQUES FOR IMPROVING MODEL PERFORMANCE

```python
def preprocess_data(
    self,
    file_path,
    feature_cols=['Open', 'High', 'Low', 'Volume'],
    train_ratio=0.8,
    seed=42
):
    """
    Preprocesses the data.

    Parameters:
        - file_path (str): Path to the CSV file.
        - feature_cols (list): List of feature
          columns.
        - train_ratio (float): Ratio of
          training data.
        - seed (int): Random seed for splitting data.

    Returns:
        - tuple: Tuple containing train and test
          DataFrames.
    """
    df = self.spark.read.csv(
        file_path,
        header=True,
        inferSchema=True
    )
    assembler = VectorAssembler(
        inputCols=feature_cols, outputCol='features')
    df = assembler.transform(df)
```

CHAPTER 9　TECHNIQUES FOR IMPROVING MODEL PERFORMANCE

```
scaler = StandardScaler(
    inputCol='features',
    outputCol='scaled_features',
    withMean=True,
    withStd=True
)
scaler_model = scaler.fit(df)
df = scaler_model.transform(df)
train_df, test_df = df.randomSplit(
    [train_ratio, 1-train_ratio],
    seed=seed
)
return train_df, test_df
```

Next, we have the convert_to_numpy method, which converts the Spark DataFrames to NumPy arrays by extracting and collecting the scaled features and labels from the Spark DataFrames into NumPy arrays:

```
[In]: def convert_to_numpy(self, train_df, test_df):
    """
    Convert Spark DataFrames to numpy arrays.

    Parameters:
        - train_df (DataFrame):
            Spark DataFrame containing training data.
        - test_df (DataFrame):
            Spark DataFrame containing test data.

    Returns:
        - tuple:
            Tuple containing train and test numpy
            arrays for
            features and labels.
    """
```

CHAPTER 9   TECHNIQUES FOR IMPROVING MODEL PERFORMANCE

```
train_features = np.array(
    train_df.select('scaled_features')
    .rdd.map(lambda x: x.scaled_features.toArray())
    .collect()
)
train_labels = np.array(
    train_df.select('Close')
    .rdd.map(lambda x: x.Close).collect())
test_features = np.array(
    test_df.select('scaled_features')
    .rdd.map(lambda x: x.scaled_features.toArray())
    .collect()
)
test_labels = np.array(
    test_df.select('Close')
    .rdd.map(lambda x: x.Close).collect()
)
return (train_features,
        train_labels,
        test_features,
        test_labels
)
```

Transitioning to the next method, we have the create_and_train_model method, which creates and trains the neural network model:

- Defines a sequential neural network model with multiple dense layers
- Compiles the model with the Adam optimizer and Mean Squared Error loss
- Trains the model on the training data

```
[In]: def create_and_train_model(
          self,
          train_features,
          train_labels,
          epochs=100,
          batch_size=32
      ):
          """
          Create and train the neural network model.

          Parameters:
              - train_features (ndarray): Training features.
              - train_labels (ndarray): Training labels.
              - epochs (int): Number of epochs for training.
              - batch_size (int): Batch size for training.

          Returns:
              - History:
                  Training history containing loss values
                  for each
                  epoch.
          """
          self.input_shape = (train_features.shape[1],)
          self.model = Sequential([
              Dense(
                  64,
                  activation='relu',
                  input_shape=self.input_shape
              ),
              Dense(
                  32,
                  activation='relu'
              ),
```

CHAPTER 9   TECHNIQUES FOR IMPROVING MODEL PERFORMANCE

```
        Dense(
            16,
            activation='relu'
        ),
        Dense(1)
    ])
    self.model.compile(optimizer='adam', loss='mse')
    history = self.model.fit(
        train_features,
        train_labels,
        epochs=epochs,
        batch_size=batch_size,
        verbose=0
    )
    return history
```

Moving on to the next step, we have the evaluate_model method. This method evaluates the trained model on the test data by computing the test loss using the model's evaluate method:

```
[In]: def evaluate_model(self, test_features, test_labels):
    """
    Evaluate the trained model.

    Parameters:
        - test_features (ndarray): Test features.
        - test_labels (ndarray): Test labels.

    Returns:
        - float: Test loss.
    """
```

CHAPTER 9 TECHNIQUES FOR IMPROVING MODEL PERFORMANCE

```
    test_loss = self.model.evaluate(
        test_features,
        test_labels
    )
    return test_loss
```

Moving on to the next method, we have the predict_and_evaluate method, which predicts the Tesla stock prices on the test data and evaluates the model by

- Making predictions on the test features
- Computing the R2 score to assess the model's performance

```
[In]: def predict_and_evaluate(
        self,
        test_features,
        test_labels
    ):
        """
        Predict and evaluate the neural network model.

        Parameters:
            - test_features (ndarray): Test features.
            - test_labels (ndarray): Test labels.

        Returns:
            - float: R2 score.
        """
        test_predictions = self.model.predict(test_features)
        r2_score_value = r2_score(test_labels, test_
        predictions)
        return r2_score_value
```

CHAPTER 9 TECHNIQUES FOR IMPROVING MODEL PERFORMANCE

In the next step, we transition to the main function, which orchestrates the entire workflow. This function, situated outside the TSLARegressor class, instantiates a TSLARegressor object and then calls methods to preprocess data, convert data to NumPy arrays, create and train the model, and evaluate the model. It also logs the test loss and R2 score and plots the training loss over epochs using Matplotlib:

```
[In]: def main(file_path):
    """
    Main function to orchestrate the entire workflow.
    Parameters:
        - file_path (str): Path to the CSV file.
    Returns:
        - None
    """
    try:
        tsla_regressor = TSLARegressor()
        train_df, test_df = tsla_regressor.
        preprocess_data(
            file_path
        )
        train_features, train_labels, test_features, \
        test_labels = tsla_regressor.convert_to_numpy(
            train_df,
            test_df
        )
        history = tsla_regressor.create_and_train_model(
            train_features,
            train_labels
        )
```

```
        test_loss = tsla_regressor.evaluate_model(
            test_features,
            test_labels
        )
        r2_score_value = tsla_regressor.predict_and_
        evaluate(
            test_features,
            test_labels
        )

        logging.info("Test Loss: {}".format(test_loss))
        logging.info("R2 Score: {}".format(r2_
        score_value))

        plt.plot(history.history['loss'])
        plt.title('Model Loss')
        plt.xlabel('Epoch')
        plt.ylabel('Loss')

        plt.show()

    except Exception as e:
        logging.error(f"An error occurred: {e}")
```

In the final step, we have the main function, which serves as the entry point to the script's workflow:

```
[In]: if __name__ == "__main__":
        logging.basicConfig(level=logging.INFO)
        file_path = "/home/ubuntu/airflow/dags/TSLA_
        stock.csv"
        main(file_path)
```

This function configures logging to the INFO level, defines the file path for the CSV data, and calls the main function with the file path if the script is executed as the main module.

The execution flow of the function is as follows:

- Data Preprocessing: Reads and processes the CSV data, scales features, and splits it into training and test sets
- Data Conversion: Converts Spark DataFrames to NumPy arrays for compatibility with TensorFlow
- Model Training: Creates a neural network model, trains it on the training data, and records the training loss
- Model Evaluation: Evaluates the trained model on the test data and calculates the R2 score
- Logging and Plotting: Logs the evaluation metrics and plots the training loss

To run the script from the /home/ubuntu/airflow/dags directory using Python 3 and considering the virtual environment named myenv, we followed these steps:

Activate the virtual environment named myenv:

[In]: source /path/to/myenv/bin/activate

Change directory to /home/ubuntu/airflow/dags:

[In]: cd /home/ubuntu/airflow/dags

Run the script using Python 3, specifying the script's name "predict_tsla_stock.py":

[In]: python3 predict_tsla_stock.py

CHAPTER 9   TECHNIQUES FOR IMPROVING MODEL PERFORMANCE

This command executes the predict_tsla_stock.py script using Python 3 from the /home/ubuntu/airflow/dags directory.

The code produces two main outputs as part of the overall TSLA stock price regression workflow. The first output is the evaluation metrics (test loss and R2 score), which are printed using the logging.info method:

```
[In]: logging.info("Test Loss: {}".format(test_loss))
[In]: logging.info("R2 Score: {}".format(r2_score_value))
```

The following is the output:

```
Test Loss: 15.5924
R2 Score: 0.9987
```

The test loss is calculated using the evaluate_model method and logged:

```
[In]: test_loss = tsla_regressor.evaluate_model(
        test_features, test_labels
    )
```

The R2 score is calculated using the predict_and_evaluate method and logged:

```
[In]: r2_score_value = tsla_regressor.predict_and_evaluate(
        test_features, test_labels
    )
```

The combination of a relatively low test loss (15.5924) and a very high R2 score (0.9987) suggests that the model performs exceptionally well on the test data. The low test loss indicates that the predictions are close to the actual values, while the high R2 score shows that the model explains nearly all the variability in the data.

CHAPTER 9   TECHNIQUES FOR IMPROVING MODEL PERFORMANCE

The second output is a plot of the training loss over the epochs, which is saved as model_loss_plot.png, using the code below:

```
[In]: plt.plot(history.history['loss'])
[In]: plt.title('Model Loss')
[In]: plt.xlabel('Epoch')
[In]: plt.ylabel('Loss')
[In]: plt.axvline(
        x=10,
        color='r',
        linestyle='--',
        label='Epoch 10 plateau'
    )
[In]: plt.savefig('model_loss_plot.png')
[In]: plt.show()
```

The training history, containing loss values for each epoch, is returned from the create_and_train_model method and used for plotting the loss:

```
[In]: history = tsla_regressor.create_and_train_model(
        train_features,
        train_labels
    )
```

The plot is shown in Figure 9-1.

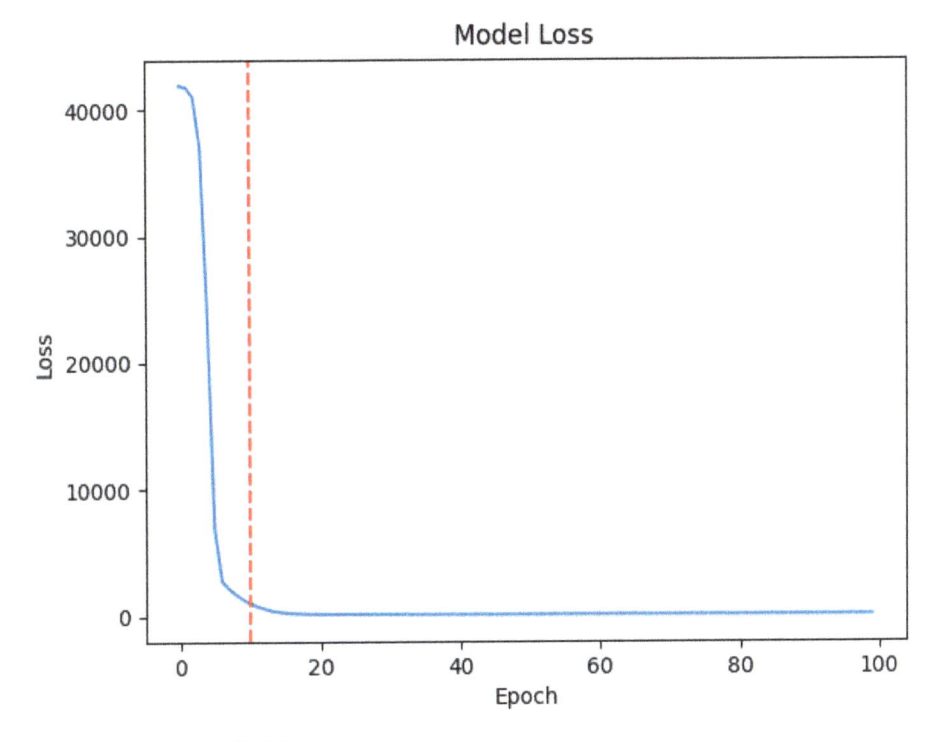

CHAPTER 9 TECHNIQUES FOR IMPROVING MODEL PERFORMANCE

Figure 9-1. Model Loss

The training loss plot shows a horizontal line beyond epoch 10 (denoted with the dashed vertical line), indicating that the training loss values do not decrease further and remain relatively constant. This typically signifies that the model has reached a plateau in its learning.

This could mean the following:

- The model has converged to a local minimum, meaning it has found a set of weights where further training does not significantly reduce the loss.

- The learning rate may be too low to escape a local minimum, or it might be just right, indicating the model has adequately learned from the data.

CHAPTER 9   TECHNIQUES FOR IMPROVING MODEL PERFORMANCE

- The model's capacity to learn from the given data might be saturated, meaning it has learned as much as it can from the training data given its architecture and hyperparameters.

In other words, the current performance level is likely the best the model can achieve with the existing architecture, training data, and hyperparameters. Alternatively, it may indicate overfitting, suggesting that the model is fitting too closely to the training data and may not generalize well to unseen data.

In our case, since the model's performance on the validation set is satisfactory, the plateau in the training loss may indicate that the model's learning process has stabilized for the training data. This can be a sign of successful training since the loss values are low and the model performs well on validation data (as judged by low test loss and high R2 Score).

In a production setting, the following actions are normally taken when the model is suspected to be overfitting:

- Review the Plateau: Confirm if both training and validation losses have plateaued, which can be visualized by plotting both losses.

- Adjust Hyperparameters: Experiment with different learning rates, batch sizes, or training durations.

- Model Simplification: Consider simplifying the model architecture by reducing the number of layers or neurons if the model seems to overfit. Alternatively, consider deeper or more complex architectures if the model seems to underfit.

- Data Augmentation: Use data augmentation techniques to artificially increase the size and variability of the training dataset.

CHAPTER 9   TECHNIQUES FOR IMPROVING MODEL PERFORMANCE

- Regularization: Implement or adjust regularization techniques such as dropout, L1 regularization, L2 regularization, or early stopping to prevent overfitting.

In the remainder of this chapter, we examine how to implement some of these strategies using the example of predicting Tesla stock price with TensorFlow.

Early Stopping

Early stopping involves monitoring the model's performance on a validation set during training and stopping the training process when the performance starts to degrade. This prevents the model from overfitting to the training data by stopping training before it begins to memorize noise in the data.

To add early stopping to the training process in our baseline model, we'll need to modify the create_and_train_model method. Specifically, we'll need to add an early stopping callback to monitor a certain metric (e.g., loss) and stop training if the monitored metric does not improve after a certain number of epochs.

Here's how we can modify the create_and_train_model method to include early stopping:

```
[In]: from tensorflow.keras.callbacks import EarlyStopping

[In]: def create_and_train_model(
          self,
          train_features,
          train_labels,
          epochs=100,
          batch_size=32
      ):
```

```python
            self.input_shape = (train_features.shape[1],)
            self.model = Sequential([
                Dense(
                    64,
                    activation='relu',
                    input_shape=self.input_shape
                ),
                Dense(
                    32,
                    activation='relu'
                ),
                Dense(
                    16,
                    activation='relu'
                ),
                Dense(1)
            ])

            early_stopping = EarlyStopping(
                monitor='val_loss',
                patience=10,
                restore_best_weights=True
            )

            self.model.compile(optimizer='adam', loss='mse')
            history = self.model.fit(
                train_features,
                train_labels,
                epochs=epochs,
                batch_size=batch_size,
                verbose=0,
```

CHAPTER 9   TECHNIQUES FOR IMPROVING MODEL PERFORMANCE

```
        callbacks=[early_stopping],
        validation_split=0.2
    )
    return history
```

In the above modification

- We imported the EarlyStopping callback from tensorflow.keras.callbacks.

- We instantiated an EarlyStopping callback with parameters:

    - monitor='val_loss': It monitors the validation loss.

    - patience=10: It waits for ten epochs before stopping if the monitored metric does not improve.

    - restore_best_weights=True: It restores the model weights from the epoch with the best validation loss.

- We added the early_stopping callback to the fit method's callbacks parameter list.

- We included a validation_split=0.2 argument in the fit method to specify that 20% of the training data should be used for validation during training. This is necessary for the early stopping callback to monitor the validation loss.

The first output we get after running the baseline model with this modification included is the early stopping information below:

```
[Out]: Epoch 87: early stopping
[Out]: Restoring model weights from the end of the best
       epoch: 77
[Out]: Epochs completed: 87
```

605

CHAPTER 9   TECHNIQUES FOR IMPROVING MODEL PERFORMANCE

Below is an explanation of this output:

- [Epoch 87: early stopping] means that early stopping was activated because the validation metric did not improve for the specified patience period. The patience period is the number of epochs to wait for an improvement in the validation metric before stopping the training. For example, a patience of ten epochs means the training will stop if the validation metric does not improve for ten consecutive epochs.

- [Restoring model weights from the end of the best epoch: 77] means that the best model (with the lowest validation loss or highest validation metric R-squared) was found at epoch 77. The model weights were restored from this epoch. Restoring weights means reverting the model to the state it was in at the specified epoch, ensuring that the model parameters used for further evaluation are those that yielded the best validation performance.

- [Epochs completed: 87] means 87 epochs were completed, including the patience period (i.e., 10 patience period + 77) before early stopping was triggered.

The second output is the loss and R-squared score at the early stopping point:

[Out]: Loss at early stopping point: 16.40
[Out]: R2 score at early stopping point: 0.9986

This means that at the early stopping point, the loss value was 16.40 and the R2 score was 0.9986.

Finally, once the best model is selected (after early stopping), the code evaluates this model on the test set to get an unbiased estimate of its performance on unseen data:

```
[Out]: Test Loss: 15.21
[Out]: R2 Score: 0.9987
```

This means that using the test dataset, the model achieved a loss value of 15.21 and an R2 score of 0.9987.

Dropout

Dropout is a technique where randomly selected neurons are ignored during training with a certain probability. This helps prevent the co-adaptation of neurons and forces the network to learn more robust features. Dropout effectively simulates training multiple neural networks with different architectures simultaneously.

To add dropout layers to our baseline neural network model, we'll need to modify the part of the code where we define the architecture of the model. Specifically, we will insert dropout layers between the dense layers in the create_and_train_model method. Dropout layers help prevent overfitting by randomly setting a fraction of input units to 0 at each update during training time, which can be beneficial for regularization.

Here's the modified create_and_train_model method with dropout layers added:

```
[In]: from tensorflow.keras.layers import Dropout

[In]: def create_and_train_model(
        self,
        train_features,
        train_labels,
```

```
        epochs=100,
        batch_size=32
    ):
        self.input_shape = (train_features.shape[1],)
        self.model = Sequential([
            Dense(
                64,
                activation='relu',
                input_shape=self.input_shape
            ),
            Dropout(0.5),
            Dense(
                32,
                activation='relu'
            ),
            Dropout(0.5),
            Dense(
                16,
                activation='relu'
            ),
            Dropout(0.5),
            Dense(1)
        ])
        self.model.compile(optimizer='adam', loss='mse')
        history = self.model.fit(
            train_features,
            train_labels,
            epochs=epochs,
            batch_size=batch_size,
            verbose=0
        )
        return history
```

CHAPTER 9   TECHNIQUES FOR IMPROVING MODEL PERFORMANCE

We have implemented the following modifications to the baseline model:

- Added from tensorflow.keras.layers import Dropout at the beginning of the file.

- Introduced Dropout(0.5) layers between the dense layers in the model architecture. The dropout rate is set to 0.5, meaning that 50% of the outputs from the previous dense layer will be randomly dropped out during training. This rate can be adjusted to other values (e.g., 0.2 or 0.3) depending on the desired extent of regularization.

After running the baseline model with the above dropout modification, we obtained the following output:

[Out]: Test Loss: 72.67
[Out]: R2 Score: 0.9939

A dropout rate of 0.5 is not uncommon, though it might be too aggressive under certain situations. Dropout rates typically range from 0.2 to 0.5, depending on the specific requirements and characteristics of the model and data.

Running the model with a dropout rate of 0.2 gave us the following output:

[Out]: Test Loss: 41.8448
[Out]: R2 Score: 0.9965

Even though we see an improvement in the metrics (lower loss and higher R2) at a dropout rate of 0.2 compared with 0.5, in a production setting, we would normally check if the model is underfitting or overfitting by comparing training with test metrics. If the model is underfitting (i.e., performing poorly on both training and test data), the dropout might

CHAPTER 9   TECHNIQUES FOR IMPROVING MODEL PERFORMANCE

be too aggressive, and reducing its rate could help. If it's overfitting (i.e., performing well on training data but poorly on test data), we might need more dropout.

L1 and L2 Regularization

L1 and L2 regularization techniques introduce a penalty term to the loss function based on the magnitudes of the model parameters. L1 regularization incorporates the absolute values of the parameters, whereas L2 regularization involves the squared values of the parameters. This incentivizes the model to maintain small parameter values, thereby mitigating overfitting.

To integrate L1 and L2 regularization into the baseline neural network model, we must adjust the create_and_train_model method. Specifically, regularization needs to be added to the dense layers where they are defined in the model architecture.

Below is the updated create_and_train_model method with L1 regularization incorporated:

```
[In]: from tensorflow.keras.regularizers import l1

[In]: def create_and_train_model(
          self,
          train_features,
          train_labels,
          epochs=100,
          batch_size=32
      ):
          self.input_shape = (train_features.shape[1],)
          self.model = Sequential([
```

```
    Dense(
        64,
        activation='relu',
        kernel_regularizer=l1(0.01),
        input_shape=self.input_shape
    ),
    Dense(
        32,
        activation='relu',
        kernel_regularizer=l1(0.01)
    ),
    Dense(
        16,
        activation='relu',
        kernel_regularizer=l1(0.01)
    ),
    Dense(1)
])

self.model.compile(optimizer='adam', loss='mse')
history = self.model.fit(
    train_features,
    train_labels,
    epochs=epochs,
    batch_size=batch_size,
    verbose=0
)
return history
```

CHAPTER 9 TECHNIQUES FOR IMPROVING MODEL PERFORMANCE

In this modification, we have added the following to the baseline model:

- from tensorflow.keras.regularizers import l1
- In each dense layer definition, we added the kernel_regularizer parameter and set it to l1(0.01). This applies L1 regularization to the weights of each dense layer with a regularization strength of 0.01. This value provides a moderate regularization effect. It is often chosen as a starting point and adjusted based on model performance during training. Typical values range from 0.001 to 0.1, with some adjustments made based on empirical observations during model training.

To add L2 regularization, we can follow a similar approach, but use the l2 function from Keras instead:

```
[In]: from tensorflow.keras.regularizers import l2

[In]: def create_and_train_model(
          self,
          train_features,
          train_labels,
          epochs=100,
          batch_size=32
      ):
          self.input_shape = (train_features.shape[1],)
          self.model = Sequential([
              Dense(
                  64,
                  activation='relu',
                  kernel_regularizer=l2(0.01),
                  input_shape=self.input_shape
              ),
```

```
        Dense(
            32,
            activation='relu',
            kernel_regularizer=l2(0.01)
        ),
        Dense(
            16,
            activation='relu',
            kernel_regularizer=l2(0.01)
        ),
        Dense(1)
    ])
    self.model.compile(optimizer='adam', loss='mse')
    history = self.model.fit(
        train_features,
        train_labels,
        epochs=epochs,
        batch_size=batch_size,
        verbose=0
    )
    return history
```

In this modification, we have added the following to the baseline model:

- from tensorflow.keras.regularizers import l2

- In each dense layer definition, we added the kernel_regularizer parameter and set it to l2(0.01). This applies L2 regularization to the weights of each dense layer with a regularization strength of 0.01. We can adjust the regularization strength as needed.

CHAPTER 9   TECHNIQUES FOR IMPROVING MODEL PERFORMANCE

After incorporating L1 and L2 regularization techniques into the baseline TensorFlow model, we evaluated the performance of the model with varying strengths of regularization. The output provides insights into how different regularization strengths impact the model's performance metrics, specifically the test loss and the R2 score.

When applying L1 regularization with a strength of 0.01, we obtained the following results:

```
Test Loss: 20.1664
R2 Score: 0.9988
```

We observed a moderate increase in the test loss and a small decrease in R2 compared with the baseline model values:

```
Test Loss (Baseline): 15.5924
R2 Score (Baseline): 0.9987
```

This increase suggests that L1 regularization didn't significantly improve the model's generalization, indicating that the model was likely not overfitting. However, the R2 score remained high at 0.9987, indicating that the model retained its predictive power even with regularization.

It's worth noting that the observed values of performance metrics can be influenced by factors such as random initialization of model weights and variations in data sampling, which should be considered when comparing baseline and regularization results.

Increasing the L1 regularization strength to 0.1 resulted in a more pronounced effect on the test loss:

```
Test Loss: 61.8990
R2 Score: 0.9989
```

This increase indicates that stronger L1 regularization led to a reduction in the model's ability to fit the training data, possibly indicating

CHAPTER 9 TECHNIQUES FOR IMPROVING MODEL PERFORMANCE

slight underfitting. However, the R2 score remained relatively high at 0.9989, indicating that the model still captured a large portion of the variance in the data.

On the other hand, applying L2 regularization with a strength of 0.01 resulted in a slightly lower test loss compared with L1 regularization:

```
Test Loss: 19.9371
R2 Score: 0.9988
```

This suggests that L2 regularization helped mitigate overfitting, if any, to a slightly greater extent compared with L1 regularization. Similarly, the R2 score remained high at 0.9988, indicating that the model maintained its predictive performance.

Increasing the L2 regularization strength to 0.1 led to a test loss similar to the result obtained with higher L1 regularization strength:

```
Test Loss: 61.4426
R2 Score: 0.9989
```

In a model that overfits, this would indicate that stronger L2 regularization had a similar effect on reducing overfitting as stronger L1 regularization. Despite the higher regularization strength, the R2 score remained high at 0.9989, indicating that the model still maintained its predictive power.

Overall, the output demonstrates the trade-off between regularization strength and model performance. While stronger regularization can help prevent overfitting and improve generalization, excessively strong regularization may lead to underfitting and a decrease in predictive performance. Therefore, it's essential to choose an appropriate regularization strength based on the specific characteristics of the dataset and the desired balance between bias and variance.

CHAPTER 9   TECHNIQUES FOR IMPROVING MODEL PERFORMANCE

Learning Rate

The learning rate is a crucial hyperparameter in training deep learning models. It determines the step size at each iteration while moving toward a minimum of the loss function. Choosing the right learning rate can significantly impact the performance and convergence speed of the model.

If the learning rate is too low, the optimization algorithm may converge very slowly or get stuck in a local minimum. On the other hand, if the learning rate is too high, the algorithm may oscillate around the minimum without converging, which can prevent the model from reaching optimal performance.

To try a different learning rate in our neural network model, we should modify the optimizer parameter when we compile the model in the create_and_train_model method. By default, TensorFlow's Adam optimizer uses a learning rate of 0.001. However, adjusting the learning rate can significantly impact the model's training process and convergence. Therefore, we can customize the learning rate to suit our specific requirements and improve model performance.

Here's the relevant part of the baseline code with the modification to include a custom learning rate:

```
[In]: from tensorflow.keras.optimizers import Adam
```

```
[In]: def create_and_train_model(
          self,
          train_features,
          train_labels,
          epochs=100,
          batch_size=32,
          learning_rate=0.001
      ):
```

```python
self.input_shape = (train_features.shape[1],)
self.model = Sequential([
    Dense(
        64,
        activation='relu',
        input_shape=self.input_shape
    ),
    Dense(
        32,
        activation='relu'
    ),
    Dense(
        16,
        activation='relu'
    ),
    Dense(1)
])
optimizer = Adam(learning_rate=learning_rate)
self.model.compile(optimizer=optimizer, loss='mse')
history = self.model.fit(
    train_features,
    train_labels,
    epochs=epochs,
    batch_size=batch_size,
    verbose=0
)
return history
```

CHAPTER 9   TECHNIQUES FOR IMPROVING MODEL PERFORMANCE

In this modified code of the create_and_train_model function

- We added a learning_rate parameter to the method signature, allowing the learning rate to be specified when the method is called. This parameter has a default value of 0.001 in the baseline model.

- We updated the optimizer instantiation to use Adam(learning_rate=learning_rate), making the learning rate customizable.

We can call the method with a different learning rate as shown below:

```
[In]: history = tsla_regressor.create_and_train_model(
        train_features,
        train_labels,
        learning_rate=0.0005
)
```

We have experimented with a moderate learning rate of 0.01. This value is higher than the default (0.001) but still within a reasonable range commonly used in neural network training. A moderate learning rate can accelerate the initial stages of training without risking instability, allowing the model to progress toward convergence more rapidly. The results obtained with this learning rate indicate promising performance:

Test Loss: 16.9786
R2 Score: 0.9986

These metrics demonstrate that the model trained with a learning rate of 0.01 achieved a relatively low test loss and a high R2 score, indicating good predictive performance and generalization capability compared with the baseline:

Baseline: 0.001
Test Loss (Baseline): 15.5924
R2 Score (Baseline): 0.9987

On the other hand, we also experimented with a more conservative learning rate of 0.0001. This value is significantly lower than the default and can be considered conservative, leading to slower but potentially more stable training progress. The results obtained with this learning rate are as follows:

```
Test Loss: 1056.1927
R2 Score: 0.9111
```

While the test loss is notably higher compared with the baseline and the model trained with a learning rate of 0.01, the R2 score remains relatively high, indicating that the model retains a significant portion of its predictive power. However, the considerably higher test loss suggests that the training progress may be slower due to the conservative learning rate, potentially requiring more epochs to achieve convergence.

Overall, these experiments illustrate the impact of different learning rates on model training and performance. While a moderate learning rate of 0.01 led to faster convergence and good performance, a conservative learning rate of 0.0001 resulted in slower progress but still maintained reasonable predictive power. The choice of learning rate should be tailored to the specific characteristics of the dataset and the desired balance between training speed and stability.

Model Capacity

The capacity of a neural network model refers to its ability to learn and represent complex patterns in the data. Insufficient capacity can lead to underfitting, where the model fails to capture the underlying structure of the data. Conversely, excessive capacity can lead to overfitting, where the model learns the noise in the training data instead of the actual patterns. Adjusting the model's capacity by varying the number of layers

CHAPTER 9   TECHNIQUES FOR IMPROVING MODEL PERFORMANCE

and neurons is a critical step in optimizing performance. In the baseline model, the neural network model defined in the TSLARegressor class consists of the following layers:

- Input Layer: Implicitly defined by the input_shape parameter passed to the first hidden layer
- First Hidden Layer: 64 neurons, ReLU activation function
- Second Hidden Layer: 32 neurons, ReLU activation function
- Third Hidden Layer: 16 neurons, ReLU activation function
- Output Layer: 1 neuron, no activation function (typical for regression tasks)

To modify the model capacity in the baseline model, we would adjust the number of neurons in the hidden layers within the create_and_train_model method. Specifically, we can modify the number of neurons in the dense layers.

Here's the relevant part of the code where we can adjust the model capacity:

```
[In]: def create_and_train_model(
         self,
         train_features,
         train_labels,
         epochs=100,
         batch_size=32
    ):
        self.input_shape = (train_features.shape[1],)
        self.model = Sequential([
```

```python
        Dense(
            128,
            activation='relu',
            input_shape=self.input_shape
        ),
        Dense(
            64,
            activation='relu'
        ),
        Dense(
            32,
            activation='relu'
        ),
        Dense(1)
    ])
    self.model.compile(optimizer='adam', loss='mse')
    history = self.model.fit(
        train_features,
        train_labels,
        epochs=epochs,
        batch_size=batch_size,
        verbose=0
    )
    return history
```

In this example, we have increased the number of neurons in each hidden layer from the baseline model's capacity. The baseline model had hidden layers with 64, 32, and 16 neurons, respectively. Now, we've augmented the model capacity to 128, 64, and 32 neurons in each corresponding hidden layer. These adjustments enable greater model complexity and flexibility in capturing intricate patterns within the data.

CHAPTER 9 TECHNIQUES FOR IMPROVING MODEL PERFORMANCE

With this modified architecture, we obtained the following results:

```
Test Loss: 10.7748
R2 Score: 0.9991
```

We also experimented with an intermediate capacity:

```
[In]: def create_and_train_model(
          self,
          train_features,
          train_labels,
          epochs=100,
          batch_size=32
      ):
          self.input_shape = (train_features.shape[1],)
          self.model = Sequential([
              Dense(
                  96,
                  activation='relu',
                  input_shape=self.input_shape
              ),
              Dense(
                  48,
                  activation='relu'
              ),
              Dense(
                  24,
                  activation='relu'
              ),
              Dense(1)
          ])
```

CHAPTER 9   TECHNIQUES FOR IMPROVING MODEL PERFORMANCE

```
        self.model.compile(optimizer='adam', loss='mse')
        history = self.model.fit(
            train_features,
            train_labels,
            epochs=epochs,
            batch_size=batch_size,
            verbose=0
        )
        return history
```

The results from the intermediate-capacity model were as follows:

```
Test Loss: 11.3233
R2 Score: 0.9990
```

As a reminder, the baseline model configuration was as follows:

```
[In]: def create_and_train_model(
            self,
            train_features,
            train_labels,
            epochs=100,
            batch_size=32
       ):
            self.input_shape = (train_features.shape[1],)
            self.model = Sequential([
                Dense(
                    64,
                    activation='relu',
                    input_shape=self.input_shape
                ),
```

CHAPTER 9 TECHNIQUES FOR IMPROVING MODEL PERFORMANCE

```
        Dense(
            32,
            activation='relu'
        ),
        Dense(
            16,
            activation='relu'
        ),
        Dense(1)
    ])
    self.model.compile(optimizer='adam', loss='mse')
    history = self.model.fit(
        train_features,
        train_labels,
        epochs=epochs,
        batch_size=batch_size,
        verbose=0
    )
    return history
```

with the following results:

```
Test Loss (Baseline): 15.5924
R2 Score (Baseline): 0.9987
```

By increasing the model capacity, we observed a significant improvement in the performance metrics. The test loss decreased from 15.59 in the baseline model to 10.77 with the higher-capacity model, and the R2 score improved from 0.9987 to 0.9991. This suggests that the increased capacity allowed the model to capture more complex patterns in the data, thereby improving its performance.

CHAPTER 9   TECHNIQUES FOR IMPROVING MODEL PERFORMANCE

For the intermediate-capacity model, the results also showed improvement over the baseline, with a test loss of 11.32 and an R2 score of 0.9990. While not as pronounced as the highest-capacity model, this intermediate configuration still demonstrated better generalization than the baseline model.

These results highlight the importance of tuning model capacity to balance complexity and generalization. While a higher-capacity model can provide better performance, it is crucial to monitor for overfitting and ensure the model is not excessively complex for the given task.

In addition to experimenting with different numbers of neurons in each layer, we also explored the impact of changing the number of layers in the neural network. Specifically, we experimented with two different architectures: a wide model and a deeper model.

A wide model has more neurons in each layer but fewer layers overall. This architecture can help us understand the impact of wide layers on model performance compared with deeper layers:

```
[In]: def create_and_train_model(
         self,
         train_features,
         train_labels,
         epochs=100,
         batch_size=32
     ):
         self.input_shape = (train_features.shape[1],)
         self.model = Sequential([
             Dense(
                 128,
                 activation='relu',
                 input_shape=self.input_shape
             ),
```

CHAPTER 9   TECHNIQUES FOR IMPROVING MODEL PERFORMANCE

```
        Dense(
            128,
            activation='relu'
        ),
        Dense(1)
    ])
    self.model.compile(optimizer='adam', loss='mse')
    history = self.model.fit(
        train_features,
        train_labels,
        epochs=epochs,
        batch_size=batch_size,
        verbose=0
    )
    return history
```

The results for this model architecture were as follows:

```
Test Loss: 12.7816
R2 Score: 0.9989
```

A deeper model has more layers with fewer neurons per layer. This architecture can help us understand the impact of deep layers on model performance compared with wider layers:

```
[In]: def create_and_train_model(
        self,
        train_features,
        train_labels,
        epochs=100,
        batch_size=32
    ):
```

```python
self.input_shape = (train_features.shape[1],)
self.model = Sequential([
    Dense(
        64,
        activation='relu',
        input_shape=self.input_shape
    ),
    Dense(
        64,
        activation='relu'
    ),
    Dense(
        32,
        activation='relu'
    ),
    Dense(
        16,
        activation='relu'
    ),
    Dense(
        8,
        activation='relu'
    ),
    Dense(1)
])
self.model.compile(optimizer='adam', loss='mse')
history = self.model.fit(
    train_features,
    train_labels,
    epochs=epochs,
```

CHAPTER 9   TECHNIQUES FOR IMPROVING MODEL PERFORMANCE

```
        batch_size=batch_size,
        verbose=0
)
return history
```

The results for the deeper model were as follows:

```
Test Loss: 10.2026
R2 Score: 0.9991
```

The wide model, with 128 neurons in each of its two hidden layers, resulted in a test loss of 12.78 and an R2 score of 0.9989. This suggests that having a larger number of neurons in fewer layers allows the model to learn and generalize well. However, the performance was not as optimal as the deeper model.

The deeper model, with five hidden layers having progressively fewer neurons (64, 64, 32, 16, and 8 neurons), achieved a test loss of 10.20 and an R2 score of 0.9991. This model outperformed the wide model, indicating that increasing the depth with moderate neuron counts per layer helped in capturing more complex patterns in the data.

Let's compare these with the baseline results:

```
Test Loss: 15.5924
R2 Score: 0.9987
```

Both the wide and deeper models showed improved performance over the baseline model. The deeper model, in particular, achieved the best results, highlighting the potential benefits of increasing model depth to capture more intricate data patterns. These experiments illustrate the importance of tuning both the number of neurons and the number of layers to optimize neural network performance.

CHAPTER 9 TECHNIQUES FOR IMPROVING MODEL PERFORMANCE

Automating Hyperparameter Optimization with Keras Tuner

In the previous sections, hyperparameters like the number of layers and the learning rate were selected through manual experimentation. While this approach can sometimes yield satisfactory results, it is inherently time-consuming and prone to suboptimal choices. As models grow more complex, the need for a systematic and automated method of hyperparameter tuning becomes evident.

Several methods exist for hyperparameter optimization, including grid search and random search. Grid search exhaustively tests every possible combination of parameters within specified ranges, which can be computationally expensive but thorough. Random search, on the other hand, samples hyperparameter combinations randomly, offering a more efficient but less comprehensive exploration of the parameter space.

In this section, we use Keras Tuner, a tool that simplifies and automates the process of hyperparameter optimization. Unlike grid or random search, Keras Tuner is specifically designed for deep learning models and integrates seamlessly with Keras, making it a natural choice for our needs. By automating the search for the best hyperparameters, Keras Tuner not only saves time but also enhances the model's performance by systematically exploring a wide range of hyperparameter combinations. We will demonstrate its use through the code example below (entire code will be provided in the "Bringing It All Together" section), which optimizes both the learning rate and the number of units in each layer of a neural network.

To incorporate Keras Tuner into our workflow, the first step is to import the necessary libraries.

Step 1: Importing Required Libraries

```
[In]: from keras_tuner import HyperModel, RandomSearch
```

CHAPTER 9   TECHNIQUES FOR IMPROVING MODEL PERFORMANCE

HyperModel allows us to define the structure of our model along with the hyperparameters to be tuned. *RandomSearch* is the tuner that will handle the exploration of different hyperparameter combinations.

Step 2: Defining the TSLAHyperModel Class

```
[In]: class TSLAHyperModel(HyperModel):
          """
          HyperModel class for TSLA stock price regression.
          Defines the model architecture and allows Keras Tuner
          to search for the best hyperparameters.
          """
          def build(self, hp):
              model = Sequential()
              model.add(
                  Dense(
                      units=hp.Int(
                          'units_1',
                          min_value=32,
                          max_value=256,
                          step=32
                      ),
                      activation='relu',
                      input_shape=(self.input_shape,)
                  )
              )
              model.add(
                  Dense(
                      units=hp.Int(
                          'units_2',
                          min_value=16,
                          max_value=128,
```

CHAPTER 9　TECHNIQUES FOR IMPROVING MODEL PERFORMANCE

```
                step=16
            ),
            activation='relu'
        )
    )
    model.add(
        Dense(
            units=hp.Int(
                'units_3',
                min_value=8,
                max_value=64,
                step=8
            ),
            activation='relu'
        )
    )
    model.add(Dense(1))

    model.compile(
        optimizer=tf.keras.optimizers.Adam(
            learning_rate=hp.Choice(
                'learning_rate',
                values=[1e-2, 1e-3, 1e-4]
            )
        ),
        loss='mse'
    )
    return model
```

In this step, we define a custom TSLAHyperModel class that extends the *HyperModel* class. This class specifies the architecture of our neural network and the hyperparameters to be tuned. The build method is where

631

CHAPTER 9 TECHNIQUES FOR IMPROVING MODEL PERFORMANCE

the model structure is defined, with hyperparameters like the number of units in each layer (units_1, units_2, units_3) and the learning rate being set dynamically using the hp (hyperparameter) object. This setup allows Keras Tuner to explore different combinations of these parameters systematically.

Step 3: Integrating the Tuner into the Model Training Process

```
[In]: def create_and_train_model(
          self,
          train_features,
          train_labels,
          test_features,
          test_labels,
          epochs=100,
          batch_size=32
      ):
          """
          Create and train the neural network model using
          Keras Tuner for hyperparameter optimization.

          Parameters:
              - train_features (ndarray): Training features.
              - train_labels (ndarray): Training labels.
              - test_features (ndarray): Test features.
              - test_labels (ndarray): Test labels.
              - epochs (int): Number of epochs for training.
              - batch_size (int): Batch size for training.

          Returns:
              - History:
                  Training history containing loss values for
                  each epoch.
          """
```

```python
self.input_shape = train_features.shape[1]
tsla_hypermodel = TSLAHyperModel()
tsla_hypermodel.input_shape = self.input_shape
tuner = RandomSearch(
    tsla_hypermodel,
    objective='val_loss',
    max_trials=5,
    executions_per_trial=3,
    directory='tuner_dir',
    project_name='tsla_regression'
)

tuner.search(
    train_features,
    train_labels,
    epochs=epochs,
    batch_size=batch_size,
    validation_data=(test_features, test_labels),
    verbose=1
)
self.model = tuner.get_best_models(num_models=1)[0]
return self.model.fit(
    train_features,
    train_labels,
    epochs=epochs,
    batch_size=batch_size,
    verbose=1,
    validation_data=(test_features, test_labels)
)
```

CHAPTER 9   TECHNIQUES FOR IMPROVING MODEL PERFORMANCE

Here, we integrate Keras Tuner into our model training process by updating the create_and_train_model method. This method now uses the TSLAHyperModel class to define the model architecture and sets up the *RandomSearch* tuner to explore different hyperparameter combinations. The tuner.search function runs the tuning process, during which the tuner trains multiple models with different hyperparameters and evaluates them using validation data. After the search is complete, the best model is selected and further trained.

In terms of results, the training process demonstrated a steady decrease in both training and validation losses over the course of 100 epochs. These results are based on the best hyperparameters selected during tuning, including the optimal number of units in each layer and the most effective learning rate. The final validation loss reached as low as 12.39, indicating a strong generalization ability, with the model maintaining a good balance between minimizing error on the training data and performing well on unseen validation data.

After training with the optimized hyperparameters, the model was evaluated on the test dataset. The final test loss was 9.16, indicating that the model performed consistently on the unseen test data. The R2 score, which measures the proportion of variance in the dependent variable that is predictable from the independent variables, was an impressive 0.9992. This high R2 score suggests that the model, with its tuned layers and learning rate, explains nearly all the variability in the test data, making it a highly accurate predictor for TSLA stock prices.

Bringing It All Together

In this section, we present an integrated version of the baseline TSLA stock price regression model with several enhancements. Throughout our exploration, we've introduced various improvements to the model architecture, including early stopping, dropout regularization, L1 and L2 regularization, and customized learning rates. We also utilized Keras Tuner for automatic hyperparameter optimization. Each of these modifications aims to enhance the model's performance, robustness, and generalization capabilities.

Below, we provide a unified code snippet that incorporates these enhancements into a single coherent model, starting with the baseline model.

Baseline Model

```
[In]: import logging
[In]: import numpy as np
[In]: import tensorflow as tf
[In]: from pyspark.sql import SparkSession
[In]: from pyspark.ml.feature import VectorAssembler,
      StandardScaler
[In]: from sklearn.metrics import r2_score
[In]: from tensorflow.keras.models import Sequential
[In]: from tensorflow.keras.layers import Dense
[In]: import matplotlib.pyplot as plt

[In]: class TSLARegressor:
          """
          Class for TSLA stock price regression.
          """
          def __init__(self):
              """
              Initialize TSLARegressor object.
              """
              self.logger = logging.getLogger(__name__)
              self.spark = SparkSession.builder \
                          .appName("TSLA_Regression") \
                          .getOrCreate()
```

CHAPTER 9   TECHNIQUES FOR IMPROVING MODEL PERFORMANCE

```python
        self.input_shape = None
        self.model = None

    def preprocess_data(
        self,
        file_path,
        feature_cols=['Open', 'High', 'Low', 'Volume'],
        train_ratio=0.8,
        seed=42
    ):
        """
        Preprocesses the data.

        Parameters:
            - file_path (str): Path to the CSV file.
            - feature_cols (list): List of feature
              columns.
            - train_ratio (float): Ratio of
              training data.
            - seed (int): Random seed for splitting data.

        Returns:
            - tuple: Tuple containing train and test
              DataFrames.
        """
        df = self.spark.read.csv(
            file_path,
            header=True,
            inferSchema=True
        )
        assembler = VectorAssembler(
            inputCols=feature_cols, outputCol='features')
```

CHAPTER 9   TECHNIQUES FOR IMPROVING MODEL PERFORMANCE

```
    df = assembler.transform(df)
    scaler = StandardScaler(
        inputCol='features',
        outputCol='scaled_features',
        withMean=True,
        withStd=True
    )
    scaler_model = scaler.fit(df)
    df = scaler_model.transform(df)
    train_df, test_df = df.randomSplit(
        [train_ratio, 1-train_ratio],
        seed=seed
    )
    return train_df, test_df

def convert_to_numpy(self, train_df, test_df):
    """
    Convert Spark DataFrames to numpy arrays.

    Parameters:
        - train_df (DataFrame):
            Spark DataFrame containing training data.
        - test_df (DataFrame):
            Spark DataFrame containing test data.

    Returns:
        - tuple:
            Tuple containing train and test numpy
            arrays for
            features and labels.
    """
```

```python
            train_features = np.array(
                train_df.select('scaled_features')
                .rdd.map(lambda x: x.scaled_features.
                toArray())
                .collect()
            )
            train_labels = np.array(
                train_df.select('Close')
                .rdd.map(lambda x: x.Close).collect())
            test_features = np.array(
                test_df.select('scaled_features')
                .rdd.map(lambda x: x.scaled_features.
                toArray())
                .collect()
            )
            test_labels = np.array(
                test_df.select('Close')
                .rdd.map(lambda x: x.Close).collect()
            )
            return (train_features,
                    train_labels,
                    test_features,
                    test_labels
            )

        def create_and_train_model(
            self,
            train_features,
            train_labels,
            epochs=100,
            batch_size=32
        ):
```

CHAPTER 9   TECHNIQUES FOR IMPROVING MODEL PERFORMANCE

```
"""
Create and train the neural network model.

Parameters:
    - train_features (ndarray): Training
      features.
    - train_labels (ndarray): Training labels.
    - epochs (int): Number of epochs for training.
    - batch_size (int): Batch size for training.

Returns:
    - History:
        Training history containing loss values
        for each
        epoch.
"""
self.input_shape = (train_features.shape[1],)
self.model = Sequential([
    Dense(
        64,
        activation='relu',
        input_shape=self.input_shape
    ),
    Dense(
        32,
        activation='relu'
    ),
    Dense(
        16,
        activation='relu'
    ),
    Dense(1)
])
```

CHAPTER 9 TECHNIQUES FOR IMPROVING MODEL PERFORMANCE

```
        self.model.compile(optimizer='adam', loss='mse')
        history = self.model.fit(
            train_features,
            train_labels,
            epochs=epochs,
            batch_size=batch_size,
            verbose=0
        )
        return history

    def evaluate_model(self, test_features, test_labels):
        """
        Evaluate the trained model.

        Parameters:
            - test_features (ndarray): Test features.
            - test_labels (ndarray): Test labels.

        Returns:
            - float: Test loss.
        """
        test_loss = self.model.evaluate(
            test_features,
            test_labels
        )
        return test_loss

    def predict_and_evaluate(
        self,
        test_features,
        test_labels
    ):
        """
```

Predict and evaluate the neural network model.

Parameters:
    - test_features (ndarray): Test features.
    - test_labels (ndarray): Test labels.

Returns:
    - float: R2 score.
"""
test_predictions = self.model.predict(test_features)
r2_score_value = r2_score(test_labels, test_predictions)
return r2_score_value

def main(file_path):
    """
    Main function to orchestrate the entire workflow.

    Parameters:
        - file_path (str): Path to the CSV file.

    Returns:
        - None
    """
    try:
        tsla_regressor = TSLARegressor()
        train_df, test_df = tsla_regressor.preprocess_data(
            file_path
)
        train_features, train_labels, test_features, \

```
            test_labels = tsla_regressor.convert_
            to_numpy(
                train_df,
                test_df
            )
            history = tsla_regressor.create_and_
            train_model(
                train_features,
                train_labels
            )
            test_loss = tsla_regressor.evaluate_model(
                test_features,
                test_labels
            )
            r2_score_value=tsla_regressor.predict_and_
            evaluate(
                test_features,
                test_labels
            )
            logging.info("Test Loss: {}".
            format(test_loss))
            logging.info("R2 Score: {}".format(r2_
            score_value))

            plt.plot(history.history['loss'])
            plt.title('Model Loss')
            plt.xlabel('Epoch')
            plt.ylabel('Loss')

            plt.show()

        except Exception as e:
            logging.error(f"An error occurred: {e}")
```

CHAPTER 9   TECHNIQUES FOR IMPROVING MODEL PERFORMANCE

```
[In]: if __name__ == "__main__":
          logging.basicConfig(level=logging.INFO)
          file_path = "/home/ubuntu/airflow/dags/TSLA_
          stock.csv"
          main(file_path)
```

Early Stopping

```
[In]: import logging
[In]: import numpy as np
[In]: import tensorflow as tf
[In]: from pyspark.sql import SparkSession
[In]: from pyspark.ml.feature import VectorAssembler,
      StandardScaler
[In]: from sklearn.metrics import r2_score
[In]: from tensorflow.keras.models import Sequential
[In]: from tensorflow.keras.layers import Input, Dense
[In]: from tensorflow.keras.callbacks import EarlyStopping

[In]: class TSLARegressor:
          """
          Class for TSLA stock price regression.
          """
          def __init__(self):
              """
              Initialize TSLARegressor object.
              """
              self.logger = logging.getLogger(__name__)
              self.spark = SparkSession.builder \
                          .appName("TSLA_Regression") \
                          .getOrCreate()
```

CHAPTER 9   TECHNIQUES FOR IMPROVING MODEL PERFORMANCE

```
            self.input_shape = None
            self.model = None

        def preprocess_data(
            self,
            file_path,
            feature_cols=['Open', 'High', 'Low', 'Volume'],
            train_ratio=0.8,
            seed=42
        ):
            """
            Preprocesses the data.

            Parameters:
                - file_path (str): Path to the CSV file.
                - feature_cols (list): List of feature
                  columns.
                - train_ratio (float): Ratio of
                  training data.
                - seed (int): Random seed for splitting data.

            Returns:
                - tuple: Tuple containing train and test
                  DataFrames.
            """
            df = self.spark.read.csv(
                file_path,
                header=True,
                inferSchema=True
            )
            assembler = VectorAssembler(
                inputCols=feature_cols, outputCol='features')
```

CHAPTER 9   TECHNIQUES FOR IMPROVING MODEL PERFORMANCE

```
    df = assembler.transform(df)
    scaler = StandardScaler(
        inputCol='features',
        outputCol='scaled_features',
        withMean=True,
        withStd=True
    )
    scaler_model = scaler.fit(df)
    df = scaler_model.transform(df)
    train_df, test_df = df.randomSplit(
        [train_ratio, 1-train_ratio],
        seed=seed
    )
    return train_df, test_df

def convert_to_numpy(self, train_df, test_df):
    """
    Convert Spark DataFrames to numpy arrays.

    Parameters:
        - train_df (DataFrame):
            Spark DataFrame containing training data.
        - test_df (DataFrame):
            Spark DataFrame containing test data.

    Returns:
        - tuple:
            Tuple containing train and test numpy
            arrays for
            features and labels.
    """
```

```python
            train_features = np.array(
                train_df.select('scaled_features')
                .rdd.map(lambda x: x.scaled_features.
                toArray())
                .collect()
            )
            train_labels = np.array(
                train_df.select('Close')
                .rdd.map(lambda x: x.Close).collect())
            test_features = np.array(
                test_df.select('scaled_features')
                .rdd.map(lambda x: x.scaled_features.
                toArray())
                .collect()
            )
            test_labels = np.array(
                test_df.select('Close')
                .rdd.map(lambda x: x.Close).collect()
            )
            return (train_features,
                    train_labels,
                    test_features,
                    test_labels
            )

    def create_and_train_model(
        self,
        train_features,
        train_labels,
        epochs=100,
        batch_size=32
    ):
```

CHAPTER 9   TECHNIQUES FOR IMPROVING MODEL PERFORMANCE

```
"""
Create and train the neural network model.

Parameters:
    - train_features (ndarray): Training features.
    - train_labels (ndarray): Training labels.
    - epochs (int): Number of epochs for training.
    - batch_size (int): Batch size for training.

Returns:
    - History:
        Training history containing loss values
        for each
        epoch.
"""
self.input_shape = (train_features.shape[1],)
self.model = Sequential([
    Dense(
        64,
        activation='relu',
        input_shape=self.input_shape
    ),
    Dense(
        32,
        activation='relu'
    ),
    Dense(
        16,
        activation='relu'
    ),
    Dense(1)
])
```

CHAPTER 9   TECHNIQUES FOR IMPROVING MODEL PERFORMANCE

```python
        self.model.compile(optimizer='adam', loss='mse')
        history = self.model.fit(
            train_features,
            train_labels,
            epochs=epochs,
            batch_size=batch_size,
            verbose=0
        )
        return history

    def evaluate_model(self, test_features, test_labels):
        """
        Evaluate the trained model.

        Parameters:
            - test_features (ndarray): Test features.
            - test_labels (ndarray): Test labels.

        Returns:
            - float: Test loss.
        """
        test_loss = self.model.evaluate(
            test_features,
            test_labels
        )
        return test_loss

    def predict_and_evaluate(
        self,
        test_features,
        test_labels
    ):
```

CHAPTER 9   TECHNIQUES FOR IMPROVING MODEL PERFORMANCE

```
"""
Predict and evaluate the neural network model.

Parameters:
    - test_features (ndarray): Test features.
    - test_labels (ndarray): Test labels.

Returns:
    - float: R2 score.
"""
test_predictions = self.model.predict(test_
features)
r2_score_value = r2_score(test_labels, test_
predictions)
return r2_score_value

def main(file_path):
    """
    Main function to orchestrate the entire workflow.

    Parameters:
        - file_path (str): Path to the CSV file.

    Returns:
        - None
    """
    try:
        tsla_regressor = TSLARegressor()
        train_df, test_df = tsla_regressor.
        preprocess_data(
            file_path
        )
```

```
            train_features, train_labels, test_features,
            \test_labels = tsla_regressor.convert_
            to_numpy(
                train_df,
                test_df
            )
            history = tsla_regressor.create_and_
            train_model(
                train_features,
                train_labels
            )
            test_loss = tsla_regressor.evaluate_model(
                test_features,
                test_labels
            )
            r2_score_value= tsla_regressor.predict_and_
            evaluate(
                test_features,
                test_labels
            )

            logging.info("Test Loss: {}".
            format(test_loss))
            logging.info("R2 Score: {}".format(r2_
            score_value))

        except Exception as e:
            logging.error(f"An error occurred: {e}")
```

```
[In]: if __name__ == "__main__":
        logging.basicConfig(level=logging.INFO)
        file_path = "/home/ubuntu/airflow/dags/TSLA_
        stock.csv"
        main(file_path)
```

Dropout

```
[In]: import logging
[In]: import numpy as np
[In]: import tensorflow as tf
[In]: from pyspark.sql import SparkSession
[In]: from pyspark.ml.feature import VectorAssembler,
      StandardScaler
[In]: from sklearn.metrics import r2_score
[In]: from tensorflow.keras.models import Sequential
[In]: from tensorflow.keras.layers import Dense
[In]: from tensorflow.keras.layers import Dropout

[In]: class TSLARegressor:
          """
          Class for TSLA stock price regression.
          """
          def __init__(self):
              """
              Initialize TSLARegressor object.
              """
              self.logger = logging.getLogger(__name__)
              self.spark = SparkSession.builder \
                          .appName("TSLA_Regression") \
                          .getOrCreate()
              self.input_shape = None
              self.model = None

          def preprocess_data(
              self,
```

CHAPTER 9 TECHNIQUES FOR IMPROVING MODEL PERFORMANCE

```
        file_path,
        feature_cols=['Open', 'High', 'Low', 'Volume'],
        train_ratio=0.8,
        seed=42
    ):
    """
    Preprocesses the data.

    Parameters:
        - file_path (str): Path to the CSV file.
        - feature_cols (list): List of feature
          columns.
        - train_ratio (float): Ratio of
          training data.
        - seed (int): Random seed for splitting data.

    Returns:
        - tuple: Tuple containing train and test
          DataFrames.
    """
    df = self.spark.read.csv(
        file_path,
        header=True,
        inferSchema=True
    )
    assembler = VectorAssembler(
        inputCols=feature_cols, outputCol='features')
    df = assembler.transform(df)
    scaler = StandardScaler(
        inputCol='features',
        outputCol='scaled_features',
        withMean=True,
```

CHAPTER 9   TECHNIQUES FOR IMPROVING MODEL PERFORMANCE

```
        withStd=True
    )
    scaler_model = scaler.fit(df)
    df = scaler_model.transform(df)
    train_df, test_df = df.randomSplit(
        [train_ratio, 1-train_ratio],
        seed=seed
    )
    return train_df, test_df

def convert_to_numpy(self, train_df, test_df):
    """
    Convert Spark DataFrames to numpy arrays.

    Parameters:
        - train_df (DataFrame):
            Spark DataFrame containing training data.
        - test_df (DataFrame):
            Spark DataFrame containing test data.

    Returns:
        - tuple:
            Tuple containing train and test numpy
            arrays for
            features and labels.
    """
    train_features = np.array(
        train_df.select('scaled_features')
        .rdd.map(lambda x: x.scaled_features.
        toArray())
        .collect()
    )
    train_labels = np.array(
```

CHAPTER 9   TECHNIQUES FOR IMPROVING MODEL PERFORMANCE

```python
            train_df.select('Close')
            .rdd.map(lambda x: x.Close).collect())
        test_features = np.array(
            test_df.select('scaled_features')
            .rdd.map(lambda x: x.scaled_features.
            toArray())
            .collect()
        )
        test_labels = np.array(
            test_df.select('Close')
            .rdd.map(lambda x: x.Close).collect()
        )
        return (train_features,
                train_labels,
                test_features,
                test_labels
        )

    def create_and_train_model(
        self,
        train_features,
        train_labels,
        epochs=100,
        batch_size=32
    ):
        """
        Create and train the neural network model.

        Parameters:
            - train_features (ndarray): Training features.
            - train_labels (ndarray): Training labels.
```

654

CHAPTER 9   TECHNIQUES FOR IMPROVING MODEL PERFORMANCE

```
    - epochs (int): Number of epochs for
      training.
    - batch_size (int): Batch size for training.

Returns:
    - History:
        Training history containing loss values
        for each
        epoch.
"""
self.input_shape = (train_features.shape[1],)
self.model = Sequential([
    Dense(
        64,
        activation='relu',
        input_shape=self.input_shape
    ),
    Dropout(0.5),
    Dense(
        32,
        activation='relu'
    ),
    Dropout(0.5),
    Dense(
        16,
        activation='relu'
    ),
    Dense(1)
])
self.model.compile(optimizer='adam', loss='mse')
history = self.model.fit(
    train_features,
```

CHAPTER 9 TECHNIQUES FOR IMPROVING MODEL PERFORMANCE

```
            train_labels,
            epochs=epochs,
            batch_size=batch_size,
            verbose=0
        )
        return history

    def evaluate_model(self, test_features, test_labels):
        """
        Evaluate the trained model.

        Parameters:
            - test_features (ndarray): Test features.
            - test_labels (ndarray): Test labels.

        Returns:
            - float: Test loss.
        """
        test_loss = self.model.evaluate(
            test_features,
            test_labels
        )
        return test_loss

    def predict_and_evaluate(
        self,
        test_features,
        test_labels
    ):
        """
        Predict and evaluate the neural network model.

        Parameters:
            - test_features (ndarray): Test features.
```

CHAPTER 9 TECHNIQUES FOR IMPROVING MODEL PERFORMANCE

```
        - test_labels (ndarray): Test labels.

    Returns:
        - float: R2 score.
    """
    test_predictions = self.model.predict(test_
    features)
    r2_score_value = r2_score(test_labels, test_
    predictions)
    return r2_score_value

def main(file_path):
    """
    Main function to orchestrate the entire workflow.

    Parameters:
        - file_path (str): Path to the CSV file.

    Returns:
        - None
    """
    try:
        tsla_regressor = TSLARegressor()
        train_df, test_df = tsla_regressor.
        preprocess_data(
            file_path
        )
        train_features, train_labels, test_
        features, \
        test_labels = tsla_regressor.convert_
        to_numpy(
            train_df,
            test_df
```

657

```
                )
                history = tsla_regressor.create_and_
                train_model(
                    train_features,
                    train_labels
                )
                test_loss = tsla_regressor.evaluate_model(
                    test_features,
                    test_labels
                )
                r2_score_value= tsla_regressor.predict_and_
                evaluate(
                    test_features,
                    test_labels
                )

                logging.info("Test Loss: {}".format(test_loss))
                logging.info("R2 Score: {}".format(r2_
                score_value))

        except Exception as e:
            logging.error(f"An error occurred: {e}")
```
```
[In]: if __name__ == "__main__":
        logging.basicConfig(level=logging.INFO)
        file_path = "/home/ubuntu/airflow/dags/TSLA_stock.csv"
        main(file_path)
```

L1 Regularization

```
[In]: import logging
[In]: import numpy as np
[In]: import tensorflow as tf
```

CHAPTER 9　TECHNIQUES FOR IMPROVING MODEL PERFORMANCE

```
[In]: from pyspark.sql import SparkSession
[In]: from pyspark.ml.feature import VectorAssembler,
      StandardScaler
[In]: from sklearn.metrics import r2_score
[In]: from tensorflow.keras.models import Sequential
[In]: from tensorflow.keras.layers import Dense
[In]: from tensorflow.keras.regularizers import l1

[In]: class TSLARegressor:
          """
          Class for TSLA stock price regression.
          """
          def __init__(self):
              """
              Initialize TSLARegressor object.
              """
              self.logger = logging.getLogger(__name__)
              self.spark = SparkSession.builder \
                          .appName("TSLA_Regression") \
                          .getOrCreate()
              self.input_shape = None
              self.model = None

          def preprocess_data(
              self,
              file_path,
              feature_cols=['Open', 'High', 'Low', 'Volume'],
              train_ratio=0.8,
              seed=42
          ):
              """
              Preprocesses the data.
```

CHAPTER 9   TECHNIQUES FOR IMPROVING MODEL PERFORMANCE

```
    Parameters:
        - file_path (str): Path to the CSV file.
        - feature_cols (list): List of feature
          columns.
        - train_ratio (float): Ratio of training data.
        - seed (int): Random seed for splitting data.
    Returns:
        - tuple: Tuple containing train and test
          DataFrames.
    """
    df = self.spark.read.csv(
        file_path,
        header=True,
        inferSchema=True
    )
    assembler = VectorAssembler(
        inputCols=feature_cols, outputCol='features')
    df = assembler.transform(df)
    scaler = StandardScaler(
        inputCol='features',
        outputCol='scaled_features',
        withMean=True,
        withStd=True
    )
    scaler_model = scaler.fit(df)
    df = scaler_model.transform(df)
    train_df, test_df = df.randomSplit(
        [train_ratio, 1-train_ratio],
        seed=seed
    )
    return train_df, test_df
```

CHAPTER 9   TECHNIQUES FOR IMPROVING MODEL PERFORMANCE

```python
def convert_to_numpy(self, train_df, test_df):
    """
    Convert Spark DataFrames to numpy arrays.

    Parameters:
        - train_df (DataFrame):
            Spark DataFrame containing training data.
        - test_df (DataFrame):
            Spark DataFrame containing test data.

    Returns:
        - tuple:
            Tuple containing train and test numpy
            arrays for
            features and labels.
    """
    train_features = np.array(
        train_df.select('scaled_features')
        .rdd.map(lambda x: x.scaled_features.
        toArray())
        .collect()
    )
    train_labels = np.array(
        train_df.select('Close')
        .rdd.map(lambda x: x.Close).collect())
    test_features = np.array(
        test_df.select('scaled_features')
        .rdd.map(lambda x: x.scaled_features.
        toArray())
        .collect()
    )
```

```
        test_labels = np.array(
            test_df.select('Close')
            .rdd.map(lambda x: x.Close).collect()
        )
        return (train_features,
                train_labels,
                test_features,
                test_labels
        )

    def create_and_train_model(
        self,
        train_features,
        train_labels,
        epochs=100,
        batch_size=32
    ):
        """
        Create and train the neural network model.

        Parameters:
            - train_features (ndarray): Training features.
            - train_labels (ndarray): Training labels.
            - epochs (int): Number of epochs for training.
            - batch_size (int): Batch size for training.

        Returns:
            - History:
                Training history containing loss values
                for each
                epoch.
        """
```

```python
self.input_shape = (train_features.shape[1],)
self.model = Sequential([
    Dense(
        64,
        activation='relu',
        input_shape=self.input_shape,
        kernel_regularizer=l1(0.01)
    ),
    Dense(
        32,
        activation='relu',
        kernel_regularizer=l1(0.01)
    ),
    Dense(
        16,
        activation='relu',
        kernel_regularizer=l1(0.01)
    ),
    Dense(1)
])
self.model.compile(optimizer='adam', loss='mse')
history = self.model.fit(
    train_features,
    train_labels,
    epochs=epochs,
    batch_size=batch_size,
    verbose=0
)
return history
```

```python
    def evaluate_model(self, test_features, test_labels):
        """
        Evaluate the trained model.

        Parameters:
            - test_features (ndarray): Test features.
            - test_labels (ndarray): Test labels.

        Returns:
            - float: Test loss.
        """
        test_loss = self.model.evaluate(
            test_features,
            test_labels
        )
        return test_loss

    def predict_and_evaluate(
        self,
        test_features,
        test_labels
    ):
        """
        Predict and evaluate the neural network model.

        Parameters:
            - test_features (ndarray): Test features.
            - test_labels (ndarray): Test labels.

        Returns:
            - float: R2 score.
        """
        test_predictions = self.model.predict(test_
        features)
```

CHAPTER 9 TECHNIQUES FOR IMPROVING MODEL PERFORMANCE

```python
    r2_score_value = r2_score(test_labels, test_
    predictions)
    return r2_score_value

def main(file_path):
    """
    Main function to orchestrate the entire workflow.

    Parameters:
        - file_path (str): Path to the CSV file.

    Returns:
        - None
    """
    try:
        tsla_regressor = TSLARegressor()
        train_df, test_df = tsla_regressor.
        preprocess_data(
            file_path
        )
        train_features, train_labels, test_
        features, \
        test_labels = tsla_regressor.convert_
        to_numpy(
            train_df,
            test_df
        )
        history = tsla_regressor.create_and_
        train_model(
            train_features,
            train_labels
        )
```

```
            test_loss = tsla_regressor.evaluate_model(
                test_features,
                test_labels
            )
            r2_score_value= tsla_regressor.predict_and_
            evaluate(
                test_features,
                test_labels
            )

            logging.info("Test Loss: {}".
            format(test_loss))
            logging.info("R2 Score: {}".format(r2_
            score_value))

        except Exception as e:
            logging.error(f"An error occurred: {e}")
```

```
[In]: if __name__ == "__main__":
        logging.basicConfig(level=logging.INFO)
        file_path = "/home/ubuntu/airflow/dags/TSLA_
        stock.csv"
        main(file_path)
```

L2 Regularization

```
[In]: import logging
[In]: import numpy as np
[In]: import tensorflow as tf
[In]: from pyspark.sql import SparkSession
[In]: from pyspark.ml.feature import VectorAssembler, StandardScaler
[In]: from sklearn.metrics import r2_score
```

CHAPTER 9   TECHNIQUES FOR IMPROVING MODEL PERFORMANCE

```
[In]: from tensorflow.keras.models import Sequential
[In]: from tensorflow.keras.layers import Dense
[In]: from tensorflow.keras.regularizers import l2

[In]: class TSLARegressor:
          """
          Class for TSLA stock price regression.
          """
          def __init__(self):
              """
              Initialize TSLARegressor object.
              """
              self.logger = logging.getLogger(__name__)
              self.spark = SparkSession.builder \
                          .appName("TSLA_Regression") \
                          .getOrCreate()
              self.input_shape = None
              self.model = None

          def preprocess_data(
              self,
              file_path,
              feature_cols=['Open', 'High', 'Low', 'Volume'],
              train_ratio=0.8,
              seed=42
          ):
              """
              Preprocesses the data.

              Parameters:
                  - file_path (str): Path to the CSV file.
                  - feature_cols (list): List of feature
                    columns.
```

CHAPTER 9   TECHNIQUES FOR IMPROVING MODEL PERFORMANCE

```
            - train_ratio (float): Ratio of
              training data.
            - seed (int): Random seed for splitting data.
        Returns:
            - tuple: Tuple containing train and test
              DataFrames.
        """
        df = self.spark.read.csv(
            file_path,
            header=True,
            inferSchema=True
        )
        assembler = VectorAssembler(
            inputCols=feature_cols, outputCol='features')
        df = assembler.transform(df)
        scaler = StandardScaler(
            inputCol='features',
            outputCol='scaled_features',
            withMean=True,
            withStd=True
        )
        scaler_model = scaler.fit(df)
        df = scaler_model.transform(df)
        train_df, test_df = df.randomSplit(
            [train_ratio, 1-train_ratio],
            seed=seed
        )
        return train_df, test_df

    def convert_to_numpy(self, train_df, test_df):
        """
        Convert Spark DataFrames to numpy arrays.
```

CHAPTER 9   TECHNIQUES FOR IMPROVING MODEL PERFORMANCE

```
Parameters:
    - train_df (DataFrame):
        Spark DataFrame containing training data.
    - test_df (DataFrame):
        Spark DataFrame containing test data.

Returns:
    - tuple:
        Tuple containing train and test numpy
        arrays for
        features and labels.
"""
train_features = np.array(
    train_df.select('scaled_features')
    .rdd.map(lambda x: x.scaled_features.
    toArray())
    .collect()
)
train_labels = np.array(
    train_df.select('Close')
    .rdd.map(lambda x: x.Close).collect())
test_features = np.array(
    test_df.select('scaled_features')
    .rdd.map(lambda x: x.scaled_features.
    toArray())
    .collect()
)
test_labels = np.array(
    test_df.select('Close')
    .rdd.map(lambda x: x.Close).collect()
)
```

CHAPTER 9   TECHNIQUES FOR IMPROVING MODEL PERFORMANCE

```python
        return (train_features,
                train_labels,
                test_features,
                test_labels
        )

    def create_and_train_model(
        self,
        train_features,
        train_labels,
        epochs=100,
        batch_size=32
    ):
        """
        Create and train the neural network model.

        Parameters:
            - train_features (ndarray): Training features.
            - train_labels (ndarray): Training labels.
            - epochs (int): Number of epochs for training.
            - batch_size (int): Batch size for training.

        Returns:
            - History:
                Training history containing loss values
                for each
                epoch.
        """
        self.input_shape = (train_features.shape[1],)
        self.model = Sequential([
            Dense(
                64,
                activation='relu',
```

CHAPTER 9   TECHNIQUES FOR IMPROVING MODEL PERFORMANCE

```python
            input_shape=self.input_shape,
            kernel_regularizer=l2(0.01)
        ),
        Dense(
            32,
            activation='relu',
            kernel_regularizer=l2(0.01)
        ),
        Dense(
            16,
            activation='relu',
            kernel_regularizer=l2(0.01)
        ),
        Dense(1)
    ])
    self.model.compile(optimizer='adam', loss='mse')
    history = self.model.fit(
        train_features,
        train_labels,
        epochs=epochs,
        batch_size=batch_size,
        verbose=0
    )
    return history

def evaluate_model(self, test_features, test_labels):
    """
    Evaluate the trained model.

    Parameters:
        - test_features (ndarray): Test features.
        - test_labels (ndarray): Test labels.
```

CHAPTER 9 TECHNIQUES FOR IMPROVING MODEL PERFORMANCE

```
        Returns:
            - float: Test loss.
        """
        test_loss = self.model.evaluate(
            test_features,
            test_labels
        )
        return test_loss

    def predict_and_evaluate(
        self,
        test_features,
        test_labels
    ):
        """
        Predict and evaluate the neural network model.

        Parameters:
            - test_features (ndarray): Test features.
            - test_labels (ndarray): Test labels.

        Returns:
            - float: R2 score.
        """
        test_predictions = self.model.predict(test_
        features)
        r2_score_value = r2_score(test_labels, test_
        predictions)
        return r2_score_value

    def main(file_path):
        """
        Main function to orchestrate the entire workflow.
```

CHAPTER 9 TECHNIQUES FOR IMPROVING MODEL PERFORMANCE

```
    Parameters:
        - file_path (str): Path to the CSV file.

    Returns:
        - None
    """
    try:
        tsla_regressor = TSLARegressor()
        train_df, test_df = tsla_regressor.
        preprocess_data(
            file_path
        )
        train_features, train_labels, test_features, \
        test_labels = tsla_regressor.convert_
        to_numpy(
            train_df,
            test_df
        )
        history = tsla_regressor.create_and_
        train_model(
            train_features,
            train_labels
        )
        test_loss = tsla_regressor.evaluate_model(
            test_features,
            test_labels
        )
        r2_score_value= tsla_regressor.predict_and_
        evaluate(
            test_features,
            test_labels
        )
```

673

CHAPTER 9   TECHNIQUES FOR IMPROVING MODEL PERFORMANCE

```
            logging.info("Test Loss: {}".
            format(test_loss))
            logging.info("R2 Score: {}".format(r2_
            score_value))

        except Exception as e:
            logging.error(f"An error occurred: {e}")
```
```
[In]: if __name__ == "__main__":
        logging.basicConfig(level=logging.INFO)
        file_path = "/home/ubuntu/airflow/dags/TSLA_
        stock.csv"
        main(file_path)
```

Learning Rate

```
[In]: import logging
[In]: import numpy as np
[In]: import tensorflow as tf
[In]: from pyspark.sql import SparkSession
[In]: from pyspark.ml.feature import VectorAssembler,
      StandardScaler
[In]: from sklearn.metrics import r2_score
[In]: from tensorflow.keras.models import Sequential
[In]: from tensorflow.keras.layers import Dense
[In]: from tensorflow.keras.optimizers import Adam

[In]: class TSLARegressor:
        """
        Class for TSLA stock price regression.
        """
```

CHAPTER 9　TECHNIQUES FOR IMPROVING MODEL PERFORMANCE

```
def __init__(self):
    """
    Initialize TSLARegressor object.
    """
    self.logger = logging.getLogger(__name__)
    self.spark = SparkSession.builder \
                .appName("TSLA_Regression") \
                .getOrCreate()
    self.input_shape = None
    self.model = None

def preprocess_data(
    self,
    file_path,
    feature_cols=['Open', 'High', 'Low', 'Volume'],
    train_ratio=0.8,
    seed=42
):
    """
    Preprocesses the data.

    Parameters:
        - file_path (str): Path to the CSV file.
        - feature_cols (list): List of feature columns.
        - train_ratio (float): Ratio of
          training data.
        - seed (int): Random seed for splitting data.

    Returns:
        - tuple: Tuple containing train and test
          DataFrames.
    """
```

CHAPTER 9   TECHNIQUES FOR IMPROVING MODEL PERFORMANCE

```
        df = self.spark.read.csv(
            file_path,
            header=True,
            inferSchema=True
        )
        assembler = VectorAssembler(
            inputCols=feature_cols, outputCol='features')
        df = assembler.transform(df)
        scaler = StandardScaler(
            inputCol='features',
            outputCol='scaled_features',
            withMean=True,
            withStd=True
        )
        scaler_model = scaler.fit(df)
        df = scaler_model.transform(df)
        train_df, test_df = df.randomSplit(
            [train_ratio, 1-train_ratio],
            seed=seed
        )
        return train_df, test_df

    def convert_to_numpy(self, train_df, test_df):
        """
        Convert Spark DataFrames to numpy arrays.

        Parameters:
            - train_df (DataFrame):
                Spark DataFrame containing training data.
            - test_df (DataFrame):
                Spark DataFrame containing test data.
```

676

```
Returns:
    - tuple:
        Tuple containing train and test numpy
        arrays for
        features and labels.
"""
train_features = np.array(
    train_df.select('scaled_features')
    .rdd.map(lambda x: x.scaled_features.
    toArray())
    .collect()
)
train_labels = np.array(
    train_df.select('Close')
    .rdd.map(lambda x: x.Close).collect())
test_features = np.array(
    test_df.select('scaled_features')
    .rdd.map(lambda x: x.scaled_features.
    toArray())
    .collect()
)
test_labels = np.array(
    test_df.select('Close')
    .rdd.map(lambda x: x.Close).collect()
)
return (train_features,
        train_labels,
        test_features,
        test_labels
)
```

```python
    def create_and_train_model(
        self,
        train_features,
        train_labels,
        epochs=100,
        batch_size=32
    ):
        """
        Create and train the neural network model.

        Parameters:
            - train_features (ndarray): Training
              features.
            - train_labels (ndarray): Training labels.
            - epochs (int): Number of epochs for
              training.
            - batch_size (int): Batch size for training.

        Returns:
            - History:
                Training history containing loss values
                for each
                epoch.
        """
        self.input_shape = (train_features.shape[1],)
        self.model = Sequential([
            Dense(
                64,
                activation='relu',
                input_shape=self.input_shape
            ),
```

CHAPTER 9   TECHNIQUES FOR IMPROVING MODEL PERFORMANCE

```python
        Dense(
            32,
            activation='relu'
        ),
        Dense(
            16,
            activation='relu'
        ),
        Dense(1)
    ])
    optimizer = Adam(learning_rate=0.01)
    self.model.compile(optimizer=optimizer,
    loss='mse')
    history = self.model.fit(
        train_features,
        train_labels,
        epochs=epochs,
        batch_size=batch_size,
        verbose=0
    )
    return history

def evaluate_model(self, test_features, test_labels):
    """
    Evaluate the trained model.

    Parameters:
        - test_features (ndarray): Test features.
        - test_labels (ndarray): Test labels.

    Returns:
        - float: Test loss.
    """
```

CHAPTER 9   TECHNIQUES FOR IMPROVING MODEL PERFORMANCE

```
        test_loss = self.model.evaluate(
            test_features,
            test_labels
        )
        return test_loss

    def predict_and_evaluate(
        self,
        test_features,
        test_labels
    ):
        """
        Predict and evaluate the neural network model.

        Parameters:
            - test_features (ndarray): Test features.
            - test_labels (ndarray): Test labels.

        Returns:
            - float: R2 score.
        """
        test_predictions = self.model.predict(test_
        features)
        r2_score_value = r2_score(test_labels, test_
        predictions)
        return r2_score_value

    def main(file_path):
        """
        Main function to orchestrate the entire workflow.

        Parameters:
            - file_path (str): Path to the CSV file.
```

CHAPTER 9   TECHNIQUES FOR IMPROVING MODEL PERFORMANCE

```
Returns:
    - None
"""
try:
    tsla_regressor = TSLARegressor()
    train_df, test_df = tsla_regressor.
    preprocess_data(
        file_path
    )
    train_features, train_labels, test_
    features, \
    test_labels = tsla_regressor.convert_
    to_numpy(
        train_df,
        test_df
    )
    history = tsla_regressor.create_and_
    train_model(
        train_features,
        train_labels
    )
    test_loss = tsla_regressor.evaluate_model(
        test_features,
        test_labels
    )
    r2_score_value= tsla_regressor.predict_and_
    evaluate(
        test_features,
        test_labels
    )
```

CHAPTER 9 TECHNIQUES FOR IMPROVING MODEL PERFORMANCE

```
                logging.info("Test Loss: {}".
                format(test_loss))
                logging.info("R2 Score: {}".format(r2_
                score_value))

                plt.plot(history.history['loss'])
                plt.title('Model Loss')
                plt.xlabel('Epoch')
                plt.ylabel('Loss')

                plt.show()

        except Exception as e:
                logging.error(f"An error occurred: {e}")
```
[In]:
```
if __name__ == "__main__":
        logging.basicConfig(level=logging.INFO)
        file_path = "/home/ubuntu/airflow/dags/TSLA_
        stock.csv"
        main(file_path)
```

Model Capacity

[In]: `import logging`
[In]: `import numpy as np`
[In]: `import tensorflow as tf`
[In]: `from pyspark.sql import SparkSession`
[In]: `from pyspark.ml.feature import VectorAssembler, StandardScaler`
[In]: `from sklearn.metrics import r2_score`
[In]: `from tensorflow.keras.models import Sequential`
[In]: `from tensorflow.keras.layers import Dense`
[In]: `import matplotlib.pyplot as plt`

CHAPTER 9   TECHNIQUES FOR IMPROVING MODEL PERFORMANCE

```
[In]: class TSLARegressor:
          """
          Class for TSLA stock price regression.
          """
          def __init__(self):
              """
              Initialize TSLARegressor object.
              """
              self.logger = logging.getLogger(__name__)
              self.spark = SparkSession.builder \
                          .appName("TSLA_Regression") \
                          .getOrCreate()
              self.input_shape = None
              self.model = None

          def preprocess_data(
              self,
              file_path,
              feature_cols=['Open', 'High', 'Low', 'Volume'],
              train_ratio=0.8,
              seed=42
          ):
              """
              Preprocesses the data.

              Parameters:
                  - file_path (str): Path to the CSV file.
                  - feature_cols (list): List of feature
                    columns.
                  - train_ratio (float): Ratio of
                    training data.
                  - seed (int): Random seed for splitting data.
```

```
Returns:
    - tuple: Tuple containing train and test
      DataFrames.
"""
df = self.spark.read.csv(
    file_path,
    header=True,
    inferSchema=True
)
assembler = VectorAssembler(
    inputCols=feature_cols, outputCol='features')
df = assembler.transform(df)
scaler = StandardScaler(
    inputCol='features',
    outputCol='scaled_features',
    withMean=True,
    withStd=True
)
scaler_model = scaler.fit(df)
df = scaler_model.transform(df)
train_df, test_df = df.randomSplit(
    [train_ratio, 1-train_ratio],
    seed=seed
)
return train_df, test_df

def convert_to_numpy(self, train_df, test_df):
    """
    Convert Spark DataFrames to numpy arrays.

    Parameters:
        - train_df (DataFrame):
            Spark DataFrame containing training data.
```

```
        - test_df (DataFrame):
            Spark DataFrame containing test data.
Returns:
    - tuple:
        Tuple containing train and test numpy
        arrays for
        features and labels.
"""
train_features = np.array(
    train_df.select('scaled_features')
    .rdd.map(lambda x: x.scaled_features.
    toArray())
    .collect()
)
train_labels = np.array(
    train_df.select('Close')
    .rdd.map(lambda x: x.Close).collect())
test_features = np.array(
    test_df.select('scaled_features')
    .rdd.map(lambda x: x.scaled_features.toArray())
    .collect()
)
test_labels = np.array(
    test_df.select('Close')
    .rdd.map(lambda x: x.Close).collect()
)
return (train_features,
        train_labels,
        test_features,
        test_labels
)
```

CHAPTER 9   TECHNIQUES FOR IMPROVING MODEL PERFORMANCE

```
def create_and_train_model(
    self,
    train_features,
    train_labels,
    epochs=100,
    batch_size=32
):
    """
    Create and train the neural network model.

    Parameters:
        - train_features (ndarray): Training
          features.
        - train_labels (ndarray): Training labels.
        - epochs (int): Number of epochs for
          training.
        - batch_size (int): Batch size for training.

    Returns:
        - History:
            Training history containing loss values
            for each
            epoch.
    """
    self.input_shape = (train_features.shape[1],)
    self.model = Sequential([
        Dense(
            128,
            activation='relu',
            input_shape=self.input_shape
        ),
```

CHAPTER 9   TECHNIQUES FOR IMPROVING MODEL PERFORMANCE

```python
        Dense(
            64,
            activation='relu'
        ),
        Dense(
            32,
            activation='relu'
        ),
        Dense(1)
    ])
    self.model.compile(optimizer='adam', loss='mse')
    history = self.model.fit(
        train_features,
        train_labels,
        epochs=epochs,
        batch_size=batch_size,
        verbose=0
    )
    return history

def evaluate_model(self, test_features, test_labels):
    """
    Evaluate the trained model.

    Parameters:
        - test_features (ndarray): Test features.
        - test_labels (ndarray): Test labels.

    Returns:
        - float: Test loss.
    """
```

CHAPTER 9   TECHNIQUES FOR IMPROVING MODEL PERFORMANCE

```python
        test_loss = self.model.evaluate(
            test_features,
            test_labels
        )
        return test_loss

    def predict_and_evaluate(
        self,
        test_features,
        test_labels
    ):
        """
        Predict and evaluate the neural network model.

        Parameters:
            - test_features (ndarray): Test features.
            - test_labels (ndarray): Test labels.

        Returns:
            - float: R2 score.
        """
        test_predictions = self.model.predict(test_features)
        r2_score_value = r2_score(test_labels, test_predictions)
        return r2_score_value

def main(file_path):
    """
    Main function to orchestrate the entire workflow.

    Parameters:
        - file_path (str): Path to the CSV file.
```

CHAPTER 9   TECHNIQUES FOR IMPROVING MODEL PERFORMANCE

Returns:
    - None
"""
try:
    tsla_regressor = TSLARegressor()
    train_df, test_df = tsla_regressor.
    preprocess_data(
        file_path
)
    train_features, train_labels, test_
    features, \
    test_labels = tsla_regressor.convert_
    to_numpy(
        train_df,
        test_df
)
    history = tsla_regressor.create_and_
    train_model(
        train_features,
        train_labels
)
    test_loss = tsla_regressor.evaluate_model(
        test_features,
        test_labels
)
    r2_score_value= tsla_regressor.predict_and_
    evaluate(
        test_features,
        test_labels
)

CHAPTER 9   TECHNIQUES FOR IMPROVING MODEL PERFORMANCE

```
            logging.info("Test Loss: {}".
            format(test_loss))
            logging.info("R2 Score: {}".format(r2_
            score_value))

            plt.plot(history.history['loss'])
            plt.title('Model Loss')
            plt.xlabel('Epoch')
            plt.ylabel('Loss')

            plt.show()

        except Exception as e:
            logging.error(f"An error occurred: {e}")
```

```
[In]: if __name__ == "__main__":
         logging.basicConfig(level=logging.INFO)
         file_path = "/home/ubuntu/airflow/dags/TSLA_
         stock.csv"
         main(file_path)
```

Automating Hyperparameter Tuning Using Keras Tuner

```
[In]: import logging
[In]: import numpy as np
[In]: import tensorflow as tf
[In]: from pyspark.sql import SparkSession
[In]: from pyspark.ml.feature import VectorAssembler,
      StandardScaler
[In]: from sklearn.metrics import r2_score
[In]: from tensorflow.keras.models import Sequential
```

CHAPTER 9　TECHNIQUES FOR IMPROVING MODEL PERFORMANCE

```
[In]: from tensorflow.keras.layers import Dense
[In]: import matplotlib.pyplot as plt
[In]: from keras_tuner import HyperModel, RandomSearch

[In]: class TSLAHyperModel(HyperModel):
          """
          HyperModel class for TSLA stock price regression.
          Defines the model architecture and allows Keras Tuner
          to search for the best hyperparameters.
          """
          def build(self, hp):
              model = Sequential()
              model.add(
                  Dense(
                      units=hp.Int(
                          'units_1',
                          min_value=32,
                          max_value=256,
                          step=32
                      ),
                      activation='relu',
                      input_shape=(self.input_shape,)
                  )
              )
              model.add(
                  Dense(
                      units=hp.Int(
                          'units_2',
                          min_value=16,
                          max_value=128,
                          step=16
                      ),
```

CHAPTER 9   TECHNIQUES FOR IMPROVING MODEL PERFORMANCE

```
                activation='relu'
            )
        )
        model.add(
            Dense(
                units=hp.Int(
                    'units_3',
                    min_value=8,
                    max_value=64,
                    step=8
                ),
                activation='relu'
            )
        )
        model.add(Dense(1))

        model.compile(
            optimizer=tf.keras.optimizers.Adam(
                learning_rate=hp.Choice(
                    'learning_rate',
                    values=[1e-2, 1e-3, 1e-4]
                )
            ),
            loss='mse'
        )
        return model
```

```
[In]: class TSLARegressor:
        """
        Class for TSLA stock price regression.
        """
```

```python
    def __init__(self):
        """
        Initialize TSLARegressor object.
        """
        self.logger = logging.getLogger(__name__)
        self.spark = SparkSession.builder \
                    .appName("TSLA_Regression") \
                    .getOrCreate()
        self.input_shape = None
        self.model = None

    def preprocess_data(
        self,
        file_path,
        feature_cols=['Open', 'High', 'Low', 'Volume'],
        train_ratio=0.8,
        seed=42
    ):
        """
        Preprocesses the data.

        Parameters:
        - file_path (str): Path to the CSV file.
        - feature_cols (list): List of feature columns.
        - train_ratio (float): Ratio of training data.
        - seed (int): Random seed for splitting data.

        Returns:
        - tuple: Tuple containing train and test
          DataFrames.
        """
```

CHAPTER 9   TECHNIQUES FOR IMPROVING MODEL PERFORMANCE

```python
        df = self.spark.read.csv(
            file_path,
            header=True,
            inferSchema=True
        )
        assembler = VectorAssembler(
            inputCols=feature_cols,
            outputCol='features'
        )
        df = assembler.transform(df)
        scaler = StandardScaler(
            inputCol='features',
            outputCol='scaled_features',
            withMean=True,
            withStd=True
        )
        scaler_model = scaler.fit(df)
        df = scaler_model.transform(df)
        train_df, test_df = df.randomSplit(
            [train_ratio, 1-train_ratio],
            seed=seed
        )
        return train_df, test_df

    def convert_to_numpy(self, train_df, test_df):
        """
        Convert Spark DataFrames to numpy arrays.

        Parameters:
        - train_df (DataFrame):
            Spark DataFrame containing training data.
        - test_df (DataFrame):
            Spark DataFrame containing test data.
```

CHAPTER 9   TECHNIQUES FOR IMPROVING MODEL PERFORMANCE

```
Returns:
- tuple:
    Tuple containing train and test numpy
    arrays for
    features and labels.
"""
train_features = np.array(
    train_df.select('scaled_features')
    .rdd.map(lambda x: x.scaled_features.
    toArray())
    .collect()
)
train_labels = np.array(
    train_df.select('Close')
    .rdd.map(lambda x: x.Close)
    .collect()
)
test_features = np.array(
    test_df.select('scaled_features')
    .rdd.map(lambda x: x.scaled_features.
    toArray())
    .collect()
)
test_labels = np.array(
    test_df.select('Close')
    .rdd.map(lambda x: x.Close)
    .collect()
)
return (
    train_features,
    train_labels,
```

CHAPTER 9   TECHNIQUES FOR IMPROVING MODEL PERFORMANCE

```
            test_features,
            test_labels
        )

    def create_and_train_model(
        self,
        train_features,
        train_labels,
        test_features,
        test_labels,
        epochs=100,
        batch_size=32
    ):
        """
        Create and train the neural network model
        using Keras
        Tuner for hyperparameter optimization.

        Parameters:
        - train_features (ndarray): Training features.
        - train_labels (ndarray): Training labels.
        - test_features (ndarray): Test features.
        - test_labels (ndarray): Test labels.
        - epochs (int): Number of epochs for training.
        - batch_size (int): Batch size for training.

        Returns:
        - History:
            Training history containing loss values
            for each
            epoch.
        """
```

```python
self.input_shape = train_features.shape[1]
tsla_hypermodel = TSLAHyperModel()
tsla_hypermodel.input_shape = self.input_shape

tuner = RandomSearch(
    tsla_hypermodel,
    objective='val_loss',
    max_trials=5,
    executions_per_trial=3,
    directory='tuner_dir',
    project_name='tsla_regression'
)

tuner.search(
    train_features,
    train_labels,
    epochs=epochs,
    batch_size=batch_size,
    validation_data=(test_features, test_labels),
    verbose=1
)
self.model = tuner.get_best_models(num_models=1)[0]
return self.model.fit(
    train_features,
    train_labels,
    epochs=epochs,
    batch_size=batch_size,
    verbose=1,
    validation_data=(test_features, test_labels)
)
```

CHAPTER 9   TECHNIQUES FOR IMPROVING MODEL PERFORMANCE

```python
    def evaluate_model(self, test_features, test_labels):
        """
        Evaluate the trained model.

        Parameters:
        - test_features (ndarray): Test features.
        - test_labels (ndarray): Test labels.

        Returns:
        - float: Test loss.
        """
        test_loss = self.model.evaluate(
            test_features,
            test_labels
        )
        return test_loss

    def predict_and_evaluate(self, test_features, test_labels):
        """
        Predict and evaluate the neural network model.

        Parameters:
        - test_features (ndarray): Test features.
        - test_labels (ndarray): Test labels.

        Returns:
        - float: R2 score.
        """
        test_predictions = self.model.predict(test_features)
        r2_score_value = r2_score(test_labels, test_predictions)
        return r2_score_value
```

CHAPTER 9   TECHNIQUES FOR IMPROVING MODEL PERFORMANCE

```
[In]: def main(file_path):
          """
          Main function to orchestrate the entire workflow.

          Parameters:
          - file_path (str): Path to the CSV file.

          Returns:
          - None
          """
          try:
              tsla_regressor = TSLARegressor()
              train_df, test_df = tsla_regressor.
              preprocess_data(
                  file_path
              )
              train_features,
              train_labels,
              test_features,
              test_labels = tsla_regressor.convert_to_numpy(
                  train_df,
                  test_df
              )
              history = tsla_regressor.create_and_train_model(
                  train_features,
                  train_labels,
                  test_features,
                  test_labels,
                  epochs=100,
                  batch_size=32
              )
```

```
            test_loss = tsla_regressor.evaluate_model(
                test_features,
                test_labels
            )
            r2_score_value = tsla_regressor.predict_and_
            evaluate(
                test_features,
                test_labels
            )

            logging.info("Test Loss: {}".format(test_loss))
            logging.info("R2 Score: {}".format(r2_
            score_value))

            plt.plot(history.history['loss'])
            plt.title('Model Loss')
            plt.xlabel('Epoch')
            plt.ylabel('Loss')
            plt.savefig('model_loss_plot.png')
            plt.show()

        except Exception as e:
            logging.error(f"An error occurred: {e}")
```

```
[In]: if __name__ == "__main__":
          logging.basicConfig(level=logging.INFO)
          file_path = "/home/ubuntu/airflow/dags/TSLA_
          stock.csv"
          main(file_path)
```

Summary

In this chapter, we examined various techniques to improve the performance of deep learning models. To deal with overfitting, we explored early stopping, dropout, and L1 and L2 regularization. These methods impose constraints on the model's parameters to prevent overfitting. Additionally, we covered hyperparameter optimization, including fine-tuning the learning rate and network architecture, to identify the best-performing model on the validation set. We demonstrated how to achieve this both manually and using Keras Tuner.

In the next chapter, which is the last chapter in the book, readers learn about best practices for deploying trained models into production environments and monitoring their performance.

CHAPTER 10

Deploying and Monitoring Deep Learning Models

Deploying machine learning models into production is a crucial step in the lifecycle of any data science project. However, this process often presents challenges related to scalability, reliability, and maintainability. Managed Workflows for Apache Airflow (MWAA), a fully managed service provided by Amazon Web Services (AWS), offers a solution to these challenges by simplifying and automating the deployment and orchestration of complex workflows, including model deployment.

In Chapter 2, we demonstrated how to set up Airflow for deep learning as a standalone service and within a Docker container. We explored the benefits and challenges of these setups, particularly in terms of flexibility and control. In Chapter 8, we built scalable data processing pipelines with standalone Apache Airflow, incorporating preprocessing steps using PySpark, and constructing deep learning models using PyTorch and TensorFlow, all orchestrated within Airflow DAGs for both regression and classification tasks. While these approaches provided significant control, they required manual installation and configuration of Apache Airflow, as well as manual triggering of DAGs, leading to increased complexity and maintenance overhead.

In this chapter, we introduce a third option: deploying deep learning models using Amazon MWAA. This service simplifies the process of setting up, managing, and scaling Apache Airflow environments in the cloud, eliminating the need for manual infrastructure management. The key advantage of using MWAA lies in its managed nature, freeing users from the burden of server management, scaling, and infrastructure maintenance. With MWAA, users can focus on building workflows and applications, confident that the underlying infrastructure is handled by AWS.

Additionally, we will explore monitoring solutions to ensure the performance and reliability of deployed models, leveraging the Amazon MWAA console for workflow monitoring and Amazon CloudWatch for comprehensive metrics and logs tracking. By monitoring deep learning models, users can detect issues such as model drift, resource bottlenecks, and latency problems, enabling timely interventions and ensuring high standards of performance, reliability, and compliance.

Steps in Deploying and Monitoring Deep Learning Models

In this section, we examine the main steps involved in deploying and monitoring deep learning models specifically using Amazon Managed Workflows for Apache Airflow (MWAA). We first provide a brief summary of these steps and then examine them in detail one by one.

Starting with the brief summary, there are six main steps:

Step 1: Set Up the Environment

- Access the Amazon MWAA console and create a new MWAA environment.

- Configure the environment by specifying an S3 bucket for DAG files, setting up networking (including VPC and subnets), and configuring security settings.

Step 2: Develop the DAG

- Begin by developing an Airflow DAG to orchestrate the deployment process. This DAG defines tasks, sets dependencies, and executes Python functions. For example, the DAG could include tasks to preprocess data, train the model, and deploy it.
- Ensure the DAG code is well-structured and includes all necessary dependencies.

Step 3: Upload to S3

- Upload the DAG file to an S3 bucket accessible by the MWAA environment. This is typically done using the AWS CLI or the AWS Management Console.

Step 4: Configure the MWAA Environment

- Configure the MWAA environment to include the uploaded DAG file. Specify any additional settings such as environment variables, connections, and plugins required for the DAG's execution. This ensures that the environment is properly set up to execute the DAG.

Step 5: Trigger DAG Execution

- Trigger the execution of the DAG manually or based on a predefined schedule within the Amazon MWAA environment. This initiates the deployment process and executes the defined tasks within the DAG, such as preprocessing data, training the model, and deploying it.

CHAPTER 10   DEPLOYING AND MONITORING DEEP LEARNING MODELS

Step 6: Monitor Execution

- Monitor the execution of the DAG from the MWAA console. Track task progress, view logs, and troubleshoot any issues that may arise during the deployment process.

After outlining the steps involved in the deployment process with Amazon MWAA, let's examine them one by one, starting with setting up the environment.

Step 1: Setting Up the Environment

Setting up an Amazon MWAA environment includes specifying an S3 bucket for the DAG files and configuring networking and security settings. Follow these steps to ensure the environment is properly configured and ready for use:

Step 1: Access the Amazon MWAA Console

- Log into the AWS Management Console using your AWS credentials.
- In the AWS Management Console, type "MWAA" in the search bar and select "Managed Apache Airflow" from the list of services.

Step 2: Create a New MWAA Environment

- On the Amazon MWAA dashboard, click the "Create environment" button.

Step 3: Specify Environment Details

- Enter a unique name for your MWAA environment.
- Choose the version of Apache Airflow you want to use.

CHAPTER 10   DEPLOYING AND MONITORING DEEP LEARNING MODELS

Step 4: Specify S3 Bucket and Folder for DAGs

- Enter the name of the S3 bucket where your DAG files are stored.

- Enter the folder path within the S3 bucket where your DAG files are located (e.g., dags/).

Step 5: Configure Execution Role

- Choose an existing AWS Identity and Access Management (IAM) role with the necessary permissions, or create a new one.

- Ensure the role has the required permissions for accessing the S3 bucket, CloudWatch logs, and other AWS services your DAGs may use.

Step 6: Configure Networking

Before creating an MWAA environment, you need to set up a Virtual Private Cloud (VPC). The VPC ensures secure interaction between your MWAA environment and other AWS resources such as S3 buckets, databases, and other services that your workflows might need. Follow these steps to set up your VPC:

- Create a VPC:

  - Navigate to the VPC Service: In the AWS Management Console, type "VPC" into the search bar at the top of the page and select VPC from the dropdown list, as shown in Figure 10-1.

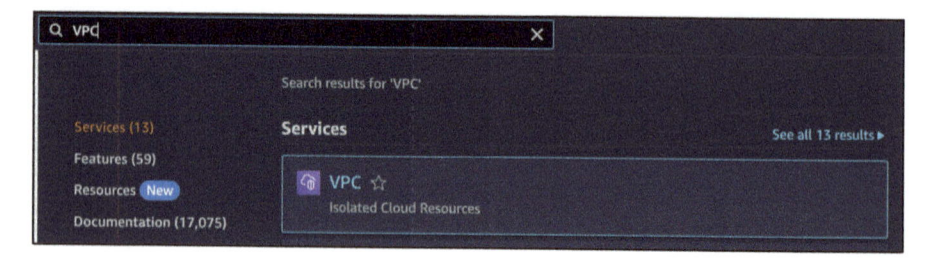

CHAPTER 10   DEPLOYING AND MONITORING DEEP LEARNING MODELS

Figure 10-1. Navigating to the VPC Service via AWS Management Console Search

- On the VPC dashboard, click Create VPC, as in Figure 10-2.

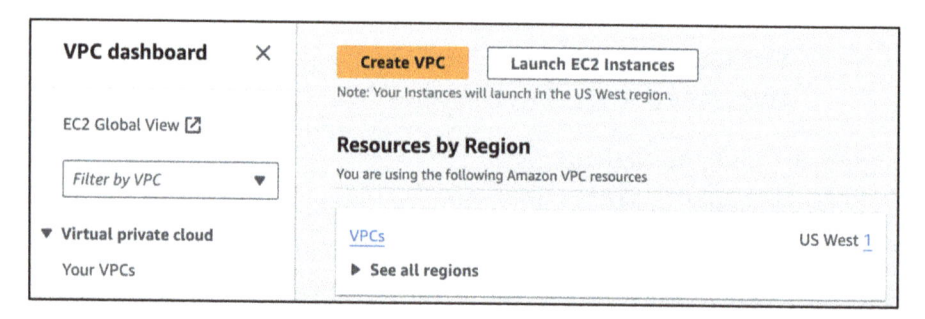

Figure 10-2. VPC Dashboard on AWS

After clicking "Create VPC," follow these steps from Figure 10-3:

- Choose the VPC with single or multiple public and private subnets.

- Specify the CIDR block (e.g., 10.0.0.0/16).

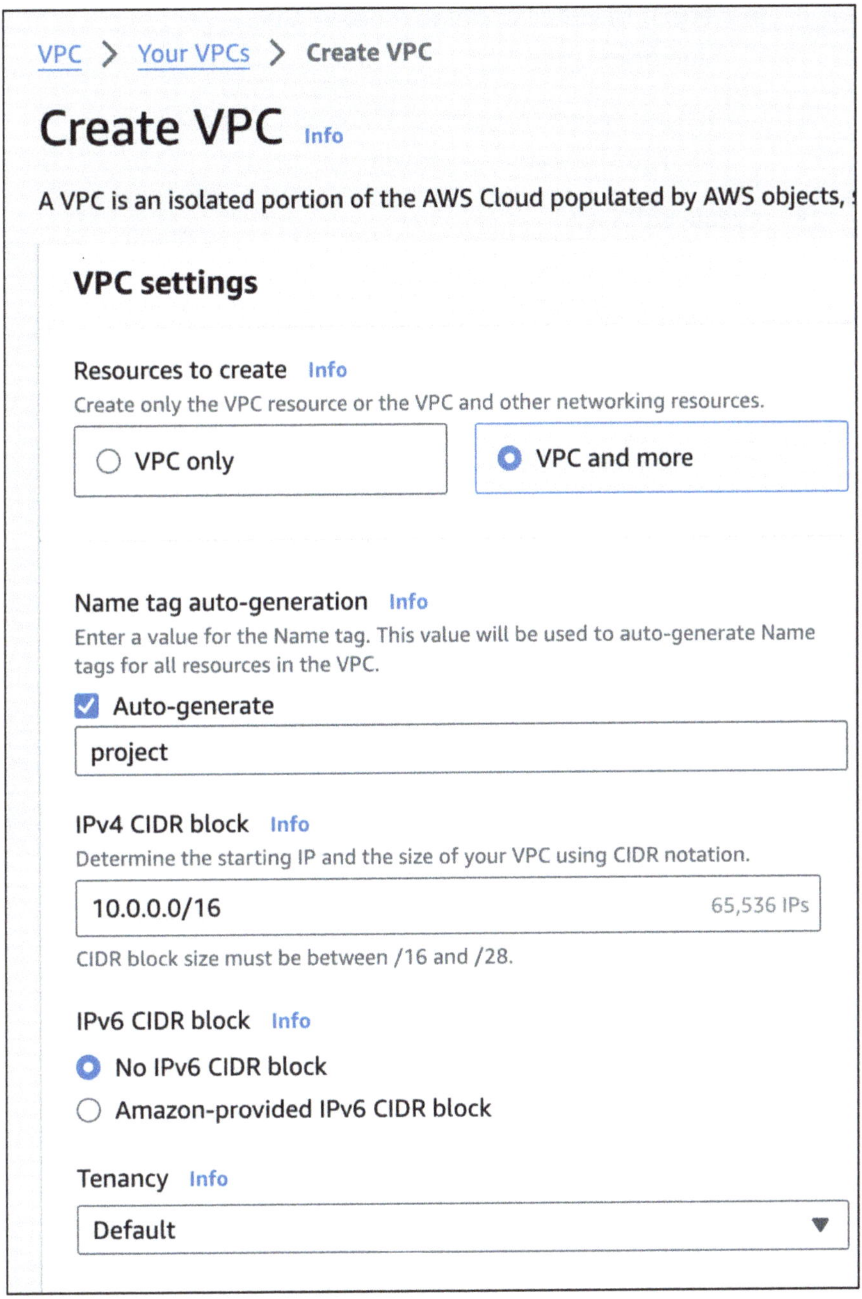

CHAPTER 10   DEPLOYING AND MONITORING DEEP LEARNING MODELS

VPC > Your VPCs > **Create VPC**

Create VPC   Info

A VPC is an isolated portion of the AWS Cloud populated by AWS objects,

VPC settings

Resources to create   Info
Create only the VPC resource or the VPC and other networking resources.

○ VPC only   ● VPC and more

Name tag auto-generation   Info
Enter a value for the Name tag. This value will be used to auto-generate Name tags for all resources in the VPC.

☑ Auto-generate

project

IPv4 CIDR block   Info
Determine the starting IP and the size of your VPC using CIDR notation.

10.0.0.0/16   65,536 IPs

CIDR block size must be between /16 and /28.

IPv6 CIDR block   Info

● No IPv6 CIDR block
○ Amazon-provided IPv6 CIDR block

Tenancy   Info

Default   ▼

Figure 10-3. Creating a VPC on AWS

709

CHAPTER 10 DEPLOYING AND MONITORING DEEP LEARNING MODELS

- For more details, refer to the AWS VPC documentation:

  https://docs.aws.amazon.com/vpc/latest/userguide/what-is-amazon-vpc.html

- Create Subnets:
  - Navigate to the Subnets section within the VPC dashboard, as shown in Figure 10-4.

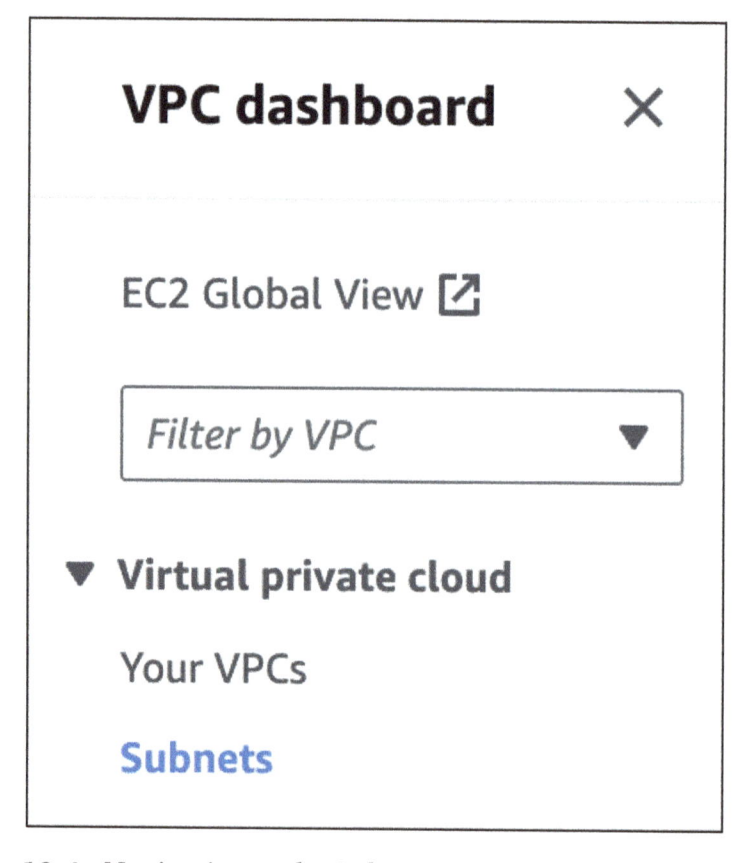

Figure 10-4. Navigating to the Subnets Section Within the VPC Dashboard on AWS

CHAPTER 10   DEPLOYING AND MONITORING DEEP LEARNING MODELS

- Click "Create subnet" as in Figure 10-5.

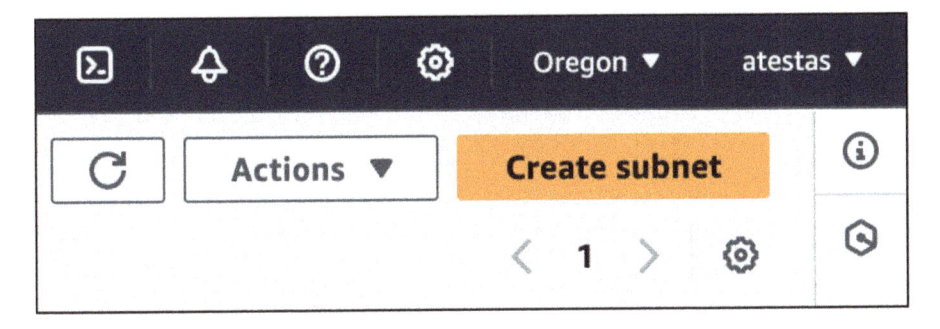

Figure 10-5. Creating a Subnet on AWS

- Create at least two private subnets in different Availability Zones.
- Name your subnets for easy identification (e.g., Private-Subnet-1 and Private-Subnet-2).
- Repeat the process to create public subnets for NAT gateways.
- For more details, refer to the AWS documentation:

  https://docs.aws.amazon.com/vpc/latest/userguide/how-it-works.html

- Create and Attach the Internet Gateway:
  - Navigate to the Internet gateways section.
  - Click "Create internet gateway."
  - Attach the Internet gateway to your VPC by selecting "Actions" and choosing "Attach to VPC."

711

CHAPTER 10 DEPLOYING AND MONITORING DEEP LEARNING MODELS

- For more details, refer to the AWS "Enable VPC internet access using internet gateways" documentation:

  https://docs.aws.amazon.com/vpc/latest/userguide/ VPC_Internet_Gateway.html#Add_IGW_Attach_Gateway

- Create the NAT Gateway:
    - Navigate to the NAT gateways section as shown in Figure 10-6.

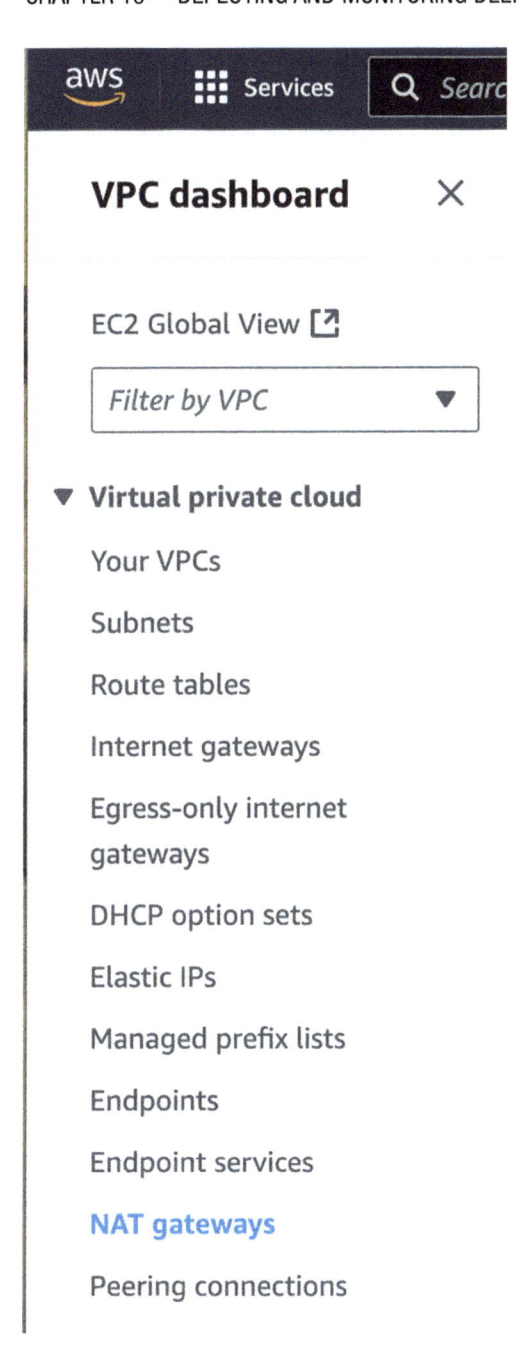

CHAPTER 10   DEPLOYING AND MONITORING DEEP LEARNING MODELS

Figure 10-6. Navigating to the NAT gateways Section on AWS

CHAPTER 10   DEPLOYING AND MONITORING DEEP LEARNING MODELS

- Click "Create NAT gateway" as in Figure 10-7.

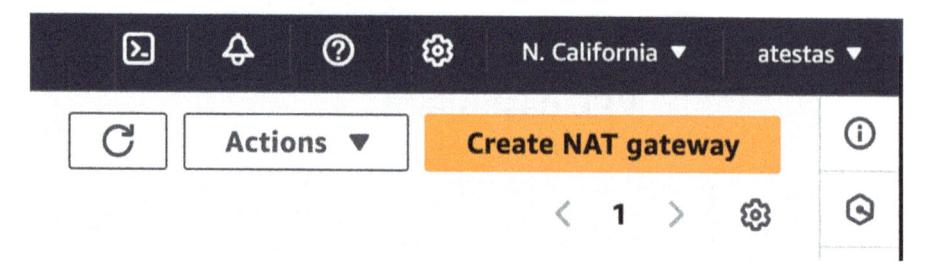

Figure 10-7. Creating a NAT Gateway on AWS

- Select the public subnet where the NAT gateway will be created.
- Allocate an Elastic IP to the NAT gateway.
- For more details, refer to the AWS NAT gateways documentation:

  https://docs.aws.amazon.com/vpc/latest/userguide/vpc-nat-gateway.html

- Update Route Tables:
  - Navigate to the Route tables section
  - Update the route tables of your private subnets to route Internet-bound traffic through the NAT gateway.
  - For more details, refer to the AWS "Configure route tables" documentation:

    https://docs.aws.amazon.com/vpc/latest/userguide/VPC_Route_Tables.html

- Configure Security Groups:

  - Navigate to Security groups as in Figure 10-8.

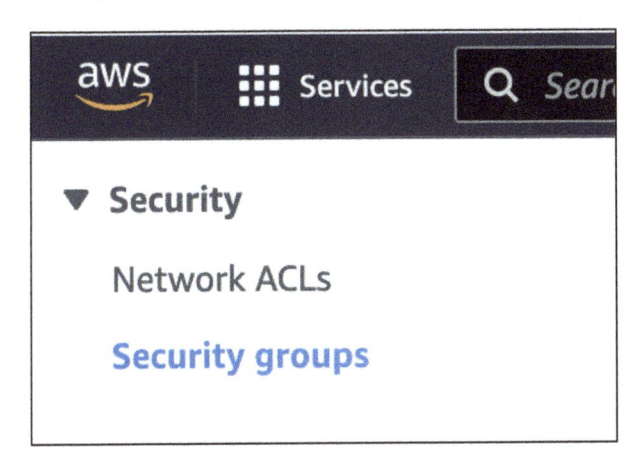

Figure 10-8. Navigating to Security groups on AWS

  - Create security groups with the necessary inbound and outbound rules for your MWAA environment.

  - Associate these security groups with your MWAA environment during its creation.

  - For more details, refer to the AWS "Control traffic to your AWS resources using security groups" documentation:

    https://docs.aws.amazon.com/vpc/latest/userguide/vpc-security-groups.html

Step 7: Configure Environment Class

- Choose the appropriate instance class for your MWAA environment. This determines the compute resources allocated to your environment.

Step 8: Additional Configurations

- Specify the maximum number of workers for your environment.
- Specify the minimum number of workers for your environment.
- Define the number of schedulers.
- Add any custom configurations for Airflow.
- Specify any custom plugins by providing the S3 path.
- Provide the S3 path to the requirements.txt file for additional Python dependencies.

Step 9: Review and Create

- Review all the configurations you have set.
- Click the "Create environment" button to create your MWAA environment.

Step 10: Monitor the Creation Process

- The creation process may take a few minutes.
- You can monitor the status of your environment on the Amazon MWAA dashboard.

Once the environment is created and the status is active, you can start deploying and managing your DAGs using the specified S3 bucket and folder.

Step 2: Developing the DAG

After setting up the Amazon MWAA environment, the next step is to develop the DAG. Let's develop a simplified DAG (simple_mwaa_dag.py) for illustration purposes. This provides a foundational understanding of

CHAPTER 10   DEPLOYING AND MONITORING DEEP LEARNING MODELS

how to define tasks, set task dependencies, and execute Python functions within Airflow.

The DAG below defines default arguments, the DAG itself, Python functions for tasks, and tasks and sets task dependencies. The DAG executes two Python tasks: one to print the current date and time and another to print a predefined message. When orchestrated by Airflow, these tasks ensure that they execute in a defined sequence:

```
[In]: from airflow import DAG
[In]: from airflow.operators.python_operator import
      PythonOperator
[In]: from datetime import datetime, timedelta

# Define default arguments
[In]: default_args = {
          'owner': 'airflow',
          'depends_on_past': False,
          'start_date': datetime(2023, 1, 1),
          'email_on_failure': False,
          'email_on_retry': False,
          'retries': 1,
          'retry_delay': timedelta(minutes=5),
      }

# Define the DAG
[In]: dag = DAG(
          'simple_mwaa_dag',
          default_args=default_args,
          description='A simple DAG for MWAA',
          schedule_interval=timedelta(days=1),
      )
```

Define Python functions for tasks
```
[In]: def print_current_date():
         print(f"Current date and time: {datetime.now()}")

[In]: def print_message():
         print("Hello, this is a message from your DAG!")
```

Define tasks
```
[In]: t1 = PythonOperator(
         task_id='print_date',
         python_callable=print_current_date,
         dag=dag,
      )

[In]: t2 = PythonOperator(
         task_id='print_message',
         python_callable=print_message,
         dag=dag,
      )
```

Set task dependencies
```
[In]: t1 >> t2
```

The DAG starts with the 'print_date' task, which utilizes a PythonOperator to execute the "print_current_date" function. Inside this function, the current date and time are retrieved using Python's datetime.now() function, and this information is printed to the Airflow logs. Once the 'print_date' task completes successfully, Airflow proceeds to the 'print_message' task. Similar to the previous task, it uses a PythonOperator to execute the "print_message" function. Inside this function, a simple message, "Hello, this is a message from your DAG!", is printed to the Airflow logs.

CHAPTER 10 DEPLOYING AND MONITORING DEEP LEARNING MODELS

These tasks are linked together with a task dependency, indicated by the >> operator, ensuring that the 'print_message' task runs only after the 'print_date' task has successfully completed. This dependency ensures a logical sequence of task execution within the DAG.

Step 3: Uploading to S3

After DAG development, the next step is to deploy the DAG to Amazon MWAA. The first task in this deployment is to upload the file containing the DAG to an Amazon S3 bucket that is accessible by the Amazon MWAA environment. The command below uploads the simple_mwaa_dag.py DAG file from /home/ubuntu/airflow/dags/ directory on Amazon EC2 to the dags folder within the instance1bucket S3 bucket:

```
[In]: aws s3 cp /home/ubuntu/airflow/dags/simple_mwaa_dag.py \
      s3://instance1bucket/dags/
```

DAG files and their dependencies are usually packaged into a zip file before loading into an S3 bucket. In this simple example, the DAG file (simple_mwaa_dag.py) does not have any external dependencies; hence, there is no need to package it into a zip file for deployment. However, as the complexity of DAGs increases and external dependencies such as custom Python modules or third-party libraries are introduced, packaging DAGs into zip files becomes essential.

A good example of a complex DAG is the `tesla_stock_prediction` DAG we developed in Chapter 8. In this chapter, we created an Airflow pipeline that included preprocessing Tesla stock data using PySpark and building, training, and evaluating a deep learning model using PyTorch to predict the Tesla stock price. The DAG has the following five modules:

- data_processing.py: Preprocesses the Tesla stock data
- model_training.py: Utilizes the PyTorch library to create and train the neural network regression model for the Tesla stock price prediction

CHAPTER 10 DEPLOYING AND MONITORING DEEP LEARNING MODELS

- model_evaluation.py: Evaluates the performance of the trained PyTorch model on a test dataset
- utils.py: Utility module that converts Spark DataFrame columns to PyTorch tensors
- tesla_stock_prediction.py: DAG that orchestrates the entire workflow

Below is the code contained in these modules:
data_processing.py:

```
[In]: from pyspark.sql import SparkSession
[In]: from pyspark.ml.feature import VectorAssembler,
      StandardScaler
[In]: from pyspark.sql import DataFrame

[In]: def load_data(file_path: str) -> DataFrame:
        """
        Load stock price data from a CSV file using
        SparkSession.
        """
        spark = SparkSession.builder \
                .appName("StockPricePrediction") \
                .getOrCreate()
        df = spark.read.csv(
                file_path,
                header=True,
                inferSchema=True
        )
        return df
```

```
[In]: def preprocess_data(df: DataFrame) -> DataFrame:
          """
          Preprocess the data by assembling feature
          vectors using
          VectorAssembler and scaling them using
          StandardScaler.
          """
          assembler = VectorAssembler(
              inputCols=['Open', 'High', 'Low', 'Volume'],
              outputCol='features'
          )
          df = assembler.transform(df)

          scaler = StandardScaler(
              inputCol="features",
              outputCol="scaled_features",
              withStd=True,
              withMean=True
          )
          scaler_model = scaler.fit(df)
          df = scaler_model.transform(df)
          df = df.select('scaled_features', 'Close')
          return df
```

model_training.py:

```
[In]: import torch
[In]: import torch.nn as nn
[In]: import torch.optim as optim
[In]: from torch.utils.data import DataLoader, TensorDataset

[In]: def create_data_loader(
          features,
          labels,
```

```
        batch_size=32
    ) -> DataLoader:
        """
        Convert the preprocessed data into PyTorch
        tensors and
        Create DataLoader objects for both the training
        and test
        sets.
        """
        dataset = TensorDataset(features, labels)
        return DataLoader(
            dataset,
            batch_size=batch_size,
            shuffle=True
        )
```

```
[In]: def train_model(
        model, train_loader, criterion, optimizer, num_epochs
    ):
        """
        Train the model on the training data using the
        DataLoader
        and the defined loss function and optimizer.
        Iterate over
        the data for a specified number of epochs,
        calculate the
        loss, and update the model parameters.
        """
        for epoch in range(num_epochs):
            for inputs, labels in train_loader:
                optimizer.zero_grad()
                outputs = model(inputs)
```

CHAPTER 10 DEPLOYING AND MONITORING DEEP LEARNING MODELS

```
            loss = criterion(outputs, labels.
            unsqueeze(1))
            loss.backward()
            optimizer.step()
    print(
        f"Epoch [{epoch + 1}/{num_epochs}], "
        f"Loss: {loss.item():.4f}"
    )
```

model_evaluation.py:

[In]: import torch

[In]: def evaluate_model(model, test_loader, criterion):
```
        """
        Evaluate the trained model on the test data to
        assess its
        performance.
        Calculate the test loss and additional
        evaluation metrics
        such as the R-squared score.
        """
        with torch.no_grad():
            model.eval()
            predictions = []
            targets = []
            test_loss = 0.0
            for inputs, labels in test_loader:
                outputs = model(inputs)
                loss = criterion(outputs, labels.unsqueeze(1))
                test_loss += loss.item() * inputs.size(0)
                predictions.extend(outputs.squeeze().tolist())
                targets.extend(labels.tolist())
```

CHAPTER 10   DEPLOYING AND MONITORING DEEP LEARNING MODELS

```
            test_loss /= len(test_loader.dataset)
            predictions = torch.tensor(predictions)
            targets = torch.tensor(targets)
            ss_res = torch.sum((targets - predictions) ** 2)
            ss_tot = torch.sum((targets - torch.
            mean(targets)) ** 2)
            r_squared = 1 - ss_res / ss_tot
            print(f"Test Loss: {test_loss:.4f}")
            print(f"R-squared Score: {r_squared:.4f}")
            return test_loss, r_squared.item()
```

utils.py:

```
[In]: import numpy as np
[In]: import torch
[In]: from typing import Tuple
[In]: from pyspark.sql import DataFrame

[In]: def spark_df_to_tensor(
        df: DataFrame
    ) -> Tuple[torch.Tensor, torch.Tensor]:
        """
        Converts Spark DataFrame columns to PyTorch tensors.
        """
        features = torch.tensor(
            np.array(
                df.rdd.map(lambda x: x.scaled_features.
                toArray())
                .collect()
            ),
            dtype=torch.float32
        )
```

CHAPTER 10   DEPLOYING AND MONITORING DEEP LEARNING MODELS

```
        labels = torch.tensor(
            np.array(
                df.rdd.map(lambda x: x.Close).collect()
            ),
            dtype=torch.float32
        )
        return features, labels
```

tesla_stock_prediction.py:

```
from datetime import datetime, timedelta
from airflow import DAG
from airflow.operators.python_operator import PythonOperator
import logging
import torch
import torch.optim as optim
import torch.nn as nn
from torch.utils.data import DataLoader
import numpy as np
from pyspark.sql import DataFrame
from typing import Tuple
from utils import spark_df_to_tensor
from data_processing import load_data, preprocess_data
from model_training import create_data_loader, train_model
from model_evaluation import evaluate_model
```
[In]: data_file_path = "/home/ubuntu/airflow/dags/TSLA_stock.csv"

[In]: logging.basicConfig(level=logging.INFO)
[In]: logger = logging.getLogger(__name__)

[In]: default_args = {
        'owner': 'airflow',
        'depends_on_past': False,
        'start_date': datetime(2024, 5, 1),

CHAPTER 10   DEPLOYING AND MONITORING DEEP LEARNING MODELS

```
            'email_on_failure': False,
            'email_on_retry': False,
            'retries': 1,
            'retry_delay': timedelta(minutes=5),
        }
```

```
[In]: dag = DAG(
            'tesla_stock_prediction',
            default_args=default_args,
            description='DAG for Tesla stock price prediction',
            schedule_interval=None,
        )
```

```
[In]: def preprocess_data_task():
            try:
                df = load_data(data_file_path)
                df = preprocess_data(df)
            except Exception as e:
                logger.error(f"Error in preprocessing data: {str(e)}")
                raise
```

```
[In]: preprocess_data_task = PythonOperator(
            task_id='preprocess_data_task',
            python_callable=preprocess_data_task,
            dag=dag,
        )
```

```
[In]: def train_model_task():
            try:
                df = load_data(data_file_path)
                df = preprocess_data(df)
                train_df, test_df = df.randomSplit([0.8, 0.2],
                seed=42)
```

```
train_features, train_labels = spark_df_
to_tensor(
    train_df
)
train_loader = create_data_loader(
    train_features,
    train_labels,
    batch_size=32
)

input_size = train_features.shape[1]
output_size = 1
model = nn.Sequential(
    nn.Linear(input_size, 64),
    nn.ReLU(),
    nn.Linear(64, 32),
    nn.ReLU(),
    nn.Linear(32, 16),
    nn.ReLU(),
    nn.Linear(16, output_size)
)
criterion = nn.MSELoss()
optimizer = optim.Adam(model.parameters(),
lr=0.001)
train_model(
    model,
    train_loader,
    criterion,
    optimizer,
    num_epochs=100
)
```

CHAPTER 10   DEPLOYING AND MONITORING DEEP LEARNING MODELS

```
    except Exception as e:
        logger.error(f"Error in training model: {str(e)}")
        raise
```

```
[In]: train_model_task = PythonOperator(
        task_id='train_model_task',
        python_callable=train_model_task,
        dag=dag,
    )
```

```
[In]: def evaluate_model_task():
        try:
            df = load_data(data_file_path)
            df = preprocess_data(df)
            train_df, test_df = df.randomSplit([0.8, 0.2], seed=42)
            test_features, test_labels = spark_df_to_tensor(test_df)
            test_loader = create_data_loader(
                test_features,
                test_labels,
                batch_size=32
            )

            input_size = test_features.shape[1]
            output_size = 1
            model = nn.Sequential(
                nn.Linear(input_size, 64),
                nn.ReLU(),
                nn.Linear(64, 32),
                nn.ReLU(),
                nn.Linear(32, 16),
```

```
            nn.ReLU(),
            nn.Linear(16, output_size)
        )
        model.load_state_dict(torch.load('trained_
        model.pth'))

        criterion = nn.MSELoss()
        evaluate_model(model, test_loader, criterion)
    except Exception as e:
        logger.error(f"Error in evaluating model:
        {str(e)}")
        raise
```

```
[In]: evaluate_model_task = PythonOperator(
        task_id='evaluate_model_task',
        python_callable=evaluate_model_task,
        dag=dag,
    )
[In]: preprocess_data_task >> train_model_task >> evaluate_
      model_task
```

The following Python code packages the five modules (data_processing.py, model_training.py, model_evaluation.py, utils.py, and tesla_stock_prediction.py) containing the DAG logic and task implementations into a zip file. Additionally, the code creates a requirements.txt file listing all the dependencies required for the DAG's execution and uploads the DAG files to Amazon S3:

```
[In]: import os
[In]: import zipfile
[In]: import boto3

# Define the directory containing the DAG files
[In]: dags_directory = '/home/ubuntu/airflow/dags'
```

CHAPTER 10 DEPLOYING AND MONITORING DEEP LEARNING MODELS

```
# Define the names of the files to be included in the zip file
[In]: files_to_zip = [
         'data_processing.py',
         'model_training.py',
         'model_evaluation.py',
         'utils.py',
         'tesla_stock_prediction.py',
         'requirements.txt'
      ]

# Define the name of the zip file
[In]: zip_file_name = 'tesla_stock_prediction.zip'

# Create a zip file containing the specified files
[In]: with zipfile.ZipFile(zip_file_name, 'w') as zip_file:
         for file_name in files_to_zip:
             file_path = os.path.join(dags_directory,
             file_name)
             zip_file.write(file_path, arcname=file_name)

# Upload the zip file to the S3 bucket
[In]: bucket_name = 'instance1bucket'
[In]: s3_client = boto3.client('s3')
[In]: s3_client.upload_file(
         zip_file_name,
         bucket_name,
         f'dags/{zip_file_name}'
      )

# Clean up: remove the zip file
[In]: os.remove(zip_file_name)
[In]: print("Files zipped and uploaded successfully.")
```

CHAPTER 10 DEPLOYING AND MONITORING DEEP LEARNING MODELS

This code prepares and uploads the DAG files to Amazon S3. First, the necessary libraries are imported. These include os for interacting with the operating system, zipfile for handling zip file operations, and boto3 for interacting with Amazon Web Services (AWS) services like S3. The directory containing the DAG files is specified as dags_directory, set to /home/ubuntu/airflow/dags. This directory holds the Python files constituting the DAG and other associated files needed for its execution. A list named files_to_zip is defined, containing the names of the files that will be included in the zip file. These comprise the Python scripts defining the DAG logic:

- data_processing.py
- model_training.py
- model_evaluation.py
- utils.py
- tesla_stock_prediction.py
- requirements.txt.

The name of the zip file is specified as zip_file_name, set to tesla_stock_prediction.zip. This file will serve as the packaged collection of DAG files ready for deployment. A zipfile.ZipFile context manager is initiated to create the zip file specified by zip_file_name. Within this context, each file listed in files_to_zip is added to the zip file. The arcname parameter ensures that the files are stored within the zip file with the same directory structure as in dags_directory.

Once the zip file is created and populated with the necessary files, it's time to upload it to an S3 bucket. The bucket name is specified as instance1bucket, and the boto3 client for S3 is initialized. The upload_file method is then called on the client, specifying the local zip file, the destination bucket, and the path within the bucket where the file should be stored.

CHAPTER 10　DEPLOYING AND MONITORING DEEP LEARNING MODELS

After the upload is successful, the local zip file is removed to clean up the workspace using os.remove(zip_file_name). Finally, a success message is printed to indicate that the files were zipped and uploaded successfully.

Packaging DAGs along with their dependencies in this way simplifies deployment processes, ensures all necessary components are included, and facilitates consistency in versioning and dependency management across different environments.

Step 4: Configuring the MWAA Environment

The next step in the deployment process after uploading the DAG files to the designated S3 bucket is to configure the MWAA environment to include the uploaded DAG and specify any additional settings such as environment variables, connections, and plugins. This step is crucial for ensuring that the MWAA environment can effectively manage and execute the DAG according to your requirements.

To configure the MWAA environment

- Start by accessing the Amazon MWAA console and navigating to the environment settings for the target MWAA environment where you intend to deploy the DAG.

- Within the environment configuration, locate the section dedicated to DAGs and specify the S3 path to the uploaded DAG file. This informs MWAA about the location of the DAG file in the S3 bucket and allows it to retrieve the file for execution.

In addition to specifying the DAG file

- Set environment variables to provide configuration parameters or secrets required by the DAG during runtime.

- Environment variables can include settings such as database connection strings, API keys, or any other sensitive information needed by the DAG for its operation.

- By configuring environment variables within the MWAA environment, you ensure that the DAG has access to the necessary resources without exposing sensitive information directly in the DAG code.

Additionally, plugins containing additional Airflow operators, hooks, or sensors may be required to support specific functionalities within the DAG:

- Ensure that the necessary plugins are installed and configured in the MWAA environment to provide the required functionality for executing the DAG.

- This includes installing any custom or third-party plugins and specifying their configurations within the MWAA environment.

Step 5: Triggering DAG Execution

After the DAG has been successfully deployed and configured in the Amazon MWAA environment, you can trigger its execution manually or based on a predefined schedule within the MWAA environment:

- Manual Execution: You can manually trigger the execution of the DAG from the MWAA console. Navigate to the DAGs view in the MWAA console, locate the DAG, and initiate its execution by clicking the "Trigger DAG" button. This will immediately start the execution of the DAG, and you can monitor its progress and view logs from the MWAA console.

- Scheduled Execution: Alternatively, you can schedule the DAG to run automatically based on a predefined schedule. In the MWAA environment configuration, specify the desired schedule interval for the DAG execution using the schedule_interval parameter. For example, to trigger the DAG execution daily, you can set `schedule_interval=timedelta(days=1)` in the DAG definition. Amazon MWAA will then automatically trigger the execution of the DAG according to the specified schedule.

Step 6: Monitoring Execution

Once triggered, Amazon MWAA will start executing the DAG tasks according to their defined dependencies and the schedule interval. The tasks will run in the specified sequence (e.g., in simple_mwaa_dag.py DAG, print_date executes first followed by print_message). You can monitor the progress of the DAG execution, view task logs, and troubleshoot any issues that may arise from the MWAA console.

Monitoring deep learning models is crucial for several reasons, particularly in production environments managed by services like Amazon MWAA. One reason is ensuring the performance and accuracy of the models over time. Models can experience performance degradation due to changes in data distribution (data drift) or shifts in the relationship between input features and target variables (concept drift). Continuous monitoring helps detect such issues early, allowing for timely retraining or adjustments to maintain accuracy.

Operational efficiency is another key reason for monitoring deep learning models. By keeping track of resource utilization—such as CPU, GPU, and memory usage—you can ensure that the model runs efficiently and identify any bottlenecks that may arise. Monitoring latency is also critical, particularly

CHAPTER 10   DEPLOYING AND MONITORING DEEP LEARNING MODELS

for real-time applications where meeting response time requirements is essential. Ensuring that models operate reliably and are available when needed is fundamental to maintaining system integrity. Monitoring can detect failures or errors in the model's operation, facilitating quick troubleshooting and minimizing downtime. Additionally, it helps in understanding and managing the scalability of the model under varying loads.

More importantly, compliance and governance are increasingly important in many industries, especially those dealing with sensitive data such as finance and healthcare. Monitoring models ensures that audit trails are maintained, which is crucial for regulatory compliance. Finally, security is another critical aspect that benefits from diligent monitoring. By detecting unusual patterns or anomalies, monitoring can alert you to potential security breaches.

You can monitor your deep learning models in two ways:

- Through the MWAA console and Apache Airflow UI for direct DAG monitoring

- Using Amazon CloudWatch for comprehensive metrics and logs tracking

The MWAA console and the Apache Airflow UI allow you to set up, manage, and monitor workflows that include tasks such as data preprocessing, model training, evaluation, and deployment. This enables you to automate and schedule these tasks while keeping an eye on their execution and performance.

To monitor DAGs running in Amazon MWAA using the MWAA console, follow the below steps:

- Log into the AWS Management Console and navigate to the Amazon MWAA service.

- Choose the MWAA environment where your DAGs are running. This will bring up the environment details page.

- In the environment details page, you will see an option to open the Airflow UI. Click the "Open Airflow UI" button, which will take you to the Apache Airflow web interface.

- Monitor DAGs:
    - DAGs View: In the Airflow UI, you can see a list of all your DAGs. This view allows you to check the status, last run, and schedule of each DAG.
    - Graph View: Click a specific DAG to view the Graph View, which shows the structure of the DAG and the status of each task.
    - Tree View: The Tree View provides a hierarchical view of DAG runs and task statuses over time.
    - Gantt Chart: The Gantt Chart view helps visualize the duration of each task in your DAG over time.
    - Code View: You can also view the code of your DAGs to ensure the logic is implemented correctly.
    - Logs: For detailed debugging and monitoring, you can access logs for each task instance to see the output and any errors that occurred during execution.

Amazon CloudWatch provides robust monitoring capabilities by tracking various metrics and logs, creating dashboards, and setting up alerts to notify you of any performance issues, failures, or other critical events. By leveraging both MWAA and CloudWatch, you can ensure that your deep learning models continue to deliver value while maintaining high standards of performance, reliability, and compliance.

CHAPTER 10   DEPLOYING AND MONITORING DEEP LEARNING MODELS

CloudWatch provides several features that can help you keep track of your MWAA environment, DAG runs, and the underlying infrastructure. Here's how you can use CloudWatch for this purpose:

- Metrics Collection:
  - MWAA Metrics: MWAA automatically publishes metrics to CloudWatch. These metrics include information about the environment's health, such as the status of the scheduler, worker, and web server, as well as task and DAG-level metrics.
  - Custom Metrics: You can also push custom metrics from your DAGs or tasks to CloudWatch using the boto3 library. For example, you can publish metrics like model inference times, accuracy scores, or resource usage.
- Logs Monitoring:
  - Task Logs: Airflow task logs can be sent to CloudWatch logs. This allows you to view and search the logs for individual tasks, which is useful for debugging and monitoring the performance of your models.
  - MWAA Environment Logs: MWAA can also send logs related to the environment itself to CloudWatch logs. This includes logs from the scheduler, web server, and other components.
- Alarms and Notifications:
  - CloudWatch Alarms: You can set up alarms based on CloudWatch metrics to get notified about specific events, such as high failure rates for tasks, increased latency, or other performance issues.

- Amazon SNS: Use Amazon Simple Notification Service (SNS) to receive notifications (email, SMS, etc.) when an alarm is triggered.
- Dashboards:
  - CloudWatch Dashboards: Create custom dashboards to visualize the metrics and logs from your MWAA environment and models. This can include graphs showing task success/failure rates, execution times, resource usage, and other relevant metrics.

Summary

In this final chapter of the book, we delved into the process of deploying and monitoring deep learning models. Using Amazon MWAA simplifies the process of setting up, managing, and scaling Apache Airflow environments in the cloud, eliminating the need for manual infrastructure management.

We examined the steps in the deployment process, including setting up the environment, developing the DAG, uploading to S3, configuring the MWAA environment, and triggering and monitoring DAG execution. Two monitoring tools were the focus of this chapter: Amazon MWAA console and Amazon CloudWatch. We also highlighted the importance of monitoring deep learning models to ensure performance and accuracy over time, maintain operational efficiency, adhere to compliance and governance standards, and enhance security.

Index

A

Access control lists (ACLs), 73
Accuracy, 29, 385–387, 395, 461, 465, 545
Adaptive Query Execution (AQE), 116, 145, 146, 210
Airflow
  Docker, 95–100
  on EC2, 87
  MWAA, 87
  standalone installation
    access airflow UI, 93–95
    activate your environment, 87
    admin user, 91, 92
    configurations and settings, 90, 91
    database, 88
    install and configure PostgreSQL, 88
    layers of abstraction, 87
    Port 8080, 94
    PostgreSQL shell, 88, 89
    PostgreSQL support, 88
    scheduler, 92
    tmux/screen, 92
    update and install dependencies, 87
    web server, 92, 94
Airflow web server, 38, 92, 94, 95, 99
Amazon CloudFormation, 28, 64
Amazon CloudWatch, 704, 736–738
Amazon Elastic Block Store (EBS), 61
Amazon Machine Images (AMIs), 26, 61, 67
Amazon S3, 2, 3, 70–75, 126, 184, 729, 731
Amazon S3 bucket, 78, 162, 164, 183, 381, 719
AmazonS3FullAccess policy, 107
Amazon Simple Storage Service (S3)
  configure bucket properties, 71, 72
  create bucket, 71
  IAM roles, 74, 75
  navigate to S3 dashboard, 71
  reliable and cost-effective, 70
  uploading data, 74
Amazon Web Services (AWS), 1, 2, 56, 361, 703, 731

© Abdelaziz Testas 2024
A. Testas, *Building Scalable Deep Learning Pipelines on AWS*,
https://doi.org/10.1007/979-8-8688-1017-6

INDEX

an __init__ method, 118
Apache Airflow, 30, 56
  active community, 30
  advantages, 48, 49
  DAG
    configuration, 34, 36, 37
    default arguments, 43
    definition, 44
    import statements, 40
    provide context, 45
    run, 37–39
    task dependencies, 45
    task functions, 41–43
    task operators, 45
  deep learning, 703
  deployment options, 39
  development, 30
  diabetes prediction (*see* Diabetes prediction)
  Docker containerized setup, 39
  features, 30
  MWAA, 39
  replacement, main() function, 47, 48
  scalable data processing pipelines, 490, 584, 703
  standalone Airflow, 39
  Tesla stock price prediction (*see* Tesla stock price prediction)
  UI, 46, 47
Application programming interface (API), 9, 85
AWS account
  AWS website, 59
  building and deploying, 59
  create a new AWS account, 59
  payment information, 60
  provide account information, 59
  review and confirm, 60
  support plan, 59
  verify your identity, 59
AWS CLI, 4–6, 59, 75, 219, 434, 705
AWS CloudFormation, 60
AWS Management Console, 58, 62, 99, 105, 106, 705–708, 735
AWS Secrets Manager, 74

B

Backpropagation, 18, 236, 237, 373, 498
Balancing model complexity, 390, 463
Baseline model, 586
  adjust hyperparameters, 602
  convert_to_numpy method, 591, 637
  create_and_train_model method, 592, 639
  current performance level, 602
  data augmentation, 602
  evaluate_model method, 594, 640
  libraries, 587
  logging, 587
  main function, 596, 641
  matplotlib.pyplot, 587

INDEX

model loss, 601
model simplification, 602
numpy, 587
overfitting, 602
performance, validation set, 602
predict_and_evaluate method, 595, 641
preprocess_data method, 588, 636
pyspark.ml.feature, 587
pyspark.sql, 587
regularization, 587, 603
review the plateau, 602
sklearn.metrics, 587
tensorflow, 587
test loss, 599, 642
training loss plot, 601
TSLARegressor class, 587, 588, 635
BCEWithLogitsLoss function, 384
Black box, 214, 273, 353
BlockDeviceMappings, 67
Boto3, 86, 117, 151, 325, 361
boto3.client('s3') method, 126
Broadcasting joins, 116, 210
Bucket Policy, 73
Built-in logging module, 362

C

cache() method, 327
Caching, 10, 116, 144, 210, 327, 567
calculate_average_price function, 34, 45, 47
calculate_average_price(taco_df) function, 34
calculate_descriptive_statistics method, 121, 132
calculate_feature_target_ correlation method(), 338–340, 344, 352
CalledProcessError exception, 220
check_for_null_values method, 117, 123, 135
Classification, 321
Cloud computing solutions, 214
Command-line interface (CLI), 4, 39, 48, 221
Compute Unified Device Architecture (CUDA), 85
Confusion matrix, 29, 30, 378, 387
convert_to_numpy() method, 279, 287, 299, 591, 637
convert_to_tensor() function, 160, 162
Convolutional Neural Networks (CNNs), 19
copy_and_print_data(), 217, 218, 221
copy_file_from_s3() function, 128, 151, 175, 183, 186, 340–343, 382
count_outcome method, 334, 335, 352
count_zeros() method, 330, 331, 347
create_and_train_model() method, 281, 288, 289, 299, 592, 600, 603, 607, 639

INDEX

create_data_loader() function, 150, 161, 162, 182, 187, 233, 234
create_taco_dataframe function, 43
create_taco_dataframe(spark) function, 33
CSV file, 4, 51, 117, 125, 176, 216, 250, 327, 435
CustomCallback class, 441, 445
Custom callback (CustomCallback), 438, 444, 453, 536
Custom Keras callback, 438, 443, 444

D

Data augmentation, 322, 389, 463, 466, 602
Databricks, 55, 83
    access S3, 103
        credentials, 108
        dbutils.fs.mount function, 110
        else:, 110
        if mount_point_exists, 109
        line of code, 109
        mounting, 105
        MOUNT_NAME, 109
        mount_point_exists = any(mount.mountPoint == MOUNT_NAME for mount in mounts), 109
        mounts = dbutils.fs.mounts(), 109
        print("S3 bucket mount failed."), 110
        print("S3 bucket was successfully mounted."), 109
        variable assignment, 104
    account creation, 102
    AWS Management Console, 105, 106
    confirm your email, 103
    DataFrame, 111–113
    environment, 105
    explore, 103
    IAM roles, 107, 108
    log into, 103
    sign up, 102
    workspace creation, 103
Data exploration, 116
    with PySpark, 191–195
DataFrame, 111–113, 120–122, 125, 221, 222, 226, 228, 258
DataFrame (self.data), 327
Data frameworks, 1, 450
Data libraries, 8
DataLoader, 225, 233, 234
Data preparation with PySpark for PyTorch, 149, 199–205
    batch processing, 165
    convert_to_tensor() function, 160, 162
    copy_file_from_s3, 162
    create_data_loader function, 161, 162

INDEX

data preparation
  process, 149
df = assembler.
  transform(df), 155
functionality, 163
is being run, 167
load_data() function, 150
logging.
  basicConfig(level=logging.
  INFO), 151
logger = logging.
  getLogger(name), 151
logging module, 150
main(), 167
NumPy, 150
parameters, 163
perform_feature_
  engineering, 156
preprocess_data
  function, 168
price_change, 171
price_range, 171
print_first_5_observations
  function, 159, 168
pyspark.sql, 150
scaled_features column, 173
scaled feature values, 173
scaling, 173
StandardScaler, 150
tensor_features, 160
Tesla stock prices, 171
VectorAssembler, 150
volume_price_
  interaction, 171

for TensorFlow, 174, 205–210
  appName method, 176
  batch method, 183
  Boto3, 175
  copy_file_from_s3 function,
    183, 186
  create_data_loader function,
    182, 187
  DataFrame from pyspark.
    sql, 175
  feature engineering
    operations, 180
  load_data function, 176
  logging information, 179
  logging module, 174
  main() function, 188
  NumPy, 175
  perform_feature_
    engineering function, 179
  preprocess_data(df)
    function, 189
  preprocess_data
    function, 177
  preprocessing step, 179
  price_change, 180
  print_first_5_observations
    function, 181, 186
  pyspark_tensorflow_
    preparation.py, 188
  s3.download_file
    method, 184
  shuffling, 183
  SparkSession from pyspark.
    sql, 174

INDEX

Data preparation with
    PySpark (*cont.*)
  StandardScaler, 175
  tensorflow as tf, 175
  tf.data.Dataset.from_tensor_
    slices method, 182
  VectorAssembler, 175, 178
  volume_price_
    interaction, 181
Data preprocessing, 2, 56, 115, 278, 321, 428
DataPreprocessor class, 364, 367, 382
DataProcessor class, 118, 121, 125, 132
DataProcessor object, 128, 129
Data representation, 90, 322, 463
Dataset, 116–137, 276–277, 321, 323
Dataset, Tesla stock, 214
  load_data, 216
  logging module, 215
  modules, 215
  pyspark.sql module, 215
  SparkSession, 216
  subprocess module, 215
Data source, S3, 3
  AWS CLI, 4
  Boto3, 5, 6
  check=True parameter, 5
  EC2, 4
  reverse process
    AWS CLI, 6
    Boto3, 6, 7
    parameters, 7

  s3_client.download_file()
    method, 6
  try/except block, 5
data_summary() method, 331–334, 349
Deeper model, 626, 628
Deep learning, 703
  neural networks, 18
  tasks, 25
  variables, 18
Deep learning
  environment on AWS
    account creation, 59, 60
    Amazon EC2 instances, 60–70
    Amazon S3, 70–75
    Databricks, 102–113
    dependencies, 83–86
    installing
      airflow, 87–100
      Boto3, 86
      JupyterLab, 100–102
    project directory, 75–79
    virtual environment, 80–83
Deep learning models, deploying
    and monitoring (MWAA), 704, 735, 738
  CloudWatch, 737, 738
  DAG development, 705, 716–718
  DAG execution, 705, 733, 734
  environment setup, 704
    access MWAA console, 706
    configurations, 716
    create MWAA
      environment, 706

INDEX

creation process, 716
DAG files, 707
environment class, 715
environment details, 706
execution role, 707
internet gateway, 711, 712
NAT gateway, 712–714
network configuration, 707
route tables, 714
S3 bucket, 707
security groups, 715
subnets, 710, 711
VPC, 707–709
monitor execution, 706
CloudWatch, 736
compliance and governance, 735
continuous monitoring, 734
DAGs running, 735, 736
failures and errors, 735
latency, 734
MWAA console, 735
operational efficiency, 734
performance and accuracy, 734
monitoring tools, 738
MWAA environment configuration, 705, 732, 733
S3 uploading, 705
command, 719
DAG files, 719, 731
DAG modules, 719, 731
DAGs packaging, 732

data_processing.py module, 720, 721
files_to_zip, 731
instance1bucket, 731
model_evaluation.py module, 723, 724
model_training.py module, 721–723
Python code packages, 729, 730
requirements.txt file, 729
tesla_stock_prediction, 719
tesla_stock_prediction.py module, 725–729
utils.py module, 724, 725
Deep learning workflow components, 2, 3
data preprocessing (*see* PySpark data preprocessing)
data source (*see* Data source, S3)
model building (*see* PyTorch; TensorFlow)
model deployment (*see* Apache Airflow)
model evaluation, 28, 29
model training, EC2 (*see* Elastic Compute Cloud (EC2))
def__init__(self), 282
Dependencies
installing and configuring
Keras, 85, 86
PySpark, 83, 84
PyTorch, 85
TensorFlow, 85, 86

745

INDEX

Development environments, 55–56, 100
DiabetesclassificationwithK-foldcross-validation, 415–428
DiabetesclassificationwithoutK-foldcross-validation, 403–415
Diabetes classifier model, 372, 376, 383, 384
DiabetesClassifierModel class, 369
diabetes_classifier_pipeline.py, 383
diabetes.csv, 326, 340, 345, 382, 434
DiabetesHyperModel class, 467, 469
DiabetesModelEvaluator class, 453, 455
DiabetesModelEvaluator.evaluate_model() method, 459
DiabetesModelTrainer class, 450, 452, 469
DiabetesModelTrainer.train_tensorflow_model() method, 459
DiabetesPedigreeFunction, 349–353, 388, 437
Diabetes prediction
   with Airflow DAG
      AWS EC2 instance, 556
      DAG object, 550
      default arguments, 548, 549
      evaluate_task, 554, 555
      execution, 582–584
      imports, 547, 548
      main() function, 546
      metrics, 558
      preprocess_task, 551, 552
      task dependencies, 556
      training loss, 557
      train_task, 553, 554
   without Airflow DAG
      columns, 544
      evaluation metrics, 545, 546
      main() function, 529, 574
      main.py module, 529, 541–543, 581
      model_evaluation.py module, 529, 538–541, 579, 580
      model_training.py module, 529, 533, 535–537, 576–579
      modules, 529
      myenv environment, 543, 544
      preprocessing.py module, 529–533, 574, 576
      script orchestrates, 543
      training loss, 544, 545
DiabetesProcessor class, 441, 442, 445, 447, 448, 458
DiabetesProcessor.preprocess_data() method, 458
Directed Acyclic Graphs (DAGs), 31, 130, 247, 489
Docker, 703
   access the airflow web interface, 99, 100
   airflow and dependencies, 95
   Airflow services, 99
   compose, 95
   Compose configuration, 98
   configuration, 95
   Docker Compose, 97

INDEX

download, 98
initialize the Airflow
    database, 99
install, 95
packages, 95
prepare the environment, 98
run a test Docker container, 97
script executable, 96
script file, 96
script with root privileges, 96
verify Docker installation, 96
Docstring, 120–122, 152, 176,
    329, 335
docstring """Class, 447, 452, 454
docstring """Custom Keras
    callback, 445
download_file method, 6, 126, 163,
    184, 382
Dropout, 585, 651–658
    create_and_train_model
        method, 607
    dropout rate, 609
    layers, 607
    modifications, baseline
        model, 609
    neurons, 607
    underfitting/overfitting, 609

E

Early stopping, 585, 643–650
    create_and_train_model
        method, 603
    EarlyStopping callback, 605

loss value, 606
output, 605, 606
performance, validation set, 603
training process, 603
TSLARegressor class, 643
Elastic Block Store (EBS), 27, 61, 67
Elastic Compute Cloud (EC2), 1, 19
    accelerated computing
        type, 23–25
    advantages, 26
    compute-optimized type, 22, 23
    general-purpose type, 19, 20
    GPUs, 25
    instance types, 19, 56
    memory-optimized type, 20, 21
    model evaluation, 28, 29
    pre-configured machine
        images, 26
    set up/provisioning, 26–28
    storage-optimized type, 21, 22
    TPUs, 25
Environment variables, 74, 98, 104,
    705, 733
Epoch/Loss table, 462
Error handling, 48, 244, 449, 453,
    455, 541
evaluate_model function, 238,
    239, 244
evaluate_model method, 291, 292,
    299, 594, 640
Exception handling, 228, 229,
    373, 378
explore_feature_distributions()
    method, 336, 344, 354

INDEX

F

Facebook's AI Research lab (FAIR), 13, 362
Feature engineering, 115, 136, 354, 448
Feedforward networks, 15, 18
FileNotFoundError exception, 220
file_path parameter, 119
Fit method, 179, 231, 453, 605
F1 score, 18, 29, 358, 386, 462, 466, 541, 558
Fundamental market dynamics, 223, 277

G

getOrCreate(), 153, 176, 217, 283
Gradient descent, 18, 233, 354
Gradients, 14, 236, 373, 498, 499, 501
Graphics processing units (GPUs), 24, 25, 85
Grid search, 629

H

handle_missing_values() method, 329, 330, 345
HVM virtualization, 61
Hyperparameter optimization, Keras Tuner, 586, 690–701
   grid search, 629
   Model Training Process, 632
   random search, 629, 630
   R2 score, 634
   test dataset, 634
   training process, 634
   TSLAHyperModel class, 630, 631, 634, 691
   tuner.search function, 634
Hyperparameter tuning, TensorFlow, 390
   DiabetesHyperModel Class, 467
   DiabetesModelTrainer, 469, 470
   imports, 467
   keras_tuner, 467
   model performance, 466
   optimizing model performance, 471

I

Identity and Access Management (IAM) roles, 74, 75, 105–108
import os, 434
Inbound rules, 62, 93, 99, 102
__init__ method, 118, 282, 372
Intermediate-capacity model, 623, 625
I/O operations per second (IOPS), 21
isNull(), 330

J

JupyterLab, 55, 76
   configuration, 101
   configure security group and access, 101, 102

INDEX

IDE, 100
install, 100
set password, 101
start, 101
JupyterLab terminal, 247, 344, 383, 436, 460
Jupyter Notebook, 55, 77, 100, 101

K

Keras, 13, 85, 86, 612
Keras's Sequential API, 452
Keras Tuner, Hyperparameter optimization, 629–634, 690–700
K-fold cross-validation, 390
   balanced evaluation, 391
   importance, 391
   importing KFold, 391
   looping through each fold, 392–395
   number of folds, 391

L

L1 and L2 regularization, 658–674
   baseline neural network model, 610
   baseline TensorFlow model, 614
   create_and_train_model method, 610
   Keras, 612
   loss function, 586, 610
   modification, baseline model, 612, 613
   neural networks, 586
   output, 615
   overfitting, 615
   R2 score, 614, 615
   squared values, parameters, 610
   squared value, weights to loss function, 586
   test loss, 614, 615
   values, parameters, 610
Learning rate, 674–682
   create_and_train_model function, 618
   high, 616
   impact, model training and performance, 619
   low, 616
   moderate learning rate, 618
   TensorFlow's Adam optimizer, 616
   test loss, 619
   training deep learning models, 616
Linear models as Lasso, 586
Linear models as Ridge, 586
load_data function, 151, 166, 174, 186, 216, 228, 493
load_data method, 118, 129, 327, 328
LocalExecutor, 90, 91
local_file_path, 128, 130, 163, 186, 219, 343
logger.error() method, 449, 453, 459

749

INDEX

Logging, 179, 362, 373, 598
logging.basicConfig(level=logging.INFO), 444
Logging module, 150, 174, 215, 279, 362, 444
Logistic regression, 323
Loss function imbalance, 358

M

Machine learning (ML), 4, 22, 55, 135, 181, 230, 247, 352, 491
main() function, 188, 254, 255, 279, 342–344, 596, 597, 641
Managed Workflows for Apache Airflow (MWAA), 1, 39, 87, 703
Matplotlib, 123, 596
Matplotlib library, 281
matplotlib.pyplot, 117, 281, 587
Matthews Correlation Coefficient (MCC), 29
Mean Absolute Error (MAE), 18, 239
Mean Squared Error (MSE), 14, 18, 28, 239
Method logs, 373
Metrics table, 385, 387, 461–463
missing_counts DataFrame, 329
Model bias, 357
Model capacity, 682–690
　baseline model, 620
　baseline model configuration, 623
　baseline results, 628
　deeper model, 626, 628
　excessive capacity, 619
　improvement, performance metrics, 624
　insufficient capacity, 619
　intermediate capacity, 622
　number of neurons, hidden layer, 620, 621
　results, 622, 623, 625
　TSLARegressor class, layers, 620
　tuning model capacity, 625
　wide model, 625, 628
Model complexity, 390, 428, 463, 466, 621
model.eval(), 239, 254
model.evaluate() method, 440
ModelEvaluator class, 363, 374, 376, 394, 453
model.fit() method, 440
Modeling steps, 278, 322, 364, 429, 438
Model learning, 358
model.load_state_dict(), 254
model_loss_plot.png, 600
ModelTrainer class, 372, 373, 394
myenv, 37, 38, 81, 221, 247, 383, 508, 598

N

__name__ variable, 187
Neural network architecture, 17, 353, 369, 443

INDEX

Neural networks, 13, 18, 86, 213, 367, 586
null_counts DataFrame, 125
NumPy, 150, 225, 362, 443
NumPy arrays, 246, 287–289, 292, 293, 296, 591
num_workers parameter, 233

O

on_epoch_end method, 445
os.system function, 434
Outbound rules, 62
Overfitting, 183, 234, 291, 309, 370, 428, 585

P, Q

Parallel processing
    in PySpark, 195–199
parentheses (), 434
Parquet format, 43, 144, 145, 449, 540
Performance metrics, 360, 363, 374, 377, 440, 527, 624
perform_feature_engineering(df) function, 189
perform_feature_engineering function, 179
Pima, 322
Pima dataset, 388–390, 396–402, 463
PimaDatasetExplorer class, 325, 326, 330, 343
PimaDatasetExplorer instance, 344
Pima datasets, PyTorch
    data augmentation, 389
    data quality, 388
    data representation, 390
    exploration, 396–402
    hyperparameter tuning, 390
    K-fold cross-validation, 390
    model complexity, 390
    resampling techniques, 389
    transfer learning, 389
Pima diabetes dataset, TensorFlow classification
    CSV file, 432
    importing necessary libraries, 433–438
Pima Indians Diabetes Dataset, 322, 323, 326
plot_actual_vs_predicted() function, 281
plt.show(), 123, 297
PostgreSQL, 87–91, 98
Precision, 29, 377, 386, 388, 461, 465
predict_and_evaluate() method, 280, 299, 595, 599, 641
Predicting diabetes with PyTorch
    additional functionality, 360, 378, 380, 381, 383, 384
    data preprocessing, 382, 383
    evaluation metrics, 385–388
    file copying from S3, 382
    observations based on loss values, 385

751

INDEX

Predicting diabetes with
    PyTorch (*cont.*)
  data preprocessing, 359
    DataPreprocessor class,
      364, 367
    randomSplit method, 367
    VectorAssembler object, 367
  features, 359
  import libraries, 361
    boto3, 361
    logging, 362
    NumPy, 362
    pyspark.ml.feature, 364
    pyspark.sql.functions, 364
    pyspark.sql module, 363
    sklearn.metrics, 363
    torch, 362
    torch.nn, 362
    torch.optim, 363
    torch.utils.data, 363
  model definition, 359
    DiabetesClassifier
      Model, 369
    forward method, 370
    hidden layers, 370
    __init__ method, 369
    layers and activation
      functions, 368
    model performance, 370
    ReLU activation
      function, 369
  model evaluation, 360
  ModelEvaluator class
    diabetes classifier model, 376
    essential components, 376
    evaluate method, 376, 377
    exception handling, 378
    __init__ method, 376
    obtaininig final
      predictions, 377
    performance metrics,
      374, 377
  model training, 360
  Model Training
    exception handling, 373
    forward pass, 373
    gradients, 373
    __init__ method, 372
    method logs, 373
    ModelTrainer class, 370,
      372, 373
  neural network model, 360
  Pima dataset (*see* Pima datasets,
    PyTorch)
Predicting diabetes with TensorFlow
  custom Keras callback, 438,
    444, 445
  data preprocessing, 445–449
  data processing, 439
  imports, 438, 441
    col, 442
    logging, 441
    numpy, 443
    Sequential, 442
    sklearn.metrics, 442
    SparkSession, 442
    tensorflow.keras.
      callbacks, 443

INDEX

tensorflow.keras.layers, 443
tensorflow.keras.
   optimizers, 443
VectorAssembler,
   StandardScaler, 442
Keras, 439
logging configuration, 438,
   443, 444
main function, 456–462
model architecture and
   training, 439
model definition and training,
   439, 450–453
model.evaluate(), 440
model evaluation, 439, 453–455
model.fit(), 440
predict_tsla_stock.py, 598, 599
preprocess_data(df) function,
   189, 190
preprocess_data() function, 150,
   168, 187, 230–232
preprocess_data() method, 280,
   328, 344, 447, 588, 636
Preprocessing, 14, 211, 230, 390
print_first_5_observations
   function, 168, 181, 186
print_first_n_rows method, 120, 131
print_top10_records.py, 436
Process ID (PID), 94
Project directory
   %%bash, 77
   design, 75
   mkdir and touch commands, 78
   mkdir-p "$project_dir", 78
   mkdir-p "$project_dir/data", 78
   project_dir="Project Directory
      Structure", 78, 79
   reproducibility, 75
   scalability and maintenance, 75
   structure, 76, 77
   structured environment, 75
   tree "$project_dir"--dirsfirst, 78
   tree command, 79
   TSLA_stock.csv, 78
Project directory structure, 56
   attributes, 50
   benefits, 49
   Tesla stock price, 50, 51
      config.yaml, 53
      data, 51
      logs, 51
      output, 52
      README.md, 53
      representation and
         overview, 53, 54
      requirements.txt, 53
      src, 52
      tests, 52
      visualization, 52
Provisioning Amazon EC2
   instances
   access your EC2 Instance, 63
   add storage, 61
   Amazon CloudFormation, 64
   AMI, 61
   associate the security group, 63
   AWSTemplateFormat
      Version, 66

INDEX

Provisioning Amazon EC2
    instances (*cont.*)
  choose an instance type, 61
  configuration, 66
  configure instance details, 61
  DeepLearningEC2
    Instance, 66, 67
  Description, 66
  EBS volume, 64–66
  InstanceSecurityGroup, 68
  JSON script, 68–70
  key pair creation, 63
  launch instance, 60
  navigate to the EC2
    Dashboard, 60
  note, 64
  Outputs section, 68
  Resources section, 66, 67
  review and apply changes, 63
  review and launch, 63
  security groups, 62
  SSH access, 64–66
  start instance, 64
  stop instance, 64
  training and inference
    tasks, 60
  YAML format, 68
PySpark, 83, 84, 116, 150, 230, 257, 329, 431
PySpark code
  calculating feature, 324
  counting outcome values, 324
  counting zero values, 324
  data summary, 324
  feature distributions, 324
  handling missing values, 323
PySpark DataFrame
  code explanation, 9–11
  creation, 8, 9
  price of tacos, 31–33
    conditional execution, 34
    defining functions, 33
    importing required
      modules, 33
    main() function, 34
  script execution, 12
PySpark data preprocessing, 7, 56
  advantages, 7
  data libraries, 8
  integration, 8
PySpark functions
  calculate_feature_target_
    correlation
    method, 338–340
  class definition, 326
  copy_file_from_s3
    function, 340–342
  correlation coefficient, 352, 353
    explore_feature_
      distributions() method, 354
    feature engineering, 354
    feature selection, 353
    model interpretability, 353
    model optimization, 353
  count_outcome method, 334, 335
  count_outcome() method, 352
  count_zeros() method, 347

INDEX

count_zeros Method definition, 330, 331
dataset's characteristics, 350
data_summary definition, 331-334
data_summary() method, 349
diabetes_data_exploration.py, 344
DiabetesPedigreeFunction and Age, 349
explore_feature_distributions method, 336-338
handle_missing_values definition, 329, 330
handle_missing_values() method, 345
histograms, 357
imports, 325
invalid readings, 349
load_data method definition, 327, 328
main function, 342-344
preprocess_data method definition, 328
skewed data distribution
　loss function imbalance, 358
　model bias, 357
　model learning, 358
　sampling strategies, 358
statistics, 350
suspicious values, 352
unbalanced dataset, 352
pyspark.ml.feature, 150, 226, 280, 364, 442

pyspark_pytorch_preparation.py, 168
PySpark show() method, 220
PySpark's parallel processing, 116, 136-149, 210
pyspark.sql, 117, 150, 174, 434
pyspark.sql.functions, 117, 118, 364, 442
pyspark.sql.functions import col, 325
pyspark.sql import SparkSession, 325
pyspark.sql module, 215, 226, 280, 325
PySpark's randomSplit() method, 449
Python 3, 12, 80, 248, 301, 345, 460, 598
Python programming, 221
PyTorch, 56, 61, 85, 439
　data loading and preprocessing, 14
　data model building, 14
　development, 13
　Evaluation Metrics, 18
　features, 13
　inference, 14
　loss function, 14, 18
　model evaluation, 14
　model training, 14
　neural network model, 15-17
　optimizer, 14
　Output Layer Activation Function, 17

755

INDEX

PyTorch (*cont.*)
   vs. TensorFlow, 305–308
   workflow, 15
PyTorch DataLoader object, 161, 162
PyTorch machine learning model, 230, 232
PyTorch multilayer algorithm, 428
PyTorch tensors, 160, 162, 202, 226, 245, 254, 492, 502
PyTorch with predict Tesla's stock prices, 223
   explore Tesla Stock dataset, 258
   fundamental market dynamics, 223
   modeling process, 224
      DataLoader creation function, 232–234
      data loading function, 227–229
      data preprocessing function, 229–232
      execution, 247
      helper function, 245–247
      imports, 224–226
      logging setup, 226, 227
      main function, 240–244
      model evaluation function, 238–240
      model training function, 234–237
      plotting actual *vs.* predicted values, 269–272

and PySpark, 260–268
Python code, 223
technical aspects, 223
trading volume, 223

R

Random search, 629, 630
randomSplit() method, 286, 367
read.csv() method, 217, 285, 448
Recall, 386, 462
Receiver Operating Characteristic Area Under the Curve (ROC-AUC), 29
Rectified linear unit (ReLU), 243, 309, 369, 452
Recurrent Neural Networks (RNNs), 19
Regression, 321
Repartitioning, 116, 137–139, 210
Resampling techniques, 389
Resilient Distributed Datasets (RDDs), 138, 245, 288
Right-skewed distribution, 357
ROC-AUC score, 360, 363, 378, 386, 394, 395
R2 score, 279, 280, 299, 302, 303, 305, 599
r2_score function, 280
R-squared score, 239, 240, 248, 249, 273, 294, 307, 501
RuntimeError, 229

S

s3_bucket_name, 128, 130, 186, 343
Scikit-Learn, 363, 440
s3.download_file() method, 126, 184
Security groups, 62, 63, 93, 101, 715
Sequential class, 280
SequentialExecutor, 90, 91
s3_file_key, 128, 130, 343
show() method, 9, 11, 34, 220, 330, 331, 335, 436
Simple Storage Service (S3), 1, 70
sklearn.metrics, 280, 363, 442, 540, 587
Software Development Kit (SDK), 6, 75
Spark DataFrame (diabetes_df), 367
Spark DataFrames, 280
spark_data_processing.py, 221
spark_df_to_tensor helper function, 245, 246
SparkSession, 117, 118, 215, 216, 228, 280, 363, 382, 434, 436, 458
SparkSession.builder API, 458
SparkSession.builder method, 10, 128, 228
SparkSession class, 9, 215, 280, 343, 363, 434
SparkSession object, 117–119, 128, 150, 152, 216, 217
SQLite, 87, 90
SSH key pair, 63, 67
StandardScaler, 115, 150, 175, 231, 280, 448, 588
StandardScaler object, 156, 179, 367, 448
stat.corr() method, 340
@staticmethod decorator, 448, 452, 455
Stochastic gradient descent (SGD), 14, 443, 453, 536
StockPricePrediction, 128, 152, 217, 494
Subprocess module, 4, 215
subprocess.run() method, 5, 243

T

TensorDataset, 150, 162, 225–226, 233, 363
tensor_features, 160
TensorFlow, 56, 61, 85, 86, 213
    computational graphs, 13
    data loading and preprocessing, 14
    development, 13
    Evaluation Metrics, 18
    inference, 14
    Keras, 13
    loss function, 14, 18
    model building, 14
    model evaluation, 14
    model training, 14
    neural network model, 15–17
    optimizer, 14

INDEX

TensorFlow (*cont.*)
  Output Layer Activation
    Function, 17
  vs. PyTorch, 305–308
  workflow, 15
tensorflow.keras.callbacks, 443, 605
tensorflow.keras.layers
    module, 281
TensorFlow (Keras) model, 450,
    453, 460
tensorflow.keras.models, 280, 442
tensorflow.keras.optimizers, 443
TensorFlow model with fixed
    hyperparameters, 472–479
TensorFlow model with
    hyperparameters
    tuning, 479–487
TensorFlow *vs.* PyTorch
  comparative analysis, 463
  evaluation metric, 465, 466
  loss value comparison,
    464, 465
TensorFlow with predict Tesla's
    stock prices, 277
  actual *vs.* predicted values, 305
  environment.yml, 310
  layers and activation
    functions, 309
  modeling steps, 278
    conditional block, 300–303
    convert_to_numpy
      method, 286–288
    create_and_train_model
      method, 288–291

evaluate_model method,
  291, 292
imports, 279–281
logging setup, 281–283
main function, 298, 299
plot_actual_vs_predicted
  function, 295–297
predict_and_evaluate
  method, 293–295
preprocess_data()
  method, 283–286
object-oriented
  methodology, 278
TensorFlow regression
  code, 310–319
TSLARegressor, 278
Tensor processing units
  (TPUs), 24, 25
Tesla stock price prediction
  main() function, 491
  modular approach, 490
  modular design, 491
  operators, 491
  PyTorch, 491
  scalability and
    performance, 491
  with Airflow DAG
    AWS EC2 instance, 525
    checkpoints, 568
    columns, 526
    DAG definition, 512, 516–518
    datasets, 567
    default arguments, 512, 515
    dynamic DAGs, 568

INDEX

evaluating model task, 512, 522–524
execution, 569, 571–573
implementation, 568
imports, 512, 513
interpretation of data, 527
load configuration, 568
main() function, 511, 567
outputs, 525
performance metrics, 527, 528
preprocessing data task, 512, 518, 519
stock_price_prediction, 525
task dependencies, 512, 524, 525
tasks, 513, 514
training loss values, 525, 526
training model task, 512, 520–522
without Airflow DAG
  columns, 509
  data_processing.py module, 492–495, 559, 560
  key metrics, 510, 511
  main() function, 507, 508
  main.py module, 492, 503, 504, 506, 564, 566, 567
  model_evaluation.py module, 492, 499–502, 562, 563
  model_training.py module, 492, 496, 498, 560–562
  modules, 492, 507
  myenv environment, 508
  output, 508, 510
  training loss, 509, 511
  utils.py module, 492, 502, 503, 563
tesla_stock_price_prediction.py, 247, 250, 300
Test loss, 599, 615, 619, 642
tf.data.Dataset.from_tensor_slices method, 182
.toPandas(), 125
toPandas method, 123
torch.load(), 254
torch.save method, 244
torch.utils.data, 150, 225, 363
Training loss, 248, 249, 301–303, 463, 509
train_model function, 235, 237, 244
Transfer learning, 322, 389, 463, 466
Transform function, 448
TSLAHyperModel Class, 630, 631, 691
tsla_regressor, 296, 299
TSLARegressor class, 278, 279, 282, 283, 291, 295, 296, 299, 587, 588, 596, 620, 635

U

User interface (UI), 38, 45, 47, 55, 92

759

V

VectorAssembler, 115, 150, 154, 155, 170, 230, 231, 280, 283, 285, 367, 448
VectorAssembler, StandardScaler, 442
Virtual environment, 54–56
  activation, 82
  bin, 81
  creation, 80
  dependencies, 80
  directories and files, 81
  EC2 instance, 80
  include, 81
  lib64, 82
  myenv, 81
  navigation, 80
  Python 3, 80
  pyvenv.cfg, 82
  reproducibility, 80
  Ubuntu, 80
Virtual Private Cloud (VPC), 27, 68, 707
visualize_data method, 117, 121, 136

W, X

Wide model, 625, 628

Y

YAML format, 68

Z

zero_counts DataFrame, 331
zero_grad method, 237

www.ingramcontent.com/pod-product-compliance
Ingram Content Group UK Ltd.
Pitfield, Milton Keynes, MK11 3LW, UK
UKHW021031090125
453260UK00005B/328